ATLAS AND CATALOGUE
OF INFRARED SOURCES IN THE MAGELLANIC CLOUDS

ATLAS AND CATALOGUE
OF INFRARED SOURCES
IN THE MAGELLANIC CLOUDS

by

P. B. W. Schwering and F. P. Israel

Sterrewacht Leiden, The Netherlands

KLUWER ACADEMIC PUBLISHERS

DORDRECHT / BOSTON / LONDON

Library of Congress Cataloging in Publication Data

Schwering, P. B. W.
    Atlas and catalogue of infrared sources in the Magellanic Clouds /
by P.B.W. Schwering and F.P. Israel.
        p.   cm.
    Includes bibliographical references.

    1. Magellanic Clouds--Atlases. 2. Magellanic Clouds--Catalogs.
3. Infrared sources--Atlases. 4. Infrared sources--Catalogs.
5. Infrared astronomy--Atlases. 6. Infrared astronomy--Catalogs.
I. Israel, F. P. II. Title.
QB858.5.M33S38 1990
522'.68--dc20   89-71712

ISBN-13: 978-94-010-6728-7   e-ISBN-13: 978-94-009-0537-5
DOI: 10.1007/978-94-009-0537-5

Published by Kluwer Academic Publishers,
P.O. Box 17, 3300 AA Dordrecht, The Netherlands.

Kluwer Academic Publishers incorporates
the publishing programmes of
D. Reidel, Martinus Nijhoff, Dr W. Junk and MTP Press.

Sold and distributed in the U.S.A. and Canada
by Kluwer Academic Publishers,
101 Philip Drive, Norwell, MA 02061, U.S.A.

In all other countries, sold and distributed
by Kluwer Academic Publishers Group,
P.O. Box 322, 3300 AH Dordrecht, The Netherlands.

Printed on acid-free paper

The Atlas and Catalogue of infrared sources in the Magellanic Clouds was prepared from data obtained with the Infrared Astronomical Satellite (IRAS). IRAS was developed and operated by the Netherlands Agency for Aerospace Programs (NIVR), the U.S. National Aeronautics and Space Administration (NASA) and the U.K. Science and Engineering Research Council (SERC).

TABLE OF CONTENTS

INTRODUCTION

1. Introduction

Around the beginning of the sixteenth century, Portuguese and Dutch sailors first ventured into southern seas. With their keen navigational interest in the skies, they noted almost immediately the continuous presence of two cloud–like features, not far from the Southern Pole. The first literature mention of these 'clouds' was in the journal written in 1520 by the Italian navigator Pigafetta on the first circumnavigation of the globe by Magalhães (*cf.* Pigafetta *et al.*, 1962). In honour of this exploit, the objects have since become known as the Magellanic Clouds, although the Dutch name 'Kaapsche Wolken' (Cape Clouds – after the Cape of Good Hope) has also been in use for centuries.

The Large and Small Magellanic Clouds are dwarf irregular galaxies, orbiting our own Milky Way Galaxy, presently at distances of 53 and 63 kpc respectively (Humphreys, 1984). They are the galaxies nearest to us: most other Local Group galaxies are of order ten times more distant. The LMC and SMC are also the prototypical blue dwarf irregulars, representatives of a class of objects in which several hundred more distant objects are now known. Their masses are a few per cent of the mass of the Milky Way Galaxy, but they are relatively gas–rich and appear to be, at the present epoch, forming stars at a more prodiguous rate than our Galaxy (*cf.* Lequeux, 1984). Compared to the Galaxy, they are poor in heavy elements (Dufour, 1984), poor in dust (Koornneef, 1984), and have a relatively weak CO signature (Israel, 1984), often taken as representative for the total abundance of molecular material (mainly H_2). In these aspects, they are similar to other blue dwarf irregulars, but different from larger spiral galaxies.

It has been difficult to quantify the dust and molecular contents of the Magellanic Clouds. Molecular measurements are only now being carried out on a significant scale with the facilities of the European Southern Observatory. In the past, several attempts have been made to gauge the dust content of the Clouds by absorption and extinction studies. These studies yielded important qualitative conclusions, but failed to produce a clear quantitative result (*cf.* Israel, 1984; Koornneef, 1984). Yet, this is a matter of importance, as dust and molecular abundances provide clues to the evolution of the Clouds. Details of dust and molecular distribution and properties also provide important clues to ongoing physical process in the Clouds, including the formation of stars. A fuller, more quantitative understanding of dust and molecular properties of the Magellanic Clouds transcends that of Cloud studies alone. The proximity of the Clouds allows analysis of the interrelation between detailed and global characteristics. In turn, this is essential to the interpretation of measurements of more distant galaxies, where only global (integrated) properties can be determined.

A more fruitful approach to the question of the dust content of the Magellanic Clouds is the measurement of thermal emission from dust particles. With the expected temperature range (15 to 1500 K), this emission will primarily take place at infrared and submillimeter wavelengths (10 microns to 1 millimeter). Prior to 1983, few such measurements existed, especially at wavelengths beyond 10 microns where the bulk of emission was to be expected.

Characteristically, the all–sky AFGL rocket survey (Price and Walker, 1976) yielded only four 20 micron sources in the LMC and none in the SMC. Succesful observations from atmospheric platforms were limited to a few sources (see Schwering, 1988, p. 12). This situation changed dramatically with the succesful IRAS mission in 1983 (IRAS, 1989). The all–sky survey, as well as the additional (pointed) observations, covered the Magellanic Clouds several times, and yielded fully sampled data at wavelengths of 12, 25, 60 and 100 microns.

The IRAS data have been used to construct maps of the infrared emission of the Magellanic Clouds at these wavelengths (Schwering, 1988; Rice *et al.*, 1988; see also Schwering and Israel, 1989; Schwering 1989). The maps presented here are based on observations obtained in the Additional Observations (AO) program, in the Deep Sky Mapping (DPM) mode. In this mode, a raster scan of 6 or 7 legs of 166 arcmin in length was made with a cross–scan step of 20 arcmin. Scanning was performed in the survey direction at a rate of 3.85 arcmin per second. The proximity of the Clouds to the Southern Galactic Pole made it possible to obtain sets of observations with almost orthogonal scan directions (approximately East–West and North–South). Because of the rectangular form of the IRAS detectors, resolution in the in–scan direction is considerably higher than in the cross–scan direction. The two datasets therefore provide maps with maximum resolution in both directions. We have chosen to present both map sets independently. In principle, it is possible to combine the orthogonal sets into full resolution maps. However, because of the non–uniform response of the IRAS detectors, the only correct way of doing this is by 'cleaning' individual detector 'streams' before combining them into a map, at a considerable cost in computer time. Attempts to combine orthogonal maps and then cleaning them in all cases created sidelobe effects and other artifacts that could not be removed satisfactorily. We therefore decided not to include those.

The main body of this publication is the Infrared Atlas of the Magellanic Clouds contained in Part 4 and the Catalogue of Infrared Sources contained in these maps in Part 3. Preceding the Atlas and Catalogue is a set of Tables correlating Catalogue sources with objects listed in other (optical) catalogues (Part 2). A series of transparent overlays facilitates identification of Catalogue sources in Atlas maps.

The purpose of this publication is to supply the IRAS infrared maps and infrared source lists, as well as source identifications in a form convenient for other workers on Magellanic Cloud studies. For a discussion of the IRAS data on the Magellanic Clouds, the reader is referred to Schwering (1988) and subsequent publications in journals such as Astronomy and Astrophysics.

References

Dufour, J.: 1984, in "Structure and Evolution of the Magellanic Clouds", IAU Symposium 108, eds. S. van den Bergh and K.S. de Boer, Reidel Publ. Co, Dordrecht, p. 353

Humphreys, R.M.: 1984, in "Structure and Evolution of the Magellanic Clouds", IAU Symposium 108, eds. S. van den Bergh and K.S. de Boer, Reidel Publ. Co, Dordrecht, p. 145

IRAS: 1989, Infrared Astronomical Satellite (IRAS) Catalogs and Atlases, Volume 1, Explanatory Supplement, eds. C.A. Beichman, G. Neugebauer, H.J. Habing, P.E. Clegg and T.J. Chester, NASA RP-1190, Washington, DC (USA)

Israel, F.P.: 1984, in "Structure and Evolution of the Magellanic Clouds", IAU Symposium 108, eds. S. van den Bergh and K.S. de Boer, Reidel Publ. Co, Dordrecht, p. 319

Lequeux, J.: 1984, in "Structure and Evolution of the Magellanic Clouds", IAU Symposium 108, eds. S. van den Bergh and K.S. de Boer, Reidel Publ. Co, Dordrecht, p. 67

Koornneef, J.: 1984, in "Structure and Evolution of the Magellanic Clouds", IAU Symposium 108, eds. S. van den Bergh and K.S. de Boer, Reidel Publ. Co, Dordrecht, p. 333

Pigafetta, A., Transylvania, M. of, Corrêa, G.: 1962, *"Magellan's Voyage around the World"*, ed. C.E. Nowell, Northwestern Univ. Press, Evanston, p. 127

Price, S.D., Walker, R.G., 1976, *"The AFGL Four Color Infrared Sky Survey: Catalog of Observations at 4.2, 11.0, 19.8 and 27.4 μm"*, AFGL–TR–76–0208, USAF Geophys. Lab., Hanscomb AFB, Mass. (USA)

Rice, W., Lonsdale, R.J., Soifer, B.T., Neugebauer, G., Kopan, E.L., Lloyd, L.A., de Jong, T., Habing, H.J.: 1988, *Astrophys. J. Suppl.* **68**, 91

Schwering, P.B.W.: 1988, Ph.D. Thesis, University of Leiden (NL)

Schwering, P.B.W., Israel, F.P.: 1989, *Astron. Astrophys. Suppl.* **79**, 79

Schwering, P.B.W.: 1989, *Astron. Astrophys. Suppl.* **79**, 105

INFRARED SOURCE

IDENTIFICATION TABLES

2. Infrared Source Identification Tables

Several infrared sources listed in the Catalogue (Part 3) have optical nebular counterparts. Hα emission nebulae (HII regions) and dark clouds are frequently associated with dusty regions emitting primarily at far–infrared wavelengths (60 and 100 microns). We have cross–correlated the Infrared Catalogue with the catalogue of Hα nebulosities by Davies *et al.* (1976) and the dark cloud catalogues by Hodge (1972, 1974). Identifications and their quality (+ : good; 0 : doubtful; – : poor or absent) are given in Tables 2.1 and 2.2.

Because of the distance of the Clouds, and the intrinsic luminosity and spectrum of the objects, only few of the Catalogue sources can reasonably be expected to be the infrared counterpart of stellar, or semi–stellar objects. We have nevertheless performed a cross–correlation of the Catalogue with the following published catalogues of such objects:

Stars:	Smithsonian Astrophysical Observatory Star Catalogue
	(SAO, 1966; version 1984)
	Radcliffe Magellanic Clouds stars (Feast *et al.*, 1960)
Star Clusters:	Kron (1956)
	Lindsay (1958)
	Westerlund and Glaspey (1971)
	Hodge and Wright (1974)
	Brück (1976)
	Hodge (1986)
	van den Bergh (1981)
Planetary Nebulae:	Sanduleak *et al.* (1978)
	Sanduleak and Pesch (1981)
	Sanduleak (1984)
Supernova Remnants:	Mathewson *et al.* (1983)

In Table 2.3 we have listed only those objects which have good infrared counterparts. Because of confusion in the infrared maps, some objects not listed may have infrared emission above the nominal detection limit. In general, for objects not listed infrared upper limits may be obtained from the Atlas maps with the aid of the transparent overlays.

A more detailed discussion of infrared source identifications in the Magellanic Clouds can be found in Schwering and Israel (1989) and Schwering (1989).

References

van den Bergh, S.: 1981, *Astron. Astrophys. Suppl.* **46**, 79

Brück, M.T.: 1976, *Occ. Rep. R. Obs. Edinburgh* **1**

Davies, R.D., Elliott, K.H., Meaburn, J.: 1976, *Mem. R. Astron. Soc.* **81**, 89 (DEM)

Feast, M.W., Thackeray, A.D., Wesselink, A.J.: 1960, *Mon. Not. R. Aston. Soc.* **121**, 337

Hodge, P.W.: 1972, *Publ. Astron. Soc. Pac.* **84**, 365

Hodge, P.W.: 1974, *Publ. Astron. Soc. Pac.* **86**, 263

Hodge, P.W., Wright, F.W.: 1974, *Astron. J.* **79**, 858

Hodge, P.W.: 1986, *Publ. Astron. Soc. Pac.* **98**, 1113

Kron, G.E.: 1956, *Publ. Astron. Soc. Pac.* **68**, 125

Lindsay, E.M.: 1958, *Mon. Not. R. Astron. Soc.* **118**, 172

Mathewson, D.S., Ford, V.L., Dopita, M.A., Tuohy, I.R., Long, K.S., Helfand, D.J.: 1983, *Astrophys. J. Suppl.* **51**, 345

Meaburn, J.: 1980, *Mon. Not. R. Astron. Soc.* **192**, 365

Sanduleak, S.N., MacDonnell, D.J., Davis Philip, A.G.: 1978, *Publ. Astron. Soc. Pac.* **90**, 621

Sanduleak, N., Pesch, P.: 1981, *Publ. Astron. Soc. Pac.* **93**, 431

Sanduleak, N.: 1984, in *"Structure and Evolution of the Magellanic Clouds"*, IAU Symposium 108, eds. S. van den Bergh, K.S. de Boer, Reidel Publ. Co., Dordrecht, p. 231

Schwering, P.B.W., Israel, F.P.: 1989, *Astron. Astrophys. Suppl.* **79**, 79

Schwering, P.B.W.: 1989, *Astron. Astrophys. Suppl.* **79**, 105

Shapley, H., Lindsay, E.M.: 1963, *Irish Astron. J.* **6**, 74

Smithsonian Astrophysical Observatory Star Catalog (4 vols.): 1966, Washington D.C., Smithsonian Institution (version 1984) USA

Westerlund, B.E., Glaspey, J.: 1971, *Astron. Astrophys.* **10**, 1

Table SMC-2.1: Infrared emission from H-alpha Emission Nebulae in the SMC (DEM, 1976).

(1) DEM Id	(2) Henize Id	(3) Intensity of H-alpha	(4) Infrared LI-SMC	(5) Detection quality	(6) Remarks
1	N 3	b	239	+	Not on DPM-map; On Co-added map only.
2		f	6	+	
3		b		0	On edge of IR extended emission.
4		b		0	On edge of weak IR extended emission.
5		vf	10	+	Offset 2'N. (Def: Offset=Pos(H-alpha)-Pos(IR)).
6		f+b	15	+	Offset 1'N.
7		b	16	+	
8		b		-	On edge of IR extended emission.
9	N 9	b	23	+	
10		f		+	Filament along edge of IR extended emission.
11	N10	b	26	+	Offset 1'N.
12		vf	24	0	Center H-alpha shell 3'NE.
13		f+b	26	+	LI-SMC 26 on N extension of H-alpha nebula; No IR peak on main body.
14		b	27	+	Offset 2'N.
15	N12 (part)	b	30	+	Offset 1'NW.
16	N13A,B	vb	29	+	
17		f+b	28	+	
18	N12B	b	30	0	Offset 2'N.
19		vf	31	+	IR coincides with southern part of nebula.
20	N12 (part)	b		0	On edge of LI-SMC 36.
21	N16	f+b	35	+	IR source on bright part of shell.
22		fb		+	Shell lining edge of IR 'hole'.
23	N12A	vb	36	+	
24		b		-	2'N of LI-SMC 35.
25	N17	b	38	+	
26		b		-	
27		fb		-	Near edge of LI-SMC 36.
28	N15	b		-	On edge of LI-SMC 29.
29		f	44,37	0	Delineating edge of extended IR emission, LI-SMC 44 coincides with E peak.
30		b		-	
31	N19	b		-	On edge of LI-SMC 49.
32		vb	43	0	IR follows nebular outline.
33		f	46	+	Nebulae on edge of IR emission.
34		fb		+	IR confused area: DEM 34, 35 and 36.
35		fb	42	+	Peak of LI-SMC 42 in between.
36		b		+	IR confused area: DEM 34, 35 and 36.
37	N22	b	45	+	IR peak LI-SMC 45 in between 37 and 38.
38	N25,N26	vb		+	IR peak LI-SMC 45 in between 37 and 38.
39		b		0	Coincides with IR 'hole'.
40	N27	b	49	+	
41		b		-	
42	N24	vb		-	On IR gradient (to LI-SMC 45 and 42).
43		b		-	On IR gradient (to LI-SMC 45).
44		f	52,56,57	+	IR peaks on East part of H-alpha nebulae.
45	N30	vb	51	+	Offset 3'NE.

Table SMC-2.1: Infrared emission from H-alpha Emission Nebulae in the SMC (continued 2).

(1)	(2)	(3)	(4)	(5)	(6)
DEM Id	Henize Id	Intensity of H-alpha	Infrared LI-SMC	Detection quality	Remarks
46		f	48	+	IR on NW part of nebula.
47		f		-	On IR gradient (to LI-SMC 63).
48	N32	b	58	+	
49		f+b	54	+	Shell borders on IR extended emission.
50		b		-	On edge of extended emission.
51		vf	59	+	On extended emission.
52		b	68	+	Offset 2'W.
53		f	70	+	IR source in S of H-alpha nebula only.
54	N36,N41	vb	65	+	Shell coincides with extended IR source.
55	N37	vb	66	+	
56		b	62	+	
57	N35	b		0	On N edge of LI-SMC 62.
58		vf		-	On edge IR extended emission, in IR-hole.
59		fb	73	+	
60	N45	b	76	+	
61		b		0	On edge of IR extended emission.
62	N46	b	77	+	
63		fb	78	0	Edge of extended emission.
64		vf		0	Bright SE component coincides with IR peak.
65		fb		+	Extended IR emission peak of LI-SMC 81.
66		fb	75,79,80,87	+	IR complex at edge of extended emission.
67		fb	84	+	Center shell offset 2'W.
68	N50	b		0	On edge of LI-SMC 92,94 IR complex/pair.
69		fb		+	Weak IR emission.
70		fb		0	Edge of IR emission.
71	N48	b		0	On edge of LI-SMC 86.
72	N51	b	86	+	
73		fb		0	3'NE of LI-SMC 86.
74		vf	88,98	+	SE peak coincides with IR peak 98, N with 88.
75		vb		-	
76		fb	93	-	Offset 2'E.
77	N52A,B	b	94	+	
78		b		+	Weak IR emission.
79		vf		0	Edge of IR emission.
80		fb	105,106	+	Main source coincides with IR 'hole', bright part with LI-SMC 105.
81		f		-	Edge of IR extended emission.
82		b		0	On edge of LI-SMC 111.
83	N57	b		0	On edge of LI-SMC 110.
84		fb		0	On edge of LI-SMC 110.
85	N58	b	110	+	
86		fb	109	+	
87		f	114	0	
88		f	111	+	
89		vf	117	+	Follows extended IR emission, close to IR 111.
90		f		+	Shell follows edge of IR extended emission.

Table SMC-2.1: Infrared emission from H-alpha Emission Nebulae in the SMC (continued 3).

(1)	(2)	(3)	(4)	(5)	(6)
DEM Id	Henize Id	Intensity of H-alpha	Infrared LI-SMC	Detection quality	Remarks
91	N59	b	119	+	Offset 1'N.
92		b		-	
93	N62	vb		+	Offset 3'SW of LI-SMC 124.
94	N63	vb	124	+	
95	N64A	vb	128	0	LI-SMC 128 in between DEM 94,95.
96		f		0	Edge of extended IR emission.
97		f		-	
98		fb		+	On IR gradient (to LI-SMC 131).
99		fb	133	+	On IR gradient (to LI-SMC 131).
100	N69	b		0	Extended IR emission.
101		f		-	
102		b		-	Edge of LI-SMC 131.
103	N66A,B,C,D	vb	131,135,137	+	LI-SMC 131 covers only SW part of very large nebula; 135 and 137 are in centre of NE part.
104		vf		-	In extended emission.
105		f		-	On edge IR extended emission.
106	N69	b		+	Weak IR emission.
107		b	138	+	
108		f	136	+	Weak extended IR source.
109	N71	vb	142	+	
110		b	144	+	
111		fb	147	+	Offset 3'SW.
112	N72	b	146	+	Offset 2'SW.
113		fb	148	+	Offset 2'SW.
114		f		+	Nebular filament delineates edge of IR 'bay'.
115	N74	b		-	On edge of LI-SMC 152.
116	N75	b		-	
117a,b	N77B,A	b,vb	152	+	DEM 117b coincides with LI-SMC 152.
118		b	153	+	LI-SMC 153 lies on arc DEM 118.
119		b	156	+	1'W of LI-SMC 156.
120	N76B	vb	156	+	1'S of LI-SMC 156.
121		b		-	Edge of LI-SMC 152.
122		f		-	On edge IR extended emission.
123	N76A	vb	160,161, 162,163	+	Complex.
124		fb		0	Shell around H-alpha corresponds to IR emission.
125		f	169	+	Confused. H-alpha follows IR emission.
126	N78A,B	vb	168	+	Offset 2'S.
127	N78D	vb	168	+	Offset 1'E.
128		f	170	+	
129		fb	173	+	Offset 1'N.
130	N78C	vb		+	Weak IR peak.
131		b		0	Edge of weak IR peak.
132		fb		-	IR extended emission.
133		f	179	+	
134		fb	174,178,181	+	Complex.
135	N80	b	184	+	LI-SMC 184 lies on brighter E part.

Table SMC-2.1: Infrared emission from H-alpha Emission Nebulae in the SMC (continued 4).

(1)	(2)	(3)	(4)	(5)	(6)
DEM Id	Henize Id	Intensity of H-alpha	Infrared LI-SMC	Detection quality	Remarks
136		f		-	Edge of extended IR emission.
137		fb	189	+	Offset 1'W.
138	N81	f+b	187	+	Offset 3'W.
139		vf	188	+	IR peak N of S-part; IR extended emission follows main source.
140		fb	190	+	Weak extended IR peak.
141		b	194	+	
142		b	193	+	3'SE of LI-SMC 193.
143		vf		0	Weak IR extended level.
144		fb	195	+	Shell at offset 2'W.
145		vf		+	Weak IR extended emission.
146		f+b		+	Weak IR peak.
147	N83A,C	vb		+	On extension of LI-SMC 199.
148	N83B	vb	199	+	
149	N84C	vb	200	+	
150	N84 (part)	b		0	Edge of LI-SMC 199, 200.
151	N84A	vb	201	+	Confused.
152	N84B,D	vb	202	+	Confused.
153		f		0	Weak extended IR emission.
154		fb	203	+	Offset 3'S.
155		b	204,205	+	
156		b		0	On extension of LI-SMC 204/205.
157		f		0	IR follows H-alpha filaments only partially. LI-SMC 206 on S filament.
158		fb	207	+	Offset 1'S.
159		b		-	Edge of IR complex.
160		fb	214	+	IR between H-alpha peaks.
161	N88	vb	215	+	
162		f		0	Weak IR emission.
163		f	216,219	0	Confused.
164	N89	b	218	+	Shell around IR peak.
165		vf		-	On edge IR extended emission.
166	N90	vb	242	+	Not on IRAS DPM-map; on Co-added map.
167		f	240,241, 245,246	+	Not on IRAS DPM-map; on Co-added map. Super giant shell SMC-1 (Meaburn, 1980), with 32' diameter. IR follows H-alpha along shell, but in the North 8'S of H-alpha.

Table SMC-2.2: Infrared emission from Hodge (1974) dark clouds in the SMC.

(1)	(2)	(3)	(4)
Number	Infrared LI-SMC	Detection quality	Remarks
1	30	+	1.8'S of LI-SMC 30; On infrared extension in SW Bar.
2	31	+	0.6' NW of LI-SMC 31; In SW Bar.
3	29	+	0.6'E of 29, 0.7'W of 32; In SW Bar.
4	35	+	1.1'W of LI-SMC 35; In SW Bar.
5		+	Infrared emission in SW Bar extended emission.
6		0	1.9'N of LI-SMC 35; On infrared gradient in SW Bar.
7	35	+	0.9'N of 35; In SW Bar.
8		0	On infrared gradient in SW Bar.
9	38	+	0.5'NE of LI-SMC 38; On infrared peak in SW Bar.
10	36	+	1.3'E of 36, Infrared extension towards peak 43; In SW Bar.
11	42	+	1.5'NW of LI-SMC 42; In SW Bar.
12	45	+	0.9'NW of 45; In SW Bar.
13	45	+	1.5'N of LI-SMC 45; In SW Bar.
14		+	1.7'E of 42; On infrared gradient to 42 and 45 in SW Bar.
15	44	0	1.0'E of 44. Weak infrared emission, on edge of SW Bar.
16		0	Weak infrared emission in Bar.
17		0	On infrared gradient to LI-SMC 51; In SW Bar.
18		+	1.1'SE of LI-SMC 58. Infrared extension near 63; In Bar.
19	63	+	1.3'NW of LI-SMC 63; In Bar.
20		+	Weak infrared extension on edge of SW Bar.
21		0	On infrared gradient SE of LI-SMC 62; In Bar.
22		0	On infrared gradient to 63 and 65 in Bar. 1.8'N of 65.
23		0	On infrared gradient to 63 and 65 in Bar. 2.7'NE of 65.
24	74	0	2.2'NW of LI-SMC 74; In extended emission of Bar.
25		0	On infrared extended emission around LI-SMC 63.
26	78	0	2.2'S of LI-SMC 78; On infrared gradient in Bar.
27		0	1.9'N of 78; On infrared gradient in Bar.
28	81	+	0.8'W of LI-SMC 81; At the edge of the Bar.
29		0	Near edge of infrared extended emission of SW Bar.
30	88	0	1.3'NE of 88. On edge of infrared extended emission of Bar.
31		–	Infrared minimum, in extended emission of Bar.
32		0	On edge of infrared extended emission; In SW Bar.
33		0	On edge of infrared extended emission; In SW Bar.
34		0	Near edge of LI-SMC 111, in extended emission of Bar.
35	108	+	1.1'S of LI-SMC 108; At the edge of the Bar.
36		0	On edge of extended emission of Bar.
37	130	+	1.0'SW of LI-SMC 130; In Bar.
38		–	7'S of N66, 3.6'SW of LI-SMC 133; In extended emission of Bar.
39		–	Infrared minimum in extended emission of Bar.
40		–	Infrared minimum in extended emission of Bar.
41		0	On extension to LI-SMC 153, edge of Bar.
42		0	Extension of extended emission in Wing.
43		–	On edge along 153, falls in infrared minimum at edge of Bar.
44		0	7'S of N76 in extended emission in NW Bar. In between LI-SMC 156, 160, 163.
45	200	+	Strong infrared emission peak of N83 of Wing; 1.0'E of 200, 1.8'E of 199, 1.9'NW of LI-SMC 201.

Table SMC-2.3: Infrared emission from SMC stellar objects.

(1)	(2)	(3)	(4)
Catalog Number	Infrared LI-SMC	Detection quality	Remarks

SAO stars .... SAO (1966).

255653	220	0	On edge of Co-added map.
255684	3	+	Faint emission at 60 μm.
255686	4	+	
255689	7	0	
255690	8	0	
255693		+	
255711	56	+	In extended emission.
255713	69	0	
255715	89	+	In extended emission.
255716	90	+	
255721	104	+	
255729	141	+	Close to extended emission.
255730	145	+	Close to extended emission.
255735	167	+	
255738	171	+	
255745		+	
255751	196	+	
255752		+	2'W of LI-SMC 198.
255767		+	
255786		+	

Star clusters .... Kron (1956), Lindsay (1958).

K 35		+	L54, NGC330. In Bar.
K 58	182	+	L85, NGC419. 1.4'N of LI-SMC 182.

Planetary nebulae .... Sanduleak et al. (1978), Saduleak and Pesch (1981).

Sk 5		0	In weak extended emission at edge of SW Bar.
Sk 6	12	0	N 6. 0.5'N of LI-SMC 12. In weak extended emission of SW-Bar.
Sk 11	50	0	N29. 0.3'N of LI-SMC 50. In SW Bar.
Sk 17		0	Weak infrared emission, far North from SMC Bar.
Sk 19	91	0	0.2'SE of LI-SMC 91. In SW Bar.
Sk 27		0	In weak extended emission of N88.

Table LMC-2.1: Infrared emission from H-alpha Emission Nebulae in the LMC (DEM, 1976).

(1)	(2)	(3)	(4)	(5)	(6)
DEM Id	Henize Id	Intensity of H-alpha	Infrared LI-LMC	Detection quality	Remarks
1		f		0	5'W of LI-LMC 40; on edge of infrared emission.
2	N 77D	b	46	+	Knot.
3	N 76	b	48	+	Knot 2'N of infrared source 48.
4	N 77E	b+f		+	Individual spots of DEM 4 on infrared peak 58.
4a	N 77C,B	b	69	+	
4b		b	58,63	+	DEM 4b is the brightest infrared part of DEM 4.
4c	N 77A	fb		+	On infrared gradient to 58.
5		f		−	On infrared gradient to source 46 (see DEM 2).
6	N 79	b	55,56, 62,67, 70	0	Extended filaments on infrared emission.
7	N 3	f	65	+	IR 65 on N-part, diffuse filaments follow IR.
8a	N 4F	b	102	+	Irregular shell; DEM 8a and 8b on LI-LMC 102.
8b	N 4A	vb	102	+	Knot; both DEM 8a and 8b on infrared peak 102.
8c	N 4B	vb		0	Two knots on IR gradient 4'N of 102.
9	N 79A,B	vf	110	+	
10	N 79A-E	vb		+	Nuclear structure and outer envelope.
10a		b	103	+	Knot DEM 10a is brightest DEM 10 spot in IR.
10b		b	107,111	+	Nuclear structure.
11	N 5	b	112	+	4'SE of infrared source LI-LMC 106.
12	N 4D,E	fb	132,146	+	Structure on IR complex; adjacent to DEM 27.
13	N 8A	b+f	122,129	+	Envelope follows IR; bright nucleus on 122, 129.
14		f	114	+	2'NE of 114.
15a		vb	117	+	Around infrared peak 121; 117 on S-part.
15b		b	121	+	On infrared peak 121.
16		vf		−	6'SE of LI-LMC 124; on infrared filament.
17		vf		−	In infrared hole, between sources 108 and 128.
18		vf		−	In infrared hole, between sources 123 and 137.
19		vf	145,147	+	IR sources on SE part of shell-like structure.
20	N 7	vb	135	+	Knot.
21		f	139	+	LI-LMC 139 corresponds to central DEM 21 blob.
22	N 83	b	148,162, 173,193	+	
22a		vb	148	+	
22b	N 83A	vb	148	+	
22c	N 83B	vb	173	+	
22d		vb	193	+	Infrared source 193 in between filaments.
23	N 87	vb	164	+	Knot close to SAO 249145.
24	N 90	vb	199	+	Knot on LI-LMC 199 in IR extended emission.
25	N185	b	131,159	+	Shell on edge of sources LI-LMC 131 and 159.
26	N 80	f+b	151	+	LI-LMC 151 coincides with bright knot on W-side.
27		vf	144,146, 185,186	0	Diffuse filaments follow IR emission. Extension of DEM 12. Bright peak LI-LMC 144 at N-side.
28		f	161,163	+	
29a	N 6	f		−	4'N of 144. In infrared extended emission.
29b		f		−	7'N of 144. In infrared extended emission.
30		f		−	Shell in IR extended emission; towards 166.

Table LMC-2.1: Infrared emission from H-alpha Emission Nebulae in the LMC (continued 2).

(1)	(2)	(3)	(4)	(5)	(6)
DEM Id	Henize Id	Intensity of H-alpha	Infrared LI-LMC	Detection quality	Remarks
31	N 9	b	166	+	IR follows blob structure, not the shell.
32	N 84	b	194	+	Diffuse shell.
33	N 86	b	191	+	At W-side of irregular filamentary shell.
34	N10,N11A-L	vb	(see Remarks)	+	Follows IR complex very well; related LI-LMC: 177,179,190,192,195,205,206,210,214,217,219, 222,226,229,243,248,259,278,284; also 244, 277,283 on separate knots close to N11.
34a	N 11L	b		0	Shell and jet.
35	N 10 (part)	fb		0	5'E of 178. On edge of broad IR peak consisting of LI-LMC 174, 176 and 178.
36	N 94A,B,C	fb	167,175, 183,213, 242,249	+	Irregular filaments on complex N11. IR peaks on brighter H-alpha part.
37	N 93	fb	230	+	Two knots in faint nebulosity.
38	N 92A,B	b	232	+	Infrared source on nucleus of DEM 38.
39	N 91A,B	vb	235,246, 262,275, 282,301	+	Filaments follow IR emission. LI-LMC 235 on nucleus at W-side.
40		f	240	+	2'S of LI-LMC 240.
41	N 11E,K	vb	251	+	Knot with extension to NW.
42	N 12A	b	266	+	Shell.
43		f	313	+	Loop follows IR emission; LI-LMC 313 on S-part.
44		f	268	0	Diffuse nebula on IR extended emission near N11.
45	N 16A	b+f		-	Shell 6'N of LI-LMC 292.
46	N 14	b	290	+	Diffuse shell 3'E of 290; edge of N11 complex.
47		b		+	Shell 3'NE of 289; at edge of infrared peak.
48	N 13	f	305,309, 316	+	Infrared sources on S-part of filaments.
49	N 15	f+b	307	+	Knot and envelope 3'E of 298.
50	N186C,D,E	b	295,317	+	Shell around IR peak 295; S-filaments on peak 317, along infrared edge.
51	N186B	vb	302	+	Knot.
52		f	315,318	+	
53		f		0	In between LI-LMC 330 and 337, on IR filament.
54		f	340	+	2'E of LI-LMC 340.
55		f	300,319, 343,358	+	Follows infrared emission. LI-LMC 300 at W-edge.
56		f	334	+	LI-LMC 334 on W-part of filaments; edge of IR.
57		f	345	+	
58		f	348,365	+	In Bar; 4'N of infrared sources 348 and 365.
59	N 17A,B	b+f	371,375, 383	+	Bright nucleus coincides with LI-LMC 371.
60		f	342	+	Diffuse nebula 4'SW on IR gradient to 342.
61		vf	381	0	Infrared peak LI-LMC 381 on E-filaments.
62		f	373,389, 390,401	+	Infrared peaks at edges of nebula structure.
63	N190	b+f	399	+	3'N of LI-LMC 399.
64a	N191B	b	409	0	Knot DEM 64a on infrared gradient, 3'NW of 409.
64b	N191A	b	409	+	Infrared source 409 coincides with knot DEM 64b.
65	N 21	b	404,416	+	Arc in Bar.

Table LMC-2.1: Infrared emission from H-alpha Emission Nebulae in the LMC (continued 3).

(1)	(2)	(3)	(4)	(5)	(6)
DEM Id	Henize Id	Intensity of H-alpha	Infrared LI-LMC	Detection quality	Remarks
66	N 23A	b	405,411	+	Shell core and curved filaments in Bar.
67	N189	b	412	+	Near centre of DEM 68.
68		f	398,417, 440	0	Diffuse filaments follow infrared structure.
69	N 20	b	414	+	
70		fb	419	+	Diffuse core and envelope in Bar.
71	in N 23	f	423	0	Circular ring in Bar.
72		fb	431	+	Sharp filaments 4'N of LI-LMC 431.
73	in N 23	f+b	429,435	+	In Bar.
74	in N 23	f+b	447	+	Core and diffuse envelope in Bar.
75		f	450,469, 479,484	0	Faint arcs on IR complex, but not followed well.
76	N100	fb	442,457	0	IR around filaments; central part in IR hole.
77		f		0	Diffuse arcs on edge of IR emission.
78		f	454	+	In Bar.
79		f+b	461,473	+	Shell outline in Bar.
80		f	448,488, 491,499	+ +	Filaments around LI-LMC 448 at W-side; on LI-LMC 488, 491, 499 at E-side.
81		f		-	Arc 4'N of infrared source LI-LMC 489.
82		f	504	0	3'S of LI-LMC 489; on infrared peak.
83		vf		-	4'SE of LI-LMC 476; at edge of IR extension.
84	N103B	b	475	0	Bright nucleus and shell in Bar, on IR gradient to LI-LMC 510.
85	N103A	b	520	+	Knot in Bar.
86	N105A	vb	534	+	Bright core and outer filament in Bar.
87		f+b	573	+	Knot in Bar; on infrared gradient to peak 534.
88	N104A	f+b	514,546	+	In Bar; on infrared gradient to peak LI-LMC 514
89		b	525	+	Irregular shell, diffuse filaments.
90		f	543	+	Shell on infrared peak.
91		vf	538	+	
92	N108	f		-	Diffuse shell in Bar; on edge of LI-LMC 551.
93	N 26	b	553	+	Knot.
94	N 27	b	564	+	Knot.
95		f		-	Knot in infrared extended emission of Bar.
96		f	586	+	Knot in Bar.
97		vf		-	Diffuse filaments on IR gradient to 599 and 594.
98		f	592	-	Shell of diffuse filaments follows IR emission; LI-LMC 592 at W-side.
99		f		-	Shell in infrared minimum.
100	N193B,E,D	b	610	+	The three knots cannot be separated on the IRAS DPM-maps; near DEM 101.
101a	N193C	b	620	+	Knot. The components of DEM 101 cannot be separated on IRAS DPM-maps.
101b	N193A	b	620	+	Knot; DEM 101 is brighter than the close nebulae DEM 100, 102.
102		f	619	+	Diffuse blob near DEM 101.
103		f		0	Diffuse nebula on infrared extension in Bar.
104	N113A-F	vb	635	+	In Bar.
105	N 30A	b	643,646	0	Shell outlines IR hole; LI-LMC 643,646 in South.

Table LMC-2.1: Infrared emission from H-alpha Emission Nebulae in the LMC (continued 4).

(1)	(2)	(3)	(4)	(5)	(6)
DEM Id	Henize Id	Intensity of H-alpha	Infrared LI-LMC	Detection quality	Remarks
106	N 30B,C,D	b	632,642	+	Cores and semicircular shell around 632 and 642.
107		b+f	641	+	Near SAO 249225 (= LI-LMC 639).
108	N113 (North)	b	640	+	Core and envelope in Bar; 7'N of DEM 104.
109	N112	b	636	+	Knot in Bar.
110		f		-	Irregular filaments outlining an IR minimum. A knot (10'E of DEM 110) corresponds to 591.
111		vf		0	Diffuse nebula on IR extension to LI-LMC 626.
112		vf		-	Diffuse filaments on gradient to IR complex 642.
113	N114A	b	665,668, 669,695	+	Arc and filaments follow IR emission in Bar.
114	N115	b		-	Knot 6'N of 660; at edge of infrared emission.
115		vf		+	Diffuse patch corresponds to IR extension.
116	N 31	b+f	667	0	Core and envelope 3'SE in IR extension of 667.
117		vf		+	On infrared extension.
118	N 34A,B,C	f	691	+	Diffuse nebula.
119		f	677	+	Irregular filaments.
120		f		0	On edge of extension in Bar.
121	N 33	vb	723	+	Knot.
122		f	734	+	In Bar.
123		f+b	716,736, 745,749, 758,759	+	Filaments on IR structure of sources in Bar.
124	N116	b		0	Knot at edge of infrared extension in Bar.
125	N 35	fb	748	+	Circular outline.
126		f	729,774	+	Extended loop around DEM 125 follows IR.
127		f		0	Semicircular diffuse nebula follows extended IR.
128		f	757	+	Diffuse nebula in Bar; 2'N of LI-LMC 757.
129	N 36	b+f	753	+	Central knot corresponds to IR source 753.
130		fb	742,754	+	Diffuse ring on infrared complex in Bar.
131	N195A,B	b	766	+	Nebula contains several neclei.
132a	N119 (part)	b	789	+	Shell on infrared complex in Bar.
132b	N119 (part)	b	807,811	+	In Bar. Apparently associated with DEM 132a.
133	N121	fb		-	In Bar, at edge of infrared emission complex.
134	N120A-D	b	804,816, 785,794	+	Diffuse partial shell on infrared peaks in Bar.
135		fb	817	+	IR 817 at N-side; S-filaments follow IR edge.
136	N 37	fb	823	+	
137		b		0	Arc and sharp filaments in infared emission.
137a		vb	851	+	Knot in DEM 137 on infrared emission peak 851.
138	N 38	b	831	+	Knot on SE-side of nebula. Nebula 3'N of 831.
139		vf		0	Diffuse nebula follows infrared edge. Near SAO 249253 (= LI-LMC 814).
140		fb	841	0	Shell-like filaments in IR complex 880 (DEM152).
141	N 41	b		0	On extension in infrared complex 880 (DEM 152).
142		b	878	+	Shell with sharp filaments on IR emission peak.
143		f	847	+	Two knots (clusters?) 3'S of LI-LMC 847.
144		f		0	On extension in infrared complex 880 (DEM 152).
145	N127B	b	872	+	Knot in Bar; close to DEM 149; 2'W of 872.

Table LMC-2.1: Infrared emission from H-alpha Emission Nebulae in the LMC (continued 5).

(1)	(2)	(3)	(4)	(5)	(6)
DEM Id	Henize Id	Intensity of H-alpha	Infrared LI-LMC	Detection quality	Remarks
146	N197	vb	864	+	
147		f	866	0	In Bar; 4'N of LI-LMC 866.
148		f	866	0	In Bar; 4'E of infrared source LI-LMC 866.
149	N127A	b	872	+	Knot in Bar; in between DEM 145 and 153.
150	N 44J	vb	855	+	Knot.
151		fb	839	0	Filamentary loop in IR complex 880 (DEM 152).
151a	N 44F	vb	861	+	
152	N 44B,C	vb	876,880, 887	+	Elliptical ring at centre of IR complex 880.
153		fb	872	+	Knot close to DEM 149; in Bar 2'E of 872.
154		fb	830,846, 852	0	Arc and filaments resemble infrared emission.
155	N 43	fb	932	+	Filaments correspond to 932; surrounding 895.
155a		b	895	+	Diffuse circular region corresponds to 895.
156	N 44I	vb	894	0	Knot 3'N of 894 (close to complex 880, DEM 152).
157	N128	f	891	0	Diffuse nebula on infrared extension.
158	N 44G,K	b	888	0	Close to IR complex 880 (DEM 152); LI-LMC 888 in between two knots.
159		fb		0	Diffuse shell on IR complex 880 (DEM 152).
160	N 44D,H	vb	911,925	+	Several nuclei near IR complex 880 (DEM 152).
161	N 45	fb	900	+	On infrared peak LI-LMC 900.
162	N 46	b	919	+	Knot.
163	N130	b	916	+	Knot.
164	N200	fb	858,934, 960	0	Shell on infrared extended emission.
165	N198	b	910,929, 966,985, 988	+	Filament around LI-LMC 929 and 910.
166a	N 44L	b	925	+	Knot; DEM 166 not separable; close to DEM 160.
166b	N 44E	b	925	+	Knot; DEM 166 not separable; close to DEM 160.
167	N 44N	b	938,946	+	Irregular filamentary shell and knot on infrared complex LI-LMC 880 (DEM 152).
168	N131	fb	921	0	Knot in Bar; 2'N of infrared source 921.
169	N 44M	b	943	+	Knot near infrared complex 880 (DEM 152).
170		f	948	+	4'SE 948; near IR complex 880 (DEM 152).
171	N132J,G	b		0	In Bar; on infrared gradient to LI-LMC 955.
172	N132A	fb	941	+	Diffuse arc in Bar.
173	N132B,C,E,I	fb	955	+	Four (non-separable) knots on IR peak 955.
174	N138D,B	b	961	+	Diffuse shell.
175		fb	952	+	On infrared gradient to complex 1015 (DEM 189).
175a	N 48E	fb	983	+	Infrared source on S-part of elliptical shell.
176a	N137A	b		0	Knot on infrared extension related to DEM 177.
176b	N137	fb		0	Knot on infrared extension related to DEM 177.
177	N140	fb	(see Remarks)	+	Extended filaments follow IR; related LI-LMC: 927,944,947,958,965,968,975,984,987,1005.
178		fb	977	+	2'S of infrared source LI-LMC 977.
179	N138C	b	978	+	Knot.
180	N138A	vb	999	+	Knot and arc.

Table LMC-2.1: Infrared emission from H-alpha Emission Nebulae in the LMC (continued 6).

(1) DEM Id	(2) Henize Id	(3) Intensity of H-alpha	(4) Infrared LI-LMC	(5) Detection quality	(6) Remarks
181		fb	996	+	Loop of filaments on IR, N of 1015 (DEM 189).
182		f	962,964, 1000,1073	+	Extended filaments around source 1073; peak in LI-LMC 962, 964, 1000.
183	N 48D	b		-	Knot on S-gradient to IR complex 1015 (DEM 189).
184		f	1101	0	IR source at N-edge of shell; shell follows IR.
185		vf	1042	0	Diffuse shell 3'W of 1042.
186	N132D,H	f		0	Knot in envelope on infrared gradient to 976.
187	N142	fb	1024	0	LI-LMC 1024 on N-part of diffuse nebula.
188	N202	vb	1016	+	Knot on infrared source 1016.
189	N 48A,B,C	f+b	1015	+	Infrared complex LI-LMC 1015.
189a	N 48B	b	1002,1009	+	The S-side of DEM 189.
189b	N 48C	b	1010,1033	+	The N-side of DEM 189.
190	N 49	vb	1022,1038	+	Knots N of infrared complex 1015 (DEM 189).
191		fb	1020	+	N of IR complex 1015 (DEM 189). 3'E of 1020.
192	N 51D	b	1003,1034, 1037,1045	0	Shell and bright filaments on IR complex LI-LMC 1090 (DEM 205).
193	N 50	fb	1017	+	Diffuse nebula.
194		f	989	+	
195		vf		0	Extended diffuse filaments on edge of infrared complex 1015 (DEM 189).
196	N 51B,E	b	1043,1052	+	Near infrared complex LI-LMC 1090 (DEM 205).
197	N143	vb	1065	0	Infrared on W-side of core and envelope nebula.
198		fb	1047,1078	0	Irregular filaments on infrared structure.
199	N144A,B	vb	1049,1064, 1076	+	Shell of filaments DEM 199 on infrared peak. Near SAO 249279.
200		f	1054	+	Infrared peak 1054 on S-blob of diffuse nebula; N-side of nebula is connected to DEM 182.
201	N 51A,C	vb	1080,1083, 1090	+	On South side of irregular nebulosity; N-blob without infrared emission.
202	N205C	b	1058,1094	+	Blob on N-side; filaments at S-side along IR. Close to DEM 206, 207.
203		fb	1023,1040, 1041,1063, 1066,1074 1129,1161	+	Filamentary shell follows IR emission.
204		b		-	Shell at edge of extended infrared emission.
205		b	1102,1090	+	Shell; infrared complex LI-LMC 1090.
206	N205A	fb	1086	+	Diffuse nebula on broad IR peak formed by 1086, 1094, 1106; close to DEM 202, 207.
207	N205A	fb	1106	+	Diffuse nebula on broad IR peak formed by 1086, 1094, 1106; close to DEM 202, 206.
208	N204	b	1098,1101, 1140,1158	+	IR peak 1098 on W-filaments; filamentary and diffuse shell follows infrared emission.
209		f+b	1099,1111	+	Diffuse and filamentary nebula follows IR.
210		f+b	1100,1110, 1118,1127, 1139,1145, 1152	+	Scattered nebulosity on infrared emission.

Table LMC-2.1: Infrared emission from H-alpha Emission Nebulae in the LMC (continued 7).

(1)	(2)	(3)	(4)	(5)	(6)
DEM Id	Henize Id	Intensity of H-alpha	Infrared LI-LMC	Detection quality	Remarks
211		vf	1093,1134, 1162	-	Centre of diffuse filaments on LI-LMC 1112.
212		vf	1131,1136, 1149,1165, 1170	0	Infrared peak LI-LMC 1136 at S-side of filament.
213		vf		-	In infrared minimum; 4'N of LI-LMC 1148.
214		f+b	1137,1141, 1176	+	Diffuse filaments follow infrared emission.
215		f	1168	+	Infrared source 1168 in centre of diffuse ring.
216		f+b	1180	0	Diffuse nebula on IR emission; 3'E of 1180.
217		vf		0	Diffuse nebula follows IR extended emission at E-side; near LI-LMC 1193.
218		fb	1166,1204, 1211	+	Diffuse filaments on infrared extension near complex LI-LMC 1249 (DEM 227).
219		f		+	Diffuse nebula on IR near complex 1274 (DEM 229).
220		f	1235	+	Blob on LI-LMC 1235; filaments along edge of IR.
221	N206A,B	f+b	(see Remarks)	+	Nebular complex follows IR. Related IR LI-LMC: 1179,1182,1183,1199,1201,1206,1211,1224,1246, 1251.
222a		f	1207	+	
222b		f	1214	+	
223		f		0	Semicircular arc on infrared filament.
224		f	1181,1192, 0 1205,1220, 1223,1265	Network	of filaments follows infrared structure.
225		b	1244	+	Knot.
226	N148I,G	b	1279	+	Filament with diffuse circle near complex LI-LMC 1249 (DEM 227).
227	N148A-F,H	b	(see Remarks)	+	Infrared complex 1249. Related LI-LMC: 1213, 1233,1247,1249,1255,1284,1293. Filaments on IR gradient. Blob at E-side on LI-LMC 1318.
228	N 55A	vb	1268,1273	+	Bright IR blob; complex nebular structure follows IR emission.
228a		vb	1253	+	Knot is extension at NW of infrared blob.
228b		vb	1256	+	Knot at N-side of infrared blob.
229	N 57A,D,E	vb	1219,1259, 1261,1274, 1292	+	Network of filaments; infrared complex LI-LMC 1274. IR peak on main nebula. LI-LMC 1261 on separate S-knot.
230		vf		-	Near infrared complex 1274 (DEM 229).
231	N 57C	b	1292	-	Diffuse ring on IR gradient to 1274 (DEM 229).
232		b	(see Remarks)	+	Network of fine filaments follows IR structure. Related LI-LMC: 1209,1218,1222,1225,1257,1275, 1277,1299,1300,1336,1357.
233	N150	b	1323	+	Two knots in DEM 233.
234	N 57,N 58	f+b	1282,1288, 1296,1305, 1315,1320, 1331	0	Diffuse filaments on infrared emission.
235	N 62B	b	1303,1326	0	Arc of fine filaments on infrared peaks.

Table LMC-2.1: Infrared emission from H-alpha Emission Nebulae in the LMC (continued 8).

(1)	(2)	(3)	(4)	(5)	(6)
DEM Id	Henize Id	Intensity of H-alpha	Infrared LI-LMC	Detection quality	Remarks
236		f		-	Diffuse nebula on extension in IR emission.
237		vf		-	Diffuse nebula on edge of LI-LMC 1376 (DEM 243).
238		f		-	Shell in infrared hole.
239	N 61 , N 62A (part)	f+b	1337,1351	+	Complex of filaments follows infrared emission.
240		b		-	Knot and envelope at edge on infrared peak.
241	N 59A,B,C	vb	1367,1392	+	Infrared peaks on nebular blobs.
242		b	1359	+	Knot on peak 1359. Filamentary structure follows infrared emission.
243	N 63A	vb	1354,1376	+	IR complex 1376; LI-LMC 1354 on N-filament.
244	N 61 (part)	f	1352,1386	0	Diffuse filaments following extended IR.
245		fb	1364,1379	+	Extended filaments; E-peak contains infrared filaments; W-side in infrared hole.
246	N154A,B	vb	1298,1341, 1353,1358, 1406	+	Complex in Greater Doradus Region (GDR).
247		f	1400	+	Diffuse loop follows IR emission; close to complex LI-LMC 1376 (DEM 243).
248		vb	1403,1425	0	Several knots on IR gradient in GDR.
249		vf	1411	0	Diffuse nebula.
250		fb		0	Diffuse shell on gradient to LI-LMC 1367 (related to DEM 241).
251	N 64A,B	f	1402,1408	+	IR along LI-LMC 1422 follows diffuse DEM 251.
252	N 64C	vb	1422,1426	+	IR peaks on central blob of complex DEM 252.
253		vb		0	Knot on infrared gradient to DEM 252.
254		vf		-	Diffuse shell on IR complex 1422 (DEM 252); close to DEM 253.
255	N 65	fb	1409,1412	0	Elliptical shell on IR sources; no clear resemblance.
256		f	1430	0	Shell on infrared gradient to complex DEM 252.
257		vf	(see Remarks)	0	Diffuse filaments follow IR structure. Related LI-LMC: 1348,1382,1396,1399,1410,1415,1420, 1423,1440,1454,1457,1492.
258	N 68	vb	1431	+	Knot.
259		f	1407	0	Diffuse circular patch 2'S of LI-LMC 1407. In Greater Doradus region.
260	N155	b	1436	+	Knot in Greater Doradus region.
261		vb	1413,1417, 1447,1470	+	Part directly South of 30 Doradus nebulosity.
262		vb	1441	0	Part of 30 Doradus. Filament in IR minimum.
263	N157A,B	vb	(see Remarks)	+	Centre of 30 Doradus nebulosity; related LI-LMC: 1342,1361,1370,1383,1384,1388,1429,1433,1448, 1458,1467,1469,1472,1482,1493,1494.
264		f		-	Diffuse nebula on IR gradient 4'N of 1444.
265	N213	vb	1471	+	Nucleus and envelope.

Table LMC-2.1: Infrared emission from H-alpha Emission Nebulae in the LMC (continued 9).

(1)	(2)	(3)	(4)	(5)	(6)
DEM Id	Henize Id	Intensity of H-alpha	Infrared LI-LMC	Detection quality	Remarks
266		f	1464	0	Diffuse nebula 3'E of 1464; on IR extension in Greater Doradus Region.
267	N171A,B	f	1486	+	4'W of bright infrared peak LI-LMC 1497.
268		f (see Remarks)		+	Filamentary structure on IR; related LI-LMC: 1368,1381,1413,1416,1443,1446,1450,1453, 1468,1481,1527.
269	N158A,B,C	vb	1483,1490, 1507,1522	+	Complex structure in Greater Doradus Region.
270		vf	1524	0	Source LI-LMC 1524 at E-side; diffuse shell follows IR at edge of emission.
271	N159A-K	vb	1501,1518	+	Several knots; in Greater Doradus Region.
272	N159L	b		-	In GDR; shell on IR gradient to 1501,1518.
273	N161	b	1519	+	In GDR; knot near complex 1469 (DEM 263).
274	N214A,B	b	1505	+	Several knots; close to DEM 276, 278.
275	N172	vb	1506	+	Knot at S-edge of Greater Doradus Region.
276	N214D	b	1521	+	Diffuse arc 1'N of 1521; close to DEM 274, 278.
277	N173	b		-	Knot at S-edge of Greater Doradus region.
278	N214E	b	1521	0	On IR gradient to DEM 276; knot 1'S of 1521.
279		b	1523	0	Knot 2'S of 1523; in Greater Doradus region.
280	N176	vb	1541	+	Knot on IR peak 1541; at S-edge of GDR.
281	N175	b	1544	+	Diffuse ring at S-edge of GDR.
282		b	1560	0	Two knots at S-edge of GDR.
283	N160F	b	1549	+	Several diffuse knots on IR gradient to DEM 271; in Greater Doradus Region.
284	N160A-E	vb (see Remarks)		+	Part of 30 Doradus nebulosity; related LI-LMC: 1503,1520,1525,1548,1552,1564,1570.
285		b	1561	+	Complex at S-edge of Greater Doradus Region.
286		f	1568	0	5'W of 1568; E- and S-filaments follow IR.
287	N214 (part)	vf	1532	0	Diffuse nebula on infrared gradient to 1521.
288		f		-	Diffuse nebula on IR close to LI-LMC 1521.
289	N214G,F	b	1572	+	Two knots.
290	N216	b	1557	+	Knot.
291		f	1550	+	
292	N214H	f	1571	+	Ring of nebulosity.
293	N214C	vb	1577	+	Close to this complex two LI-LMCs are related to nebular emission: 1600 on a filament NE and 1529 on filaments SW of DEM 293.
294		f	1551	+	Diffuse nebula.
295	N177	f	1567	0	At S-edge of Greater Doradus Region.
296		f		0	Diffuse nebula at edge of IR extended emission.
297		b	1573	+	Two arcs; in Greater Doradus Region.
298	N164	vb	1586,1594	+	Complex; in Greater Doradus Region.
299	N165	f+b	1597	+	Shell with second shell on NW; in GDR.
300	N163	vb	1609	+	Core and envelope; at edge of GDR.
301	N 70	b	1605	+	Shell with fine filaments.
302		vb	1613	+	Knot on infrared peak LI-LMC 1613.
303	N 71	vb	1602	+	Knot.
304	N 72	b+f	1629	+	Group of filaments; at E-edge of GDR.
305	N 73	f	1621	+	Knot.

Table LMC-2.1: Infrared emission from H-alpha Emission Nebulae in the LMC (continued 10).

(1)	(2)	(3)	(4)	(5)	(6)
DEM Id	Henize Id	Intensity of H-alpha	Infrared LI-LMC	Detection quality	Remarks
306		fb	(see Remarks)	+	Irregular shell follows IR emission. Related LI-LMC: 1591,1614,1634,1635,1671,1674.
307	N167	b	1633	+	Diffuse shell on IR gradient to LI-LMC 1643.
308		f	1650,1655, 1697	+	Diffuse filaments follow infrared emission.
309	N 74A,B	b	1667,1673	+	Filaments and shell in SE on IR complex.
310		fb	(see Remarks)	+	Network of filaments (E of 30 Doradus) follows IR outline; connected to GDR. Related LI-LMC: 1658,1662,1671,1677,1685,1693,1694,1700,1706, 1707,1729.
311	N168	vb	1675,1692	+	Diffuse shell on LI-LMC 1675; 1692 at N-tip.
312	N169C	b		0	Diffuse nebula on IR extension in GDR.
313	N169B	vb		0	Knot on infrared extension in GDR.
314	N169A	b	1696	+	Knot on infrared peak LI-LMC 1696.
315	N 74 (part)	f	1698	+	Shell on infrared extension.
316		fb		-	Two diffuse shells on IR gradient of GDR.
317		vf		0	Diffuse nebula 4'S of 1704; on IR extension.
318	N179C	b	1733	+	Knot on LI-LMC 1733. DEM 318-321 non-separable.
319	N179B	b	1733	+	Knot on LI-LMC 1733. DEM 318-321 non-separable.
320	N179A,D	vb	1733	+	Knot on LI-LMC 1733. DEM 318-321 non-separable.
321		f	1733	+	On LI-LMC 1733. DEM 318-321 non-separable.
322	N180C	vb	1740	+	Diffuse nebula near IR complex 1744 (DEM 323).
323	N180A,B	vb	1743,1744	+	Circular filaments; IR complex 1744.
324		f		-	Diffuse nebula on edge on IR extended emission.
325		b		0	Diffuse shell N of DEM 323, on IR gradient to LI-LMC 1744.
326	N180	fb	1741,1749	+	Shell surrounds DEM 323 and follows infrared outline around infrared complex 1744.
327		vf	1747	+	Diffuse nebula on edge of IR extended emission.
328		vf	1753,1778	+	IR follows arc nebula. At E-end of arc: LI-LMC 1791 also in nebulae.
329		f	1767	+	Galaxy?

Table LMC-2.2: Infrared emission from Hodge (1972) dark clouds in the LMC.

(1)	(2)	(3)	(4)
Number	Infrared-Id LI-LMC	Detection quality	Remarks
1		0	On infrared extension in Bar (2.7'E of 438; 2.9'NW of 465).
2	478	+	0.4'E of LI-LMC 478. Infrared peak in LMC Bar.
3	530	+	1.9'SW of 530. On infrared emission in Bar, peaking at 502.
4	541,536	+	1.4'W of 541; 1.8'N of 536. On infrared peak in Bar.
5		+	On infared gradient at 2.6'E of peak LI-LMC 582.
6	591	+	0.7'SE of LI-LMC 591. Infrared peak in Bar. See DEM 110.
7		−	Infrared minimum in Bar (2.8'N of 608).
8	601	+	1.6'NE of LI-LMC 601. On infrared gradient to 634.
9		+	On infrared gradient at 2.8'W of peak 635.
10	659	0	1.7'S of LI-LMC 659. On infrared extension in Bar.
11	660	+	0.7'SE of 660. In infrared peak, close to Bar. See DEM 114.
12	728	0	1.6'SE of LI-LMC 728. At edge of infrared hole.
13	743	+	2.0'N of LI-LMC 743. On infrared extension in Bar.
14	768	+	0.5'S of LI-LMC 768. On infrared extension in Bar.
15		0	At edge of extension in Bar (2.5'N of 795; 2.7'NE of 786).
16	825	+	1.8'NW of LI-LMC 825. Close to infrared peak in Bar.
17		+	On infrared gradient of LI-LMC 816 in Bar.
18	854	+	1.5'W of LI-LMC 854. Infrared peak in Bar.
19		+	On infrared gradient at 2.8'SE of LI-LMC 866.
20		+	(2.9'W of LI-LMC 897).
21		+	(4.4'NW of LI-LMC 897).
22	876	+	0.2'SE of 876. IR peak close to main peak of complex. DEM 152.
23	899	+	0.6'W of LI-LMC 899. Infrared peak in Bar.
24		+	On infrared gradient in between peaks.
25	984,965	0	1.2'NW of 984; 1.4'S of 965. On IR peak in Bar. See DEM 177.
26	1070	+	2.0'NE of LI-LMC 1070. On infrared gradient to 1076.
27		+	In between infrared peaks: 2.3'NW of 1080; 2.7'SW of 1083.
28		+	On infrared gradient at 3.4'SE of LI-LMC 1076.
29	1115	0	1.1'SE of LI-LMC 1115. At SE end of Bar.
30	1142	0	1.9'N of LI-LMC 1142. At SE end of Bar.
31	1171	0	1.8'SW of LI-LMC 1171. At SE end of Bar.
32		+	On infrared extension at SE end of Bar (2.7'SE of 1180).
33	1195	0	1.1'E of LI-LMC 1195. On infrared extension near Bar.
34		0	In infrared extended emission, near GDR (3.3'W of 1218).
35	1222,1209	0	1.5'NW of 1222; 1.8'N of 1209. IR extension to 30 Doradus. See DEM 232.
36		0	IR from edge of 30 Doradus (2.6'SE of 1217; 2.7'NW of 1252).
37		0	IR from edge of 30 Doradus (2.1'N of 1252; 2.5'NW of 1270).
38		0	In infrared extended emission, near GDR.
39		−	At edge of infrared extended emission, near GDR.
40	1332	0	2.0'S of LI-LMC 1232.
41	1335	+	1.5'NE of LI-LMC 1335. At infrared edge of GDR.
42		+	On extension to infrared peak LI-LMC 1323.
43		0	At edge of Greater Doradus region (2.4'NW of 1374).
44		+	In Greater Doradus region (2.6'SE of 1361).
45		+	On infrared extension in Greater Doradus region (GDR).

Table LMC-2.2: Infrared emission from Hodge (1972) dark clouds in the LMC (continued 2).

(1)	(2)	(3)	(4)
Number	Infrared-Id LI-LMC	Detection quality	Remarks
46		+	On infrared extension in GDR (2.6'SW of 1433).
47		+	On infrared extension in GDR (2.6'NW of 1436).
48	1435	+	1.4'W of LI-LMC 1435. On infrared extension in GDR.
49		0	In Greater Doradus region.
50	1446	+	0.8'E of LI-LMC 1446. On infrared extension in GDR. See DEM 268.
51	1467	+	1.4'W of LI-LMC 1467. On infrared extension in GDR. See DEM 263.
52		+	In Greater Doradus region (2.8'SE of 1451).
53		+	Infrared peak in Greater Doradus region (2.6'N of 1486).
54	1487	0	1.7'NE of 1487. On infrared extension South of N159, in GDR.
55		0	In infrared extended emission of GDR (3.0'NE of 1483).
56		0	In 30 Doradus outskirts.
57	1525	+	0.7'S of LI-LMC 1525. Infrared peak in GDR. See DEM 284.
58	1518	+	1.8'SE of LI-LMC 1518. Infrared peak in GDR. See DEM 271.
59	1539	+	0.7'NE of LI-LMC 1539. In GDR.
60		0	In 30 Doradus outskirts.
61	1553	+	1.7'SE of LI-LMC 1553. In GDR.
62		+	On infrared extension in GDR (2.6'W of 1588).
63	1586	+	1.4'N of LI-LMC 1586. Infrared peak in GDR. See DEM 298.
64	1629	0	1.7'S of LI-LMC 1629. Infrared peak in GDR. See DEM 304.
65		+	On infrared extension to 1643 (2.3'S of 1662).
66	1671	+	1.1'S of LI-LMC 1671. On IR extension to 1643. See DEM 306.
67		+	On infrared extension to 1696 (2.9'NW of 1692).
68	1741	+	1.7'NW of LI-LMC 1741. On IR gradient to 1743. See DEM 326.

Table LMC-2.3: Infrared emission from LMC stellar objects.

(1)	(2)	(3)	(4)
Catalog Number	Infrared LI-LMC	Detection quality	Remarks

SAO stars .... SAO (1966).

249044	1828	+	
249050	1831	+	
249051	1833	+	
249055	1835	+	
249071	1847	+	
249074	1848	+	
249075	1851	+	
249084	7	+	
249099	19	+	
249100	20	+	
249104	1864	+	
249109	25	+	
249123	78	+	Faint 60 μm emission.
249125	87	+	
249126	93	+	
249138	130	+	
249150	1870	+	
249165	1872	+	
249172	327	+	
249182	1876	+	
249201	500	+	
249212	532	+	
249217	585	+	Also 60 μm emission present.
249218	1866	+	Near edge of IRAS DPM-map.
249225	639	+	Near DEM 107.
249241	750	+	
249242	761	+	
249253	814	+	Near DEM 139.
249260	893	+	
249281	1081	+	Also faint 60 μm emission.
249286	1117	+	
249293	1135	+	
249316	1346	+	
249320	1395	+	
249322	1418	+	0.3'S of LI-LMC 1418, 0.8'NE of LI-LMC 1414.
249329	1469:	0	30 Doradus centre (DEM 263); also 60 and 100 μm. 0.7'NE of LI-LMC 1469.
249334	1558	+	
249336	1565	+	
249337	1575	+	
249339	1587	+	
249341	1618	+	Faint, probably unrelated 60 and 100 μm emission.
249346	1644	+	Faint 60 μm emission.
249361	1728	+	
249362	1891	+	
249363	1731	+	

Table LMC-2.3: Infrared emission from LMC stellar objects (continued 2).

(1)	(2)	(3)	(4)
Catalog Number	Infrared LI-LMC	Detection quality	Remarks
249368	1748	+	
249372	1752	+	
249396	1788	+	
249398	1789	+	
249408	1804	+	
256122	14	+	
256123	17	0	Near edge of IRAS DPM-map.
256129	30	+	
256139	142	+	
256146	233	+	
256154	359	+	
256158	424	+	
256161	470	+	
256169	670	+	
256173	762	+	Probably unrelated 100 μm emission.
256174	763	+	Near edge of IRAS DPM-map.
256178	838	+	
256187	1075	+	
256207	1380	+	
256221	1530	+	
256237	1725	+	
256239	1736	+	Near edge of IRAS DPM-map.
256257	1805	+	
256260	1814	+	
256261	1816	+	

Radcliffe stars .... Feast et al. (1960)

R 66	225	+	S73. 0.4'SW of LI-LMC 225; dust-shell emission.
R 71	346	+	S155. 0.2'NW of LI-LMC 346; dust-shell emission.
R 76	423	+	S171 (DEM 71). Visually brightest known LMC member. 0.5'SE of LI-LMC 423. Only 25 and 60 μm emission.

Star clusters .... van den Bergh (1981), Shapley and Lindsay (1963).

NGC 1978	1126	+	SL501. 1.4'W of LI-LMC 1126.
NGC 2002	1186	+	SL517. 1.2'NW of LI-LMC 1186.

Supernova remnants .... Mathewson et al. (1983).

0505-67.9	423	+	DEM 71; 0.3'NE of LI-LMC 423.
0509-68.7	510	+	N103B; 1.0'N of LI-LMC 510.
0525-66.1	1022	+	N49 (DEM 190); 0.5'SW of LI-LMC 1022.
0535-66.0	1376	+	N63A (DEM 243); 0.2'NW of LI-LMC 1376.
0538-69.1	1448	+	N157B (DEM 263); 0.9'N of LI-LMC 1448. In extended infrared emission in 30 Doradus outskirts.

Table LMC-2.3: Infrared emission from LMC stellar objects (continued 3).

(1)	(2)	(3)	(4)
Catalog Number	Infrared LI-LMC	Detection quality	Remarks

Planetary nebulae .... Sanduleak et al. (1978), Sanduleak (1984).

Sk 1	2	0	N182. Near edge of DPM-map, 1.3'NE of LI-LMC 2.
Sk 11	89	0	0.7'N of LI-LMC 89.
Sk 28	481	0	In IR extended emission of LMC-Bar. 0.4'SE of LI-LMC 481.
Sk 29	483	0	N102. In LMC-Bar. 0.6'NE of LI-LMC 483.
Sk 31	513	0	N25. 0.5'NW of LI-LMC 513.
Sk 36	557	0	N107. 0.4'SW of LI-LMC 557.
Sk 51	850	0	N125. In extended emission; 0.5'E of LI-LMC 850.
Sk 58	982	0	N133. In extended emission. 0.4'N of LI-LMC 982.
Sk 62	1014	0	N201. 0.5'N of LI-LMC 1014.
Sk 64	1100	0	N145 (DEM 210). 0.8'S of LI-LMC 1100.
Sk 75	1327	0	N151. 0.7'E of LI-LMC 1327.
Sk 78	1336	0	N153 (DEM 232). 1.2'N of LI-LMC 1336.
Sk 85	1513	0	N69. 0.3'N of LI-LMC 1513.

THE LEIDEN–IRAS

MAGELLANIC CLOUDS

INFRARED SOURCE CATALOGUES

3.   The Leiden–IRAS Magellanic Clouds Infrared Source Catalogues

The LMC and SMC infrared source catalogues presented in this Part contain sources determined by a visual inspection of the two–dimensional DPM and Co–Add maps presented in the Atlas (Part 4). By comparison, the official IRAS Point Source Catalogue (PSC – see IRAS, 1989a) was derived from template fitting to individual one–dimensional survey scans. This method is unreliable in confused areas, such as the Clouds, mainly resulting in a higher degree of incompleteness. We have searched the different map sets in all four bands for both resolved and unresolved sources down to four times the median noise level (*i.e.* 5, 2, 1 and 0.5 x 10^{-8} Watt m^{-2} sr^{-1} at 12, 25, 60 and 100 microns respectively). At any given wavelength, derived fluxes were roughly equal (dispersion about 15 per cent). The source lists thus obtained were merged, first over all map sets, then over all bands. Whenever an unambiguous identification with a PSC entry could be made, PSC positions are listed as these are the more accurate. For each source, we first list the peak intensity as well as the background emission level near the source. This is followed by a size estimate, based on the observed size and the nominal gaussian resolution in each band. Finally, 'true' source flux densities are given, taking into account source size and source background level. A more detailed discussion of the Catalogue source lists is given by Schwering (1988, 1989) and Schwering and Israel (1989).

The Leiden IRAS – Magellanic Clouds Catalogues contain the following information:

Column 1: Sequential number. We recommend the name LI–SMC for the SMC sources and LI–LMC for the LMC sources.

Column 2: The position of the source (1950.0). Right Ascension is given in hours, minutes and seconds. If given in 0.1^s the position and error are that of the IRAS PSC entry given in Column 12, otherwise they are taken from the maps with an error of 12^s. Declination is given in degrees, arc–minutes and arc–seconds. If given in 1$''$ the position and error are that of the IRAS PSC entry given in Column 12, otherwise they are taken from the DPM or Co–Add maps and have an error of 1$'$. The positional errors in the IRAS PSC, have been estimated with the use of stellar infrared observations at ESO–La Silla (Chile) and of the positions of SAO stars. The errors are approximately $10'' - 15''$. The errors in positions of sources directly taken from the infrared maps are approximately $1'$.

Column 3: 12 μm intensity peak and background level (in 10^{-8} Watt m^{-2} sr^{-1}). A dash indicates that the source is below the intensity search level.

Column 4: The same as Column 3 for the 25 μm band.

Column 5: The same as Column 3 for the 60 μm band.

Column 6: The same as Column 3 for the 100 μm band.

Column 7: The size of the source in $\alpha(') \times \delta(')$. p denotes a source that cannot be discerned from a point source response. A colon denotes an uncertain size. A dash indicates that no reliable size could be estimated because no FWHM could be determined. The size in these cases often agrees with a point source. The error in the source size is inversely proportional to the size itself, but no numerical value is derived.

Column 8: Flux density of the source at 12 μm in Jy, assuming an intrinsic source spectrum $f_\nu \propto \nu^{-1}$ (the same as for the released IRAS products; IRAS, 1989a); see also Column 14. C denotes confusion with either background or other discrete sources and a colon indicates an uncertain flux density. Flux densities are calculated using the size of Column 7. When the size could not be measured the point source response was assumed. In section 4 the Point Source Conversion Factors are given to convert (Peak–Background) Watt m^{-2} sr^{-1} to Jy. The error in flux densities is approximately 10 – 20 %, and depends on the intensity peak and background level. The flux densities have been obtained from maps produced with the November 1984 IRAS calibration (of the official IRAS products).

Column 9: The same as Column 8 for the 25 μm band.

Column 10: The same as Column 8 for the 60 μm band.

Column 11: The same as Column 8 for the 100 μm band.

Column 12: If the source is present in the IRAS Point Source Catalog (PSC; IRAS, 1989a) and/or in the IRAS Small Scale Structure Catalogue (SSS; IRAS, 1989b) its PSC/SSS designation is given. A colon indicates that the associated PSC or SSS entry is at a large distance from the source. An asterisk indicates that the association has been made for more than one source in the Catalogue.

Column 13: The fields of the Cloud in which the object is present. Only the two most important field numbers are given. Field numbers of the DPM–fields and of the Co–Added fields are given.

Column 14: The spectral type of the source. Spectrum type C is a 'cool dust' spectrum (typical $T_d \approx 30$ K), peaking beyond 100 μm; colour correction factors are of order 1.00, 0.95, 0.99, 1.00. Type W is a 'warm dust' spectrum (typical $T_d \approx 70$ K), peaking between 12 and 100 μm; colour correction factors are of order 1.03, 1.00, 1.00, 1.04. Type S is a 'stellar spectrum' (typical blackbody of about 5000 K); colour correction factors are of order 1.43, 1.40, 1.32, 1.09. A semi–colon indicates that the infrared spectrum is uncertain. Actual, colour-corrected flux densities can be calculated from the quoted ones by *dividing* the latter by these colour correction factors (see IRAS, 1989a).

The Atlas section (Part 4) contains, for each LMC and SMC field, transparent overlays listing positions (marked by +) and Catalogue source numbers allowing immediate comparison of map structures and Catalogue entries. As these overlays also contain the positions of all SAO stars in the field (marked by *), comparison of infrared sources with optical images is likewise possible.

References

IRAS: 1989a, Infrared Astronomical Satellite (IRAS) Catalogs and Atlases, Volume 1, Explanatory Supplement, eds. C.A. Beichman, G. Neugebauer, H.J. Habing, P.E. Clegg and T.J. Chester, NASA RP–1190, Washington, DC (USA).

IRAS: 1989b, Infrared Astronomical Satellite (IRAS) Catalogs and Atlases, Volume 7, The Small Scale Structure Catalog, eds. C.A. Beichman, G. Neugebauer, H.J. Habing, P.E. Clegg and T.J. Chester, NASA RP–1190, Washington, DC (USA).

Infrared Sources in the SMC, Based on IRAS Additional DPM Observations.

Peak / Bg columns (3)–(6) are in units of 1.0E-8 Watt / (m·m sr).

(1) Number LI-SMC	(2) RA(1950) h m s	(2) DEC(1950) ° ' "	(3) 12μm Peak	(3) Bg	(4) 25μm Peak	(4) Bg	(5) 60μm Peak	(5) Bg	(6) 100μm Peak	(6) Bg	(7) Size arcmin	(8) F 12μm Jy	(9) F 25μm Jy	(10) F 60μm Jy	(11) F 100μm Jy	(12) IRAS-Id	(13) DPM field	(14) Spectrum
1	00 33 04.3	-73 25 06	-	-	-	-	1	1	1	-	4x9	-	-	1.9	5.7	00330-7325	4	C
																X0032-734		
2	00 33 43.2	-73 37 49	2	-	4	-	2	-	1	0.5	p	0.07:	0.44	0.8	1.0	00337-7337	4	C
3	00 33 47.8	-74 09 09	23	-	2	-	-	-	-	-	p	0.85	0.22	-	C	00337-7409	4	S
4	00 34 04.0	-73 08 03	12	-	-	-	-	-	-	-	p	0.44	-	-	-	00340-7308	4	S
5	00 35 04.2	-74 36 17	8	-	2	-	-	-	-	-	p	0.30	0.22	-	-	00350-7436	4	S
6	00 35 10.5	-73 16 19	-	-	1	-	4	1	2	1	4x2	-	0.36:	2.5	3.0	00351-7316	4	C
7	00 36 24	-74 14	3	-	-	-	-	-	-	-	-	0.11:	-	-	-		4	S
8	00 36 44	-73 25	3	-	-	-	<3	-	-	-	-	0.11:	-	-	C		4	C
9	00 37 46.8	-73 18 55	-	-	-	-	4	2	4	2	-	-	-	0.8	4.2	00377-7318	4	C
10	00 38 55.4	-73 53 40	-	-	-	-	2	1	2	1	-	-	-	0.4	2.1	00389-7353	4	C
11	00 39 33.5	-73 17 35	-	-	1	-	8	4	-	-	p	-	0.11:	1.7	C	00395-7317	4	M
12	00 39 33.7	-74 03 45	-	-	2	-	1	-	-	-	-	-	0.22:	0.4:	C	00395-7403	4	M
13	00 40 00	-73 58	5	-	-	-	<2	-	-	-	-	0.19	-	-	C		4	S
14	00 40 16.3	-73 47 27	-	-	2	-	7	4	4	3	p	-	0.22	1.2	2.1	00402-7347	4	C
15	00 40 20.7	-73 16 28	-	-	2	-	13	6	9	6	p	-	0.22	2.9	6.3	00403-7316	4	C
16	00 40 25.8	-74 00 47	-	-	-	-	5	2	2	1	p:	-	-	1.2	2.1	00404-7400	4	C
17	00 40 39.3	-74 45 25	-	-	-	-	3	-	10	0.3	p	-	-	1.2	1.5	00406-7445	4	C
18	00 40 42	-73 31	-	-	-	-	11	6	10	5	-	-	-	2.1	10.0		4	C
19	00 41 01.0	-73 39 45	3	-	-	-	-	-	-	-	-	0.11:	-	C	C	00410-7339	4	S
20	00 41 10.0	-73 36 35	3	-	-	-	-	-	-	-	-	0.11:	-	C	C	00411-7336	4	S
21	00 41 20.6	-73 16 38	3	-	6	-	18	6	12	5	pxl	0.14:	0.80	5.3	15.0	00413-7316	4	C
																X0040-732		
22	00 41 45.2	-74 00 29	12	-	-	-	-	-	-	-	-	0.44	-	C	C	00417-7400	4	S
23	00 41 46.3	-73 18 34	3	-	2	-	14	11	-	-	p	0.11:	0.22	1.2	C	00417-7318	4	M
24	00 42 51.1	-74 17 36	-	-	-	-	2	-	2	0.5	14x14	-	-	14.0	11.0	00428-7417	4	M
25	00 42 59.9	-73 13 58	-	-	6	-	13	9	-	-	p	-	0.67	1.7	C	00429-7313	4	M
26	00 43 03.7	-73 26 45	15	-	30	3	60	20	-	-	p	0.41	3.00	17.0	C	00430-7326	4	M
27	00 43 13.8	-73 32 52	12	-	15	3	72	20	46	22	p	0.26	1.33	21.0	50.0	00432-7332	4	C
28	00 43 30	-73 29	14	-	9	5	72	30	45	22	p	0.30	0.44	17.0	48.0:		4	C
29	00 43 32.2	-73 39 10	22	-	21	4	88	20	48	22	p:	0.63	1.89	28.0	54.0	00435-7339	4	C
30	00 43 37.1	-73 21 32	9	-	10	2	55	20	35	22	p:	0.22	0.89	14.0	27.0	00436-7321	4	C

Type	4/2	ID					Pat									Dec	RA	No.
N	4	00437-7334	C	C	0.22	0.11	-	-	-	-	-	4	6	6	9	-73 34 33	00 43 47.1	31
S	4		C	C	0.22	0.33	p:	-	-	-	-	4	6	6	17	-73 39	00 43 51	32
S	2		-	-	-	0.41	p:	-	<4	-	<3	-	-	-	11	-72 52	00 44 32	33
S	4		C	-	-	0.33:	p	-	-	-	<2	-	-	-	9	-74 08	00 44 36	34
C	4	00446-7339	46.0	19.0	1.60	0.60	1x2	22	42	30	65	4	12	5	13	-73 39 02	00 44 38.5	35
C	4	00447-7322	42.0	21.0	1.78	0.52	p	40	60	50	100	5	21	6	20	-73 22 29	00 44 47.0	36
N	4		C	2.1	0.22	0.22	-	-	-	20	25	2	4	3	9	-73 44	00 44 51	37
C	4	00449-7347	30.0	13.0	1.40	0.51	1x2	10	23	12	36	1	8	-	7	-73 47 35	00 44 55.0	38
S:	4		C	C	0.53	0.57	2xp:	-	-	-	-	3	6	4	13	-73 39	00 45 09	39
S	2		-	-	-	0.30	p:	-	<4	-	<3	3	-	-	8	-72 57	00 45 36	40
N	4	00456-7354	C	0.8:	-	0.19:	pxl	-	-	6	8	-	-	-	5	-73 54 38	00 45 38.1	41
N	4	X0045-735	C	11.0	2.79	0.64	p:	60	70	125	125	7	28	7	21	-73 34	00 46 00	42
C	4	00462-7339	21.0	10.0	0.44	0.19	-	-	-	100	125	11	15	10	15	-73 24	00 46 12	43
N	4	00462-7331	C	4.1	0.11:	0.11:	p	22	83	25	35	1	2	-	3	-73 39 56	00 46 15.6	44
C	4		128.0	56.0	9.77	1.07		-	-	25	160	7	95	5	34	-73 31 37	00 46 17.3	45
N	4	00463-7352	C	0.8	-	-	-	-	-	-	9	-	-	-	-	-73 52 11	00 46 21.7	46
C	2	00463-7238	20.0	10.0	-	-	5x4	4	9	7	13	-	-	-	-	-72 38 22	00 46 23.8	47
S	4/2	00466-7322	C	C	0.11:	0.19	-	-	-	5	-	-	1	-	5	-73 01	00 46 34	48
C	4	X0046-733	149.0	52.0	2.78	0.67	p	22	93	30	155	8	33	5	23	-73 22 10	00 46 37.6	49
N	4	00467-7314	4.2	4.5	1.00	0.19	p	25	27	25	36	2	11	-	5	-73 14 30	00 46 47.2	50
C	4	00472-7325:	C	21.0	1.33	0.33	p:	45	63	50	100	8	20	5	14	-73 26	00 46 54	51
N	4	00469-7341:	38.0	2.1:	0.56	0.11:	p:	-	-	15	20	2	7	-	3	-73 43	00 47 06	52
N	4	00474-7350	C	0.4:	0.11:	0.11:	-	-	-	8	9	-	1	-	3	-73 50 07	00 47 26.1	53
S	4	00474-7330	C	C	C	0.19:	-	-	-	-	-	-	-	3	8	-73 30 45	00 47 26.9	54
N	4	00477-7328:	C	10.0	0.78	0.19:	p	-	-	30	55	5	12	-	5	-73 27	00 47 30	55
S	4	00477-7343	C	C	-	0.44	p	12	25	16	50	-	-	-	12	-73 45	00 47 37	56
C	4	00478-7305	27.0	14.0	1.11	0.19	p	-	-	30	50	2	12	5	5	-73 43 04	00 47 42.8	57
N	4		C	8.3	1.22	0.19:	p:	-	-	8	36	2	13	-	5	-73 05 08	00 47 53.2	58
N	4		C	2.5	0.11:	0.19:	-	-	<3	30	<3	1	3	3	5	-73 19	00 47 57	59
S	2		-	-	-	0.19:	-	-	-	-	-	-	-	-	5	-72 25	00 48 03	60
N:	4	00483-7347	C	0.9:	0.53	0.78	1xp	-	-	8	10	1	5	-	17	-73 47 48	00 48 22.1	61
C	2	00483-7250	19.0	12.0	0.87:	-	3x4	7	13	9	21	1	2	-	-	-72 50 11	00 48 23.9	62
C	4		80.0	61.0	1.31	0.83	3x4	18	43	25	88	7	10	-	5	-73 09	00 48 25	63
C	2	00486-7237	4.2	1.7	-	0.11:	p	4	6	6	10	-	-	-	3	-72 37 39	00 48 39.5	64
N	4	X0048-731	C	19.0	0.89	0.44	p	-	-	25	70	6	14	4	16	-73 08	00 48 45	65
N	4/2	00489-7302	C	4.1	1.11	0.19:	p	-	-	30	40	3	13	-	5	-73 02 59	00 48 57.5	66
C	2	00489-7235	6.8	1.0	-	-	2xp	4	7	9	11	-	-	-	-	-72 35 48	00 48 59.8	67
C	4	00489-7336	4.3	2.8	0.15:	-	1x1	14	16	18	24	2	3	-	3	-73 36 26	00 49 00.0	68
S	2	00490-7125	-	-	-	0.11:	-	-	-	17	-	-	-	-	-	-71 25 36	00 49 00.3	69
N	4	00491-7340	2.1	2.1	0.22	0.11:	p:	15	16	17	22	2	4	-	3	-73 40 54	00 49 07.3	70

Infrared Sources in the SMC (continued 2).

Peak/Bg surface brightness columns (3)–(6) in units of 1.0E-8 Watt / (m*m sr).

(1) Number LI-SMC	(2) RA(1950) h m s	(2) DEC(1950) ° ′ ″	(3) 12 μm Peak	(3) 12 μm Bg	(4) 25 μm Peak	(4) 25 μm Bg	(5) 60 μm Peak	(5) 60 μm Bg	(6) 100 μm Peak	(6) 100 μm Bg	(7) Size arcmin	(8) F 12μm Jy	(9) F 25μm Jy	(10) F 60μm Jy	(11) F 100μm Jy	(12) IRAS-Id	(13) DPM field	(14) Spectrum
71	00 49 07.4	-72 46 43	-	-	2	-	15	11	10	8	1x2	-	0.40	2.2	4.6	00491-7246	2	C
72	00 49 18	-73 27	5	-	4	2	20	15	-	-	-	0.19	0.22	C	C		4	M
73	00 49 28.5	-73 47 29	3	-	2	-	-	-	16	12	p:	0.11:	0.22	2.1	8.4	00494-7347	4	C
74	00 49 30	-73 00	5	-	3	1	-	-	-	-	p:	0.19	0.22:	C	C		2	M:
75	00 49 35.7	-72 16 23	-	-	-	-	7	3	6	3	4x5	-	-	5.0	12.0	00495-7216	2	C
76	00 49 54.5	-73 30 05	5	-	6	2	38	17	25	17	1x1	0.28	0.62	9.9	17.0	00499-7330	4	C
77	00 50 03.0	-73 06 55	5	-	7	2	50	30	-	-	1xp	0.23	0.66	8.8	C	00500-7306	4	M
78	00 50 09	-72 57	3	-	4	2	-	-	-	-	-	0.11:	0.22	C	C		2	M
79	00 50 09.7	-72 22 14	-	-	-	-	8	4	4	3	-	-	-	1.7	2.1	00501-7222	2	C
80	00 50 18.2	-72 20 02	-	-	-	-	6	3	6	3	-	-	-	1.2	6.3	00503-7220	2	C
81	00 50 22	-72 35	-	-	-	-	17	9	14	6	4x4	-	-	8.8	28.0		2	C
82	00 50 25.9	-73 53 09	-	-	2	-	15	6	10	6	2x2	-	0.50	5.6	9.8	00504-7353	4	C
83	00 50 36	-72 57	-	-	4	2	44	15	25	10	5x4	-	1.18	37.0	59.0		2	C
84	00 50 38.1	-72 07 39	-	-	-	-	4	2	4	2	10x3	-	-	3.8	11.0	00506-7207	2	C
85	00 50 46.4	-72 45 56	-	-	1	-	12	9	10	8	-	-	0.11:	1.2	4.2	00507-7245	2	C
86	00 50 54.7	-73 42 47	7	-	10	1	30	15	21	13	p	0.26	1.00	6.2	17.0	00509-7342	4	C
87	00 51 15.2	-72 29 46	-	-	-	-	17	9	11	6	3x3	-	-	6.7	15.0	00512-7229	2	C
88	00 51 18	-73 29	3	-	6	2	36	17	30	18	2xp	0.19:	0.71	9.7	27.0		4	C
89	00 51 23	-73 23	5	-	-	-	-	-	-	-	p:	0.19	-	C	C		4	S
90	00 51 24	-74 56	10	-	-	-	-	-	-	-	p	0.37	-	-	-	00515-7455:	4	S
91	00 51 24	-73 01	-	-	6	2	-	-	-	-	p	-	0.44	C	C		4/2	M
92	00 51 38.4	-72 59 12	5	-	5	2	36	20	23	20	p:	0.19:	0.33	6.6	6.3	00516-7259	2/4	M
93	00 51 47.9	-72 40 19	-	-	1	-	17	10	10	7	3x5	-	0.51:	7.7	11.0:	00517-7240	2	C
94	00 51 56.0	-72 55 35	5	-	6	2	42	20	24	12	2x3	0.57	1.22	16.0	32.0	00519-7255	2	C
95	00 52 00	-73 10	-	-	2	-	17	10	-	-	-	-	0.22	2.9	C	X0051-729	4	M
96	00 52 04	-72 59	5	-	-	-	-	-	-	-	-	0.19:	-	C	C		2	S
97	00 52 06	-73 00	5	-	2	-	-	-	-	-	-	0.19	0.22:	C	C		4	M:
98	00 52 12	-73 36	7	-	11	2	30	17	-	-	p	0.26	1.00	5.4	C	X0052-735	4	M
99	00 52 17.7	-73 05 43	-	-	2	-	25	16	16	12	p:	-	0.22	3.7	8.4	00522-7305	4	C
100	00 52 21	-73 38	3	-	4	-	-	-	-	-	p	0.11:	0.44	C	C	00524-7335:	4	M

No.	Type	N	Name	(1)	(2)	(3)	(4)	(5)	(6)	(7)	(8)	(9)	(10)	(11)	(12)	(13)	RA	Dec
101	S	2	00524-7153	-	-	-	0.48	p	-	<1	-	-	-	-	-	13	00 52 25.2	-71 53 26
102	C	2	00525-7245	11.0	5.2	1.86	0.58	4x1	10	14	15	22	-	6	-	5	00 52 36.0	-72 45 15
103	M	2	00532-7244	-	1.2	0.22:	-	-	-	-	15	18	-	2	-	-	00 53 16.3	-72 44 11
104	C	4	00533-7434	-	-	-	0.67	p	-	-	-	-	-	-	-	18	00 53 21.4	-74 34 35
105	C	2	00535-7255	13.0	8.6	0.87:	-	3x4	12	16	16	25	-	2	-	-	00 53 31.0	-72 55 00
106	S	2		C	C	-	0.19	-	-	-	-	-	-	-	-	5	00 53 46	-72 58
107	C	4	00542-7334	C	C	0.56	1.15	p	-	-	-	-	-	5	-	31	00 54 12.7	-73 34 43
108	M	2	00542-7220	6.4	2.9	-	-	4x3	6	8	7	10	-	-	-	-	00 54 16.7	-72 21 00
109	C	2		C	4.5	0.22:	0.11:	p:	-	-	20	31	-	2	-	3	00 54 18	-72 37
110	C	2		22.0	18.0	0.87	0.82	4x3	15	22	13	32	-	2	-	5	00 54 28	-72 34
111	M	4/2	00544-7303	C	7.5	0.66	0.23:	px1	-	-	15	32	1	6	-	5	00 54 28.6	-73 03 04
112	C	2		C	C	0.22	-	p:	-	-	-	-	-	2	-	-	00 54 32	-73 23
113	C	2		13.0	4.1	0.22	0.11:	p:	15	21	17	27	1	3	-	-	00 54 49	-72 39
114	C	2		6.3	2.1	0.11	0.23:	p:	13	16	17	22	1	2	-	3	00 55 00	-72 47
115	C	2	X0055-730	27.0	10.0	0.61	-	2x3	12	22	15	29	-	2	-	2	00 55 00	-73 03
116	M	2		C	C	0.22	-	5x1	12	23	15	27	-	2	-	3	00 55 03	-72 56
117	C	4		34.0	10.0	0.74	0.41:	-	6	8	5	9	-	2	-	-	00 55 10	-73 04
118	C	4	00553-7317	4.2	1.7	-	0.44	p:	-	-	-	-	-	-	-	12	00 55 18.6	-73 17 02
119	S	4/2	00554-7351	C	C	0.22	0.11:	-	-	-	-	-	-	2	-	3	00 55 24.8	-73 51 29
120	S	2		C	C	-	-	p	-	-	-	-	-	-	-	-	00 55 30	-73 02
121	M	2	00557-7248	6.3	7.0	1.00	0.19:	p	15	18	17	34	1	10	-	5	00 55 47.8	-72 48 41
122	M	4		C	C	0.22	0.19	-	-	-	-	-	2	2	-	5	00 55 57	-73 10
123	C	2		C	1.2:	-	-	-	-	-	10	13	-	-	-	-	00 56 00	-72 16
124	C	2	00562-7255	17.0	9.1	0.78	0.19	p	13	21	13	35	1	8	-	5	00 56 17.1	-72 55 23
125	C	2	00563-7220	13.0	3.7	0.33	0.11:	p	10	16	12	21	2	5	-	3	00 56 22.7	-72 20 22
126	C	2		20.0	3.6	0.35:	0.19:	px2	15	24	20	27	-	2	-	-	00 56 34	-72 42
127	M	2		C	C	1.44:	0.19	-	-	-	-	-	2	15	-	5	00 56 40	-72 27
128	C	2		C	9.1	0.67	-	p:	-	-	13	35	-	6	-	5	00 56 41	-72 56
129	C	2		8.4	2.9	0.22:	0.37:	-	12	16	16	23	-	2	-	-	00 56 52	-72 52
130	M	2	00569-7243	C	7.4	2.00	-	p	-	-	18	36	2	20	-	10	00 56 59.9	-72 43 43
131	C	2	00574-7226	242.0	200.0	43.50	5.99	3x3	10	93	20	260	2	120	-	43	00 57 26.5	-72 26 36
132	M	4/3	X0057-724	C	1.2	-	-	-	-	-	9	12	-	-	-	-	00 57 28.4	-73 09 12
133	M	2/1	00574-7309	C	C	0.44:	0.19:	-	-	-	-	-	2	6	-	5	00 57 42	-72 31
134	C	1/2		15.0	5.8	0.55:	-	3x3	5	10	6	13	-	1.5	-	-	00 57 54	-72 03
135	M	2		C	21.0	0.22	0.19:	-	-	-	20	70	10	12	-	5	00 58 06	-72 25
136	C	1		4.5	2.0	0.27:	-	px2	5	7	3	7	-	1.5	-	-	00 58 12	-71 47
137	M	1		C	19.0	0.89	-	p	-	-	15	60	4	12	-	-	00 58 12	-72 24
138	C	1		11.0	5.0	0.61:	-	2x3	9	13	10	17	-	2	-	-	00 58 36	-72 05
139	M	3		C	0.4	0.11:	0.19	p:	-	-	6	6	-	1	-	5	00 59 06	-73 07
140	S	1		C	C	-	0.19	p:	-	-	-	-	-	-	-	5	00 59 10	-71 55

Infrared Sources in the SMC (continued 3).

Peak and Bg values for 12, 25, 60 and 100 μm are in units of 1.0E-8 Watt / (m² m sr). Fluxes F are in Jy.

(1) Number LI-SMC	(2) RA(1950) h m s	(2) DEC(1950) ° ′ ″	(3) 12 μm Peak	(3) 12 μm Bg	(4) 25 μm Peak	(4) 25 μm Bg	(5) 60 μm Peak	(5) 60 μm Bg	(6) 100 μm Peak	(6) 100 μm Bg	(7) Size arcmin	(8) F 12 μm	(9) F 25 μm	(10) F 60 μm	(11) F 100 μm	(12) IRAS-Id	(13) DPM field	(14) Spectrum
141	00 59 13.2	-72 58 17	18	-	1.5	-	-	-	-	-	p	0.67	0.17:	-	-	00592-7258	1	S
142	00 59 18.4	-71 51 24	5	-	9	1	26	5	14	5	p	0.19	0.89	8.7	19.0	00593-7151	1	C
143	00 59 25	-72 04	-	-	2	-	-	-	4	1	-	-	0.22	C	C		1	W
144	00 59 27.9	-71 44 06	-	-	-	-	8	4	4	2	p	-	-	1.7	4.2	00594-7144	1	C
145	00 59 51.8	-71 49 03	9	-	-	-	-	-	-	-	p	0.33	-	-	-	00598-7149	1	S
146	00 59 52.2	-72 06 50	-	-	8	1	26	13	16	10	p	-	0.78	5.4	13.0	00598-7206	1	C
147	00 59 56.1	-72 12 16	-	-	1.5	1	20	13	16	15	-	-	0.06	2.9	2.1	00599-7212	1	W
148	01 00 04.8	-72 03 07	-	-	-	-	13	10	6	5	p:	-	-	1.2	C	01000-7203	1	W
149	01 00 05.5	-71 55 32	-	-	1.5	-	7	4	6	5	p:	-	0.17:	1.2	2.1	01000-7155	3	C
150	01 00 27	-74 17	5	-	-	-	-	-	-	-	-	0.19:	-	-	-		3	S
151	01 01 00	-73 13	-	-	-	-	6	4	5	3	-	-	-	0.8	4.2		3	C
152	01 01 09.3	-72 09 42	10	-	11	2	42	16	29	16	p	0.37	1.00	11.0	27.0	01011-7209	1	C
153	01 01 12.8	-72 41 34	-	-	2	2	17	7	11	7	3x2	-	0.61:	7.2	11.0	01012-7241	1	C
154	01 01 18	-72 14	5	-	5	2	30	20	-	-	p	0.19	0.33	4.1	C		1	W
155	01 01 19	-72 18	11	4	10	3	-	-	-	-	p	0.26	0.78	C	C		1	M
156	01 01 31.0	-72 22 16	10	-	8	2	-	-	-	-	p	0.37	0.67	C	C	01015-7222	1	M
157	01 01 32.0	-72 56 42	-	-	-	-	7	5	6	5	-	-	-	1.2	2.1:	01015-7256	1	C
158	01 01 32.8	-73 06 59	18	-	-	-	-	-	-	-	p	0.67	-	-	-	01015-7106	1	S
159	01 01 38	-73 30	5	-	-	-	4	2	3	2	-	0.19:	-	0.8	2.1		3	C
160	01 01 41.9	-72 28 06	-	-	6	2	25	14	16	15	p	-	0.44	4.5	2.1	01016-7228	1	M
161	01 01 42	-72 20	10	-	13	3	75	20	46	15	p:	0.37	1.11	23.0	65.0		1	C
162	01 02 11.2	-72 19 19	5	-	17	3	85	20	-	-	p	0.19	1.55	27.0	C	01021-7219	1	W
163	01 02 13.4	-72 24 53	-	-	4	1.5	26	17	18	15	p	-	0.28	3.7	6.3	01022-7224	1	C
164	01 02 27.8	-73 43 47	5	-	-	-	-	-	-	-	-	0.19:	-	-	-	01024-7343	3	S
165	01 02 44.9	-72 07 39	3	-	2	-	18	15	16	14	-	0.11:	0.22:	1.2	4.2	01027-7207	1	C
166	01 02 51.4	-73 10 15	-	-	1	-	5	3	5	3	-	-	0.11:	1.2	4.2	01028-7310	3	C
167	01 03 30.1	-72 00 08	12	-	1.5	-	-	-	-	-	p	0.44	0.17:	C	C	01035-7200	1	S
168	01 03 30.3	-72 15 28	37	-	100	2	128	20	43	15	p	1.37	10.90	45.0	59.0	01035-7215	1	C
169	01 03 36.3	-72 40 39	-	-	-	-	9	5	7	5	p	-	-	1.7	4.2	01036-7240	1	C
170	01 03 44	-72 25	-	-	2	-	20	15	13	12	p	-	0.22	2.1	2.1		1	W

No.	Type	n	Identification	D1	D2	R1	R2	P	n1	n2	n3	n4	n5	n6	n7	n8	Dec (1950)	RA (1950)
171	S	1	01038-7112	-	-	-	0.37:	P	-	-	-	-	-	-	-	10	-71 12 03	01 03 48.5
172	M	1	-	C	0.8	0.28	0.19	P	-	-	18	20	1.5	4	-	-	-73 12 12	01 03 50
173	C	3	01039-7305	6.3	3.7	1.00	-	P	3	6	4	13	-	9	-	5	-73 05 59	01 03 56.9
174	C	1	01039-7246	4.9	3.1	-	-	2x2	5	7	5	10	-	-	-	-	-72 46 34	01 03 59.2
175	C	1	-	6.3	5.4	0.22	0.11:	P	17	20	17	30	2	4	-	3	-72 20	01 04 06
176	M	1	01042-7215	C	8.3	1.11	0.19	P	-	-	15	35	2	12	-	5	-72 15 51	01 04 13.9
177	C	3	-	2.1	1.2	-	-	2x2	3	4	2	5	-	-	-	-	-73 24	01 05 06
178	C	1	-	9.8	3.7	0.22:	-	-	5	9	5	11	-	-	-	-	-72 51	01 05 24
179	C	3	01054-7307	6.3	1.7	-	0.19:	P:	3	6	3	7	-	2	-	5	-73 07 22	01 05 26.4
180	S	3	-	-	-	-	0.26:	P	-	-	-	-	-	7	-	7	-74 04	01 05 45
181	M	1	01060-7250	2.1:	2.5	0.17:	-	p:	7	8	5	11	1.5	-	-	-	-72 50 21	01 06 01.5
182	S	3	-	C	C	-	0.19	p:	-	-	-	-	-	-	-	5	-73 10	01 06 41
183	C	1	01067-7228	2.1	0.8	1.48	-	p:	4	5	4	6	-	-	-	-	-72 28 09	01 06 46.1
184	C	1	01069-7215	35.0	17.0	0.44	-	3x3	6	18	6	26	1	5	-	-	-72 15 46	01 06 58.3
185	M	1	01074-7140	-	-	-	0.41	P	-	-	-	-	1	4	-	11	-71 40 06	01 07 27.8
186	S	1	01075-7254	C	0.4	0.78	2.59	P	-	16	3	4	-	7	-	70	-72 54 39	01 07 33.8
187	C	3	01077-7327	27.0	18.0	2.55	0.44	P	3	16	2	46	-	23	-	12	-73 27 40	01 07 43.9
188	M	1	01077-7245	-	0.8	-	-	-	3	5	3	5	-	-	-	-	-72 45 35	01 07 45.2
189	C	1	01080-7237	2.5	1.9	-	-	2x2:	3	4	3	6	1	-	-	-	-72 37 25	01 08 03.6
190	C	1/3	-	7.1	3.3	-	-	4x4	3	5	3	6	-	-	-	-	-72 58	01 09 18
191	S	1	01094-7152	-	-	0.17:	1.00	P	-	-	-	-	1.5	1.5	-	27	-71 52 26	01 09 27.6
192	C	3	-	4.2	0.8	-	-	-	3	16	2	4	-	-	-	-	-73 21	01 09 30
193	C	1	01095-7225	4.2	1.7	0.33	-	P	3	5	3	7	-	3	-	-	-72 25 33	01 09 31.4
194	C	1	01098-7238 / X0110-724 *	8.7	4.7	0.93	-	1x1	3	7	3	13	-	6	-	-	-72 38 47	01 09 50.3
195	C	1/3	01106-7300	7.4	3.1	-	-	2x2	3	6	3	8	-	-	-	-	-73 00 12	01 10 41.3
196	S	3	-	-	-	-	0.19:	p:	-	-	1	-	-	-	-	5	-74 11	01 10 44
197	C	1	01116-7226 / X0110-724 *	2.1	0.8	-	-	p:	1	2	2	4	2	-	2	-	-72 26 36	01 11 39.0
198	S	1	01121-7108	-	-	0.30	0.30	P	-	-	-	-	2	-	-	8	-71 08 07	01 12 10.9
199	C	3	01124-7332 / X0112-735 *	117.0	46.0	2.22	0.52	P	4	60	6	117	2	22	3	17	-73 32 49	01 12 29.2
200	M	3	01126-7332 / X0112-735 *	C	C	4.00	0.70	p:	-	-	-	-	2	38	3	22	-73 32 42	01 12 41.2
201	M	3	01133-7333 / X0112-735 *	C	C	2.00	0.30	P	-	-	-	-	2	20	2	10	-73 33 42	01 13 19.1
202	C	3	01133-7336	88.0	32.0	2.22	0.33	p:	6	48	6	83	2	22	2	11	-73 36 33	01 13 23.1
203	C	1	01139-7234	2.1	0.4	0.33	-	-	1	2	1	2	-	-	-	-	-72 34 44	01 13 56.3
204	C	3	01142-7327	10.0	6.2	0.44	0.19:	p:	9	14	6	21	1	4	-	5	-73 27 38	01 14 17.3
205	M	3	01143-7326	C	C	-	-	P	-	-	-	-	1	4	-	-	-73 26 04	01 14 18.1
206	C	3	X0115-737	2.1	0.8	-	-	-	2	3	2	4	-	-	-	-	-73 45 14	01 15 00
207	M	3	01153-7324	C	0.8	-	-	P	1	8	6	8	-	-	-	7	-73 24 14	01 15 21.1
208	S	3	-	C	C	0.11:	0.26	P	-	-	-	-	-	1	-	-	-73 59	01 16 06
209	C	3	01164-7311	2.1	0.4	-	-	-	2	3	2	3	1	-	-	-	-73 11 09	01 16 28.7
210	C	3	01196-7357	2.1	0.4	-	-	-	1	2	1	2	-	-	-	-	-73 37 13	01 19 36.8

Infrared Sources in the SMC (continued 4).

(1) Number LI-SMC	(2) RA(1950) h m s	(2) DEC(1950) o ' "	(3) 12 μm Peak Bg	(4) 25 μm Peak Bg	(5) 60 μm Peak Bg	(6) 100 μm Peak Bg	(7) Size arcmin	(8) F 12μm Jy	(9) F 25μm Jy	(10) F 60μm Jy	(11) F 100μm Jy	(12) IRAS-Id	(13) DPM field	(14) Spectrum
						1.0E-8 Watt / (m²m sr)							
211	01 20 00	-74 15	-	-	1 -	1 -	-	-	-	0.4:	2.1		3	C
212	01 20 12	-73 20	-	-	2 1	2 1	-	-	-	0.4	2.1		3	C
213	01 21 37.9	-74 50 50	-	-	1 -	<1 -	-	-	-	0.4:	-	01216-7450	3	M
214	01 22 24.0	-73 38 54	-	-	5 2	4 2	1x2	-	-	1.6	4.6	01224-7338	3	C
215	01 22 52.8	-73 24 45	42 2	150 2	120 2	24 3	1x1	2.21	22.90	55.0	46.0	01228-7324	3	M
216	01 22 56.2	-73 29 43	2 -	1 -	7 4	6 5	-	0.07:	0.11:	1.2	2.1	01229-7329	3	C
217	01 23 24.3	-73 53 31	-	2 -	2 -	1 -	p:	-	0.22:	0.8	2.1	01234-7353	3	C
218	01 24 10.3	-73 40 01	-	-	8 2	6 3	p:	-	-	2.5	6.3	01241-7340	3	C
219	01 24 12.7	-73 30 50	2 -	3 -	9 2	5 3	p:	0.07:	0.33:	2.9	4.2	X0123-736 01242-7330	3	C

Infrared Sources in the SMC (continued 5), Additions from Co-added IRAS Survey data.

(1)	(2)			(3)	(4)	(5)	(6)	(7)	(8)	(9)	(10)	(11)	(12)	(13)	(14)
Number LI-SMC	Position RA(1950) h m s	DEC(1950) o ′ ″		12 μm Peak Bg	25 μm Peak Bg 1.0E-8 Watt / (m×m sr)	60 μm Peak Bg	100 μm Peak Bg	Size arcmin	F 12μm Jy	F 25μm Jy	F 60μm Jy	F 100μm Jy	IRAS-Id	CoAdd field	Spectrum
220	00 14 58	-74 14		5	2	-	-	-	0.19:	0.22:	-	-	-	6	S:
221	00 15 35.0	-73 56 10		5	-	-	0.5	p	0.19	-	1.2	1.0	00155-7356	6	s
222	00 16 02.3	-73 25 51		-	2	3	-	p:	-	0.22	-	-	00160-7325	6	W
223	00 16 21	-73 28		6	-	-	-	p	0.22	-	-	-	-	6	s
224	00 16 21	-74 03		-	3	-	-	p:	-	0.33	-	-	-	6	W
225	00 16 35.2	-74 18 54		14	3	-	-	p	0.52	0.33	-	-	00165-7418	6	s
226	00 18 05.8	-73 37 30		-	-	1.5	-	p:	0.19:	-	0.6	-	00180-7337	6	W
227	00 18 49.3	-74 52 38		5	-	1.5	-	-	-	-	0.6:	-	00188-7452	6	W:
228	00 19 56.4	-74 26 10		-	4	12	3	p:	-	0.44	4.9	6.2	00199-7426	6	C
229	00 21 53.8	-73 54 00		-	2	-	-	-	-	0.22	-	-	00218-7353	6	W
230	00 22 07	-74 34		-	-	-	0.5	-	-	-	-	1.0	-	6	C
231	00 22 31	-74 21		-	-	-	0.5	-	-	-	-	1.0	-	6	C
232	00 25 03	-74 35		-	-	-	0.7	-	-	-	-	1.5	-	6	C
233	00 26 03.0	-73 15 18		6	20	16	1.5	p	0.22	2.20	6.6	3.1	00260-7315	6	W
234	00 26 28.0	-74 37 47		-	-	1.5	0.5	p:	-	-	0.6	1.0	00264-7437	6	C
235	00 26 37	-73 56		-	2	-	-	p	-	0.22	-	-	-	6	W
236	00 27 20	-74 12		-	-	1.5	1	-	-	-	0.6	2.1	-	6	C
237	00 27 40	-74 20		-	-	1	0.5	-	-	-	0.4	1.0	-	6	C
238	00 28 05	-74 29		5	-	-	-	p	0.19	-	-	-	-	6	s
239	00 29 37.0	-74 04 17		-	2	4	1.5	p	-	0.22	1.6	3.1	00296-7404	6	C
240	01 26 16	-73 31		5	-	2	1	p:	0.19	-	0.8	2.1	-	5	C
241	01 27 47	-73 30		-	-	1	1	-	-	-	0.4	2.1	01283-7349	5	C
242	01 28 23.1	-73 49 17		-	6	22	12	2xp	-	1.05	11.1	27.2	X0128-738	5	C
243	01 28 43	-74 10		-	-	-	0.5	-	-	-	-	1.0	-	5	C
244	01 29 07.7	-73 25 38		6	2	-	-	p:	0.22	0.22	-	-	01291-7325	5	S:
245	01 29 08	-73 28		-	-	-	0.5	-	-	-	-	1.0	-	5	C
246	01 31 19.7	-73 42 30		-	-	1	1.5	p:	-	-	0.4	3.1	01313-7342	5	C
247	01 33 06.0	-73 21 00		5	-	-	-	p:	0.19	-	-	C	01331-7321	5	s
248	01 41 26	-73 33		-	-	-	0.5	-	-	-	-	1.0	-	5	C
249	01 41 38.5	-73 43 12		-	-	-	0.5	p:	-	-	-	1.0	01416-7343	5	C

Infrared Sources in the LMC, Based on IRAS Additional DPM Observations.

(1)	(2) RA(1950) h m s	(2) DEC(1950) ° ′ ″	(3) 12 μm Peak Bg		(4) 25 μm Peak Bg		(5) 60 μm Peak Bg		(6) 100 μm Peak Bg		(7) Size arcmin	(8) F 12μm Jy	(9) F 25μm Jy	(10) F 60μm Jy	(11) F 100μm Jy	(12) IRAS-Id	(13) DPM field	(14) Spec-trum
						1.0E-8 Watt / (m²m sr)											
1	04 38 36	-70 53	13	-	-	-	-	-	-	-	P	0.48	-	-	-		16/12	S
2	04 38 49.4	-70 42 47	3	-	3	-	1	0.5	1	0.5	P	0.11:	0.33	0.2:	1.0:	04388-7042	12	C
3	04 40 24	-71 04	-	-	-	-	-	-	2	1	-	-	-	-	2.1:		16	C
4	04 40 46.7	-70 00 54	22	-	6	-	-	-	-	-	P	0.81	0.67	-	-	04407-7000	12	S
5	04 41 03.9	-69 13 00	-	-	-	-	1	-	1	-	-	-	-	0.4:	2.1:	04410-6912	12	C
6	04 41 36	-71 28	6	-	-	-	-	-	-	-	P	0.22	-	-	-		16	S
7	04 41 41.6	-68 42 08	5	-	-	-	1	-	-	3	-	0.19	-	0.4:	-	04416-6842	08	S
8	04 42 15	-70 50 50	-	-	-	-	3	-	4	3	-	-	-	1.2	2.1		12/16	C
9	04 42 59.0	-69 33 07	-	-	-	-	-	-	4	2	-	-	-	-	4.2	04429-6933	12	C
10	04 43 00	-71 35	3	-	-	-	-	-	-	-	-	0.11:	-	-	-		16	S
11	04 43 10	-70 43	-	-	-	-	4	-	5	3	-	-	-	1.7:	4.2		12	C
12	04 43 26.9	-70 39 36	5	-	10	-	18	-	9	4	P	0.19:	1.11	7.5	10.4	04434-7039	12	C
13	04 43 30	-70 58	-	-	-	-	-	-	4	3	-	-	-	-	2.1		16	C
14	04 43 33	-71 01	4	-	-	-	-	-	-	-	-	0.15:	-	-	-		16	S
15	04 43 56.4	-68 46 53	4	-	-	-	2	-	1	-	P	0.15:	-	0.8	2.1:	04439-6846	08	C
16	04 44 31.0	-68 12 56	-	-	2	-	-	-	3	-	-	0.33	0.22:	-	6.2:	04445-6812	08	C
17	04 44 33.6	-72 13 35	9	-	1	-	-	-	-	-	-	0.26	0.11:	-	-	04445-7213	16	S
18	04 45 03.0	-70 48 24	7	-	1	-	-	-	-	-	-	0.19	0.11:	-	-	04450-7048	12	S
19	04 45 06	-68 29	5	-	-	-	-	-	-	-	P:	0.19	-	-	-		08	S
20	04 45 11	-68 07	5	-	-	-	-	-	-	-	-	-	-	-	-		08	S
21	04 45 45	-69 53	2	-	1	-	1	-	2	1	-	0.07:	-	0.4:	2.1	04459-6955:	12	C
22	04 46 10.8	-68 51 48	-	-	-	-	6	5	-	-	-	-	0.11:	0.4	C	04461-6851	08	W
23	04 46 12.3	-68 23 01	-	-	-	-	4	2	2	-	-	-	-	0.8	4.2:	04462-6823	08	C
24	04 46 40.8	-71 00 21	-	-	-	-	3	1	4	2	-	-	-	0.8	4.2	04466-7100	16	C
25	04 46 46	-68 38	6	-	-	-	-	-	-	-	-	0.22	-	-	-		08	S
26	04 47 00	-70 50	2	-	-	-	3	-	5	4	-	0.07:	-	1.2	2.1		12	C
27	04 47 00	-71 22	-	-	-	-	1	-	2	1.5	-	-	-	0.4	1.0	04470-6712	16	C
28	04 47 01.0	-67 12 17	5	-	3	-	10	-	9	5	P:	0.19	0.33	2.5	8.3	X0447-672*	08	C
29	04 47 04.8	-71 09 08	-	-	-	-	3	1	4	2	-	-	-	0.8	4.2	04470-7109	16	C
30	04 47 10	-71 12	5	-	-	-	1	-	-	-	-	0.19	-	-	-		16	S

No.	RA (1950)	Dec (1950)									P					Name	n	T
31	04 47 22.0	−68 29 43	8	−	2	−	−	−	−	−	P	−	−	0.22:	0.30	04473-6829	08	S
32	04 47 25	−67 19 25	8	−	2	−	6	4	7	6	−	2.1:	0.8	0.22	0.30	X0447-672*	08	C
33	04 47 30.5	−69 14 41	10	5	13	3	20	14	21	16	P	10.4	2.5	1.11	0.19	04475-6914	12	C
34	04 47 50	−70 40 50	3	−	1	−	5	−	8	5	−	6.2	2.1	0.11:	0.11:		12	C
35	04 47 58	−69 48 58	5	−	1	−	4	2	3	2	−	2.1:	0.8:	0.11:	0.19	04481-7037:	12	C
36	04 48 00	−67 55 55	−	−	−	−	4	3	5	3	−	4.2	0.4	−	−		08	C
37	04 48 08	−68 51 51	4	−	−	−	−	−	−	−	−	−	−	−	0.15:		08	S
38	04 48 10.0	−68 24 01	−	−	−	−	4	2	5	3	−	4.2:	0.8:	−	−	04481-6824	08	C
39	04 48 15	−68 55 55	4	2	1.5	−	13	10	13	11	−	4.2	1.2	0.17	0.07:		06/12	C
40	04 48 30	−69 24 24	6	4	4	2	26	18	25	20	−	10.4	3.3	0.22:	0.07:		12	C
41	04 48 30	−71 54 30	4	−	−	−	1	−	2	1	−	3.1	0.4	0.11:	0.15:	04487-7024	16	C
42	04 48 44.7	−70 24 06	2	−	1	−	2	−	5	4	−	2.1	0.8	0.11:	0.07:		12	C
43	04 49 00	−70 30 00	2	−	1	1	3	1	5	4	−	2.1:	1.2:	−	0.07:		12	C
44	04 49 00	−71 11 00	−	2	−	−	30	25	4	2	−	2.1	0.4	−	−		16	C
45	04 49 06.2	−69 26 02	12	7	7	5	−	−	−	−	−	C	2.1:	0.22	0.19	04491-6926	12	M
46	04 49 07.7	−69 15 01	34	15	33	6	84	40	15	14	P	C	18.2	3.00	0.70	04491-6915	12	M
47	04 49 09.2	−69 01 16	6	4	4	1.5	16	13	18	14	−	2.1:	1.2	0.28	0.07:	04491-6901	12/08	C
48	04 49 14.5	−68 29 22	10	3	6	1.5	18	9	11	10	−	16.6	3.7	0.50	0.26	04492-6829	08	C
49	04 49 15	−68 42 15	−	−	3	1.5	10	7	10	8	−	6.2	1.2	0.17	−		08	C
50	04 49 17.1	−70 20 35	7	−	1	−	−	−	−	−	P:	−	−	0.11:	0.26	04492-7020	12	S
51	04 49 20.3	−66 55 02	66	−	7	5	35	27	13		P	C	3.3:	0.78	2.44	04495-6655	04/08	S
52	04 49 30	−69 14 30	17	13	7	5	12	8	10		−	−	1.7	0.22	0.15:		12	M
53	04 49 30.1	−68 35 29	8	4	3	1.5	6	4	11		−	6.2	0.8	0.17	0.15:	04495-6835	08	C
54	04 49 33.2	−68 12 56	3	−	2	−	9	6			−	10.4	1.2	0.22	0.11	04495-6812	08	C
55	04 49 35	−69 46 35	5	−	2	−	−	−	8		−	6.2	−	0.22	0.19		12	C
56	04 49 37.5	−69 29 34	20	9	9	5	40	30	−	−	P:	C	4.1:	0.44	0.41	04496-6929	12	M
57	04 49 38.4	−69 58 17	10	−	3	8	200	50	38		P	−	62.1	0.33	0.37	04496-6958	12	S
58	04 49 40.5	−69 17 07	62	15	80	1	8	5	5	5	−	139.4	1.2	7.99	1.74	04496-6917	12	C
59	04 49 47	−66 56 47	6	1	−	−	−	−	−	−	P	−	−	0.22	0.19		04/08	S
60	04 49 50.3	−68 42 53	34	2	11	1	−	−	−	−	−	C	−	1.11	1.18	04498-6842	08	M
61	04 49 52.2	−71 21 22	4	−	1	−	10.5	0.5	−	−	−	C	0.2	0.11:	0.15:	04498-7121	16	M
62	04 49 53.0	−69 25 09	15	12	14	7	40	35	10		−	C	2.1:	0.78	0.11:	04498-6925	12	M
63	04 49 55	−69 17 55	30	17	22	9	13	8	13		P	C	C	1.44	0.48		12	C
64	04 50 14.9	−68 30 21	9	4	4	1	14	3	11		−	6.2	2.1	0.33	0.19	04502-6830	08	C
65	04 50 15	−67 44 15	2	−	2	−	−	−	4		−	14.6	4.6	0.22	0.07:		08	C
66	04 50 22.7	−69 45 32	10	3	2	−	9	7	−	−	−	−	0.8:	0.22:	0.26	04503-6945	12	M
67	04 50 29.8	−69 34 47	25	10	15	7	74	38	61	38	−	47.8	14.9	0.89	0.56		12	C
68	04 50 30	−66 51 30	2	−	1	−	5	5	9	6	−	6.2	1.7	0.11:	0.07:		04	C
69	04 50 30	−69 17 30	19	11	7	5	45	35	50	40	−	20.8:	4.1:	0.22:	0.30	04504-6934	12	C
70	04 50 30	−69 27 30	24	15	15	9	75	65	70	55	−	31.2	4.1	0.67	0.33		12	C

Infrared Sources in the LMC (continued 2).

(1) Number LI-LMC	(2) RA(1950) h m s	(2) DEC(1950) o ' "	(3) 12 µm Peak	(3) 12 µm Bg	(4) 25 µm Peak	(4) 25 µm Bg	(5) 60 µm Peak	(5) 60 µm Bg	(6) 100 µm Peak	(6) 100 µm Bg	(7) Size arcmin	(8) F 12µm Jy	(9) F 25µm Jy	(10) F 60µm Jy	(11) F 100µm Jy	(12) IRAS-Id	(13) DPM field	(14) Spectrum
									1.0E-8 Watt / (m²·m·sr)									
71	04 50 30	-69 37	20	10	9	6	50	40	50	40	-	0.37	0.33	4.1	20.8		12	C
72	04 50 30.0	-69 38 45	12	8	8	3	20	15	-	-	-	0.15:	0.56	2.1:	C	04505-6938	12	M
73	04 50 30.7	-72 02 33	-	-	-	-	1	1	0.5	-	-	-	-	0.4:	1.0:	04505-7202	16	C
74	04 50 31.1	-71 01 36	3	-	1	-	4	1	6	3	-	0.11:	0.11:	1.2	6.2	04505-7101	16/12	C
75	04 50 31.2	-70 52 32	4	-	2	-	6	1	7	4	-	0.15:	0.22	2.1	6.2	04505-7052	12	C
76	04 50 45	-70 22	2	-	1	-	2	-	2	1	-	0.07:	0.11:	0.8:	2.1:		12	C
77	04 50 55.7	-69 22 32	36	16	17	8	-	-	-	-	P	0.74	1.00	C	C	04509-6922	12	M
78	04 51 04.0	-69 54 49	135	-	14	-	2	-	-	-	P	4.99	1.55	0.8:	-	04510-6954	12	S
79	04 51 14.7	-69 05 55	17	9	6	4	29	18	32	20	P	0.30	0.22	4.6	25.0	04512-6905	12	C
80	04 51 16.7	-69 24 34	23	15	15	9	65	50	-	-	P	0.30	0.67	6.2:	C	04512-6924	12	M
81	04 51 19.6	-70 27 07	10	-	5	-	17	-	10	3	p:	0.37:	0.56	7.0	14.6	04513-7027	12	C
82	04 51 20	-67 01	11	5	7	4	20	12	-	-	-	0.22	0.33	3.3:	C		08/04	M
83	04 51 20	-69 11	12	8	5	3	31	26	30	28	-	0.15	0.22:	2.1	4.2:		12	C
84	04 51 22.0	-68 14 33	3	-	2	-	10	7	11	8	-	0.11:	0.22:	1.2	6.2	04513-6814	08	C
85	04 51 22.3	-68 32 39	5	3	1.5	-	9	7	10	9	-	0.07:	0.17	0.8	2.1	04513-6832	08	C
86	04 51 27.7	-69 31 36	15	6	35	6	83	50	-	-	P	0.33	3.22	13.7	C	04514-6931	12	M
87	04 51 28.0	-68 09 00	11	-	3	1.5	-	-	-	-	P	0.41	0.17:	-	-	04514-6808	08	S
88	04 51 30	-68 47	6	-	1	-	-	-	-	-	-	0.22	0.11:	-	-		08	S
89	04 51 35.4	-67 10 14	17	8	8	3	-	-	-	-	P	0.33	0.56	C	C	04515-6710	08	M
90	04 51 39.0	-69 19 12	15	10	6	4	35	30	37	33	-	0.19:	0.22	2.1:	8.3:	04516-6919	12	C
91	04 51 40	-67 26	4	-	2	-	9	7	10	8	-	0.15	0.22	0.8	4.2	X04451-690	08	C
92	04 51 41.3	-69 02 49	24	7	9	2	17	14	-	-	P	0.63	0.78	1.2:	C	04516-6902	12/08	M
93	04 51 41.5	-68 10 38	16	4	3	1.5	35	20	35	25	P	0.44	0.17:	-	-	04516-6810	08	S
94	04 51 45	-67 07	17	8	7	4	20	-	-	-	-	0.33	0.33	6.2	20.8		08	C
95	04 51 46.5	-65 51 32	-	-	-	-	3	-	-	-	-	-	-	1.2:	-	04517-6551	04	M
96	04 51 50	-67 04	17	7	7	-	-	-	-	-	-	0.37	0.33	C	C		08	S
97	04 51 50	-70 30	7	-	-	-	10	4	10	6	p:	0.26	-	-	-		12	S
98	04 51 50.6	-67 34 15	6	-	2	-	-	-	-	-	-	0.22	0.22	2.5	8.3	04518-6734	08	C
99	04 51 51.1	-68 52 23	10	-	2	-	-	-	-	-	P	0.37	0.22	-	-	04518-6852	08	S
100	04 51 55	-67 15	13	6	2	1	-	-	-	-	-	0.26	0.11:	C	C		08	S

No.	Cl	N	IRAS	a	b	c	d	v	e	f	g	h	i	j	k	l	RA	Dec
101	C	16		1.0	0.4	–	–	I	4	4.5	1	2	–	–	–	–	04 52 00	-71 22
102	C	08/04	04520-6700	191.4	116.7	13.99	3.63	P	23	115	18	300	4	130	12	110	04 52 04.8	-67 00 09
103	C	12	04521-6928	353.6	343.6	65.16	9.10	P	60	230	70	900	13	600	14	260	04 52 09.5	-69 28 21
104	C	12	04521-6913	6.2:	2.9	0.11:	0.04:	–	29	32	28	35	4	5	8	9	04 52 11.0	-69 13 02
105	C	12	04521-6945	8.3	2.5	0.17	0.11:	–	12	16	9	15	1.5	3	4	7	04 52 11.3	-69 45 29
106	C	08	04521-6720	45.8:	10.3:	0.89	0.37	–	10	32	8	33	2	10	4	14	04 52 11.7	-67 20 03
107	M	12	04522-6925	C	62.1	4.22:	0.93:	–	–	–	70	220	12	50	20	45	04 52 17.9	-69 25 22
108	C	12	04523-7043	10.4	2.5	0.44	0.52	P:	7	12	3	9	–	4	–	14	04 52 19.5	-70 43 23
109	C	08			–	0.22:	0.22	–	–	–	–	–	–	2	–	6	04 52 20	-67 27
110	C	08		4.2	1.2	0.22:	0.22	–	8	10	7	10	–	2	–	6	04 52 25	-68 27
111	C	16	04524-7235	3.1:	0.2:	–	–	P:	–	1.5	–	0.5	–	–	–	–	04 52 25.9	-72 35 27
112	M	08	04524-6721	C	10.3:	0.78	0.19:	–	–	–	8	33	2	9	4	9	04 52 27.0	-67 21 43
113	C	04	04526-6951	8.3:	1.2	0.22	0.19	–	7	11	7	10	–	2	–	5	04 52 36.3	-69 51 47
114	C	08/12	04526-6859	18.7	4.6	0.44	0.19	–	–	22	13	24	2	6	5	10	04 52 41.4	-68 59 24
115	M	12	04527-6925		18.6:	4.77	1.37	P	13	–	80	125	12	55	15	52	04 52 42.8	-69 25 45
116	M	08/04		C	2.9	0.22	0.22	P:	–	–	22	29	2	4	6	12	04 52 45	-67 02
117	C	12		C	C	0.33	0.26	–	–	–	–	–	14	17	15	22	04 52 45	-69 19
118	C	04		4.2:	0.8	0.11:	–	–	6	8	6	8	1	1	–	–	04 52 50	-66 40
119	C	04		4.2	2.1	0.22:	0.19	–	8	10	8	13	1	3	–	–	04 53 00	-66 50
120	M	08		C	C	0.22		–	–	–	1	–	2	4	4	9	04 53 00	-68 12
121	C	12	04530-6916	39.5	24.8	5.11	2.07	P	65	84	65	125	12	58	17	73	04 53 00.4	-69 16 43
122	C	08	04531-6808	60.3	35.6	4.16	1.00	P	10	39	8	94	1.5	39	3	30	04 53 07.7	-68 08 41
123	M	08	X0453-681	C	6.6	0.50	0.37	–	–	–	12	28	6	6	–	10	04 53 10	-67 10
124	C	08		6.2	1.7	0.22:	0.15:	–	9	12	9	13	1.5	2	–	4	04 53 10	-68 50
125	M	12		C	C	0.22	0.19	–	–	–	–	–	4	6	9	14	04 53 10	-69 32
126	C	12	04531-6935	10.4:	4.1:	0.22	0.11:	–	25	30	20	30	4	6	6	9	04 53 11.6	-69 35 46
127	M	16		4.2	0.8	–	0.07:	–	3	5	1	3	–	–	–	2	04 53 12	-71 06 02
128	C	12		4.2:	0.8	0.22	0.15:	–	6	8	5	7	1	2	–	4	04 53 13.8	-70 51 02
129	M	08		C	0.8:	0.22	0.33	–	–	–	10	12	2	4	–	9	04 53	-68 09
130	S	04	04534-6645	–	–	1.55	4.33	–	–	–	–	–	1	14	3	120	04 53 25.2	-66 45 22
131	C	12	04534-7003	4.2	0.8	0.11:	–	–	2	4	3	5	–	1	–	7	04 53 29.7	-70 03 19
132	C	04/08		10.4:	5.4	0.33	0.22	–	14	19	15	28	2	5	1	9	04 53 30	-66 58
133	C	08		10.4	3.3	0.17	0.07	–	10	15	16	18	1.5	3	4	6	04 53 30	-68 37
134	C	12		10.4	2.5	0.22	0.07	–	31	36	34	40	4	6	10	12	04 53 30	-69 35
135	C	08	04535-6728	4.2:	1.2	0.22	–	–	4	6	3	6	–	2	–	–	04 53 30.5	-67 28 16
136	M	04	04535-6616	–	0.4:	0.33	0.19	P:	–	–	4	5	–	3	–	5	04 53 35.3	-66 16 31
137	C	08	04536-6704	16.6	5.8	0.22	0.19	–	16	24	12	26	2	20	4	18	04 53 37.8	-67 04 04
138	M	12	04537-6922	C	C	1.11	0.22	–	–	–	–	–	10	2	12	5	04 53 46.0	-69 22 36
139	C	12	04538-7040	6.2	1.7	0.22	0.19	–	6	9	5	9	–	2	–	5	04 53 49.6	-70 40 30
140	C	12	04538-6952	8.3	2.1	0.22	0.19	–	9	13	7	12	–		–		04 53 52.0	-69 52 39

Infrared Sources in the LMC (continued 3).

(1)	(2) Position		(3) 12 μm		(4) 25 μm		(5) 60 μm		(6) 100 μm		(7)	(8)	(9)	(10)	(11)	(12)	(13)	(14)
Number LI-LMC	RA(1950) h m s	DEC(1950) o ' "	Peak	Bg	Peak	Bg	Peak	Bg	Peak	Bg	Size arcmin	F 12μm Jy	F 25μm Jy	F 60μm Jy	F 100μm Jy	IRAS-Id	DPM field	Spec-trum
							1.0E-8 Watt / (m² m sr)											
141	04 53 54.2	-68 21 11	9	4	2	-	-	-	-	-	p:	0.19	0.22	-	-	04539-6821	08	M
142	04 53 55.1	-72 29 20	10	2	-	-	-	-	-	-	-	0.30:	-	-	-	04539-7229	16	S
143	04 53 58	-69 03	9	6	3	1.5	33	8	23	9	-	0.11	0.17:	10.3	29.1		08/12	M
144	04 54 00.8	-66 50 39	14	-	17	2	27	8	15	9	p:	0.52	1.66	7.9	12.5	04540-6650	04	C
145	04 54 03.1	-67 21 05	4	-	20	1.5	15	10	-	-	-	0.15	2.05	2.1	C	04540-6721	08	C
146	04 54 10	-66 57	4	-	2	-	-	-	-	-	-	0.15:	0.22:	-	-		04	M
147	04 54 15	-67 22	23	3	5	2	500	70	200	65	P	0.74	0.33	178.0	280.8	04544-6722:	08	S
148	04 54 17.0	-69 16 23	140	25	170	15	20	16	20	18	P	4.25	17.20	1.7:	4.2:	04542-6916	12	C
149	04 54 19.4	-69 08 26	-	-	4	2	7	6	9	8	-	-	0.22:	0.4:	2.1:	04543-6908	12	C
150	04 54 20	-70 54	5	-	1	-	34	9	28	13	-	0.19	0.11:	10.3	31.2		12	C
151	04 54 20.8	-68 27 03	15	5	10	1.5	30	20	-	-	-	0.37	0.94	4.1	C	04543-6827	08	C
152	04 54 22.5	-66 29 49	8	3	4	1	-	-	-	-	-	0.19	0.33	-	-	04543-6629	04	M
153	04 54 24.6	-68 49 02	11	3	1.5	-	74	55	54	39	p:	0.30	0.17:	7.9	31.2	04544-6849	08	S
154	04 54 25.2	-69 25 08	22	15	22	10	12	9	-	-	-	0.26	2.33	1.2:	C	04544-6925	12	C
155	04 54 30	-66 40	6	-	3	1	20	15	20	15	p:	0.22	0.22	2.1	10.4		04	M
156	04 54 30	-68 40	9	6	5	2	52	44	-	-	-	0.11	0.33	3.3:	C		08	C
157	04 54 30	-69 29	18	12	10	5	9	7	10	8	-	0.22	0.56	0.8:	4.2:		12	M
158	04 54 30.1	-69 46 33	4	-	1	-	5	3	6	4	-	0.15:	0.11:	0.8:	4.2:	04545-6946	12	C
159	04 54 32.0	-70 00 44	12	-	8	-	9	7	10	8	-	0.44	0.89	0.8	4.2:	04545-7000	12	C
160	04 54 34.6	-66 44 35	-	-	1	-	9	7	-	-	P	-	0.11:	0.8:	C	04545-6644	04	C
161	04 54 40	-65 56	-	-	-	-	200	100	-	-	-	-	-	41.4:	C		04	M
162	04 54 40.6	-69 15 39	58	20	125	10	10	6	9	4	P	1.41	12.76	1.7	10.4	04546-6915	12	M
163	04 54 41.6	-65 58 00	5	-	2	-	37	30	34	28	-	0.19	0.22	2.9	12.5	04546-6558	04	C
164	04 54 42.1	-69 34 23	19	9	9	6	9	7	8	6	P	0.37	0.33	0.8:	4.2:	04547-6934	12	C
165	04 54 43.8	-67 24 15	2	-	1	-	32	12	25	12	-	0.07:	0.11:	8.3	27.0	04547-6724	08	C
166	04 54 45	-67 17	10	-	6	1.5	-	-	-	-	-	0.37	0.50	C	C		08	C
167	04 54 50.2	-69 31 14	12	10	5	4	6	4	7	5	-	0.07:	0.11:	0.8	4.2	04548-6931	12	M
168	04 54 55	-69 54	5	-	1	-	-	-	-	-	-	0.19	0.11:	-	-		12	C
169	04 55 00	-65 48	-	-	-	-	4	-	4	-	10x10:	-	-	16.3	43.0		04	C
170	04 55 00	-70 24	5	-	-	-	-	-	-	-	-	0.19	-	-	-		12	S

No.	RA (1950)	Dec (1950)	Name	F1	F2	F3	F4	flag	n1	n2	n3	n4	n5	n6	n7	n8	Type	Q
171	04 55 00	−70 58	04551−6607	4.2	0.4:	0.11:	−	−	7	9	7	6	−	1	−	−	C	12
172	04 55 00	−71 18	04551−6928	2.1	0.8	−	−	−	2	3	3	1	−	−	−	−	M	16
173	04 55 05	−69 19		C	C	0.33	0.30	−	−	−	1	1	7	10	12	20	M	12
174	04 55 10.0	−66 07 57	04552−6605	6.2	0.8	0.33	0.22	−	10	13	11	9	7	3	−	6	C	04
175	04 55 10.1	−69 28 41	04552−6636	C	1.2:	0.22:	0.33:	−	−	−	30	27	5	7	6	15	M	12
176	04 55 13.1	−66 05 58		C	0.8	0.33	0.15:	−	−	−	11	9	−	3	4	4	M	04
177	04 55 13.6	−66 36 14	04552−6536	C	C	0.78	0.11:	P:	7	11	−	−	9	16	18	18	M	04
178	04 55 15	−66 03		8.3:	2.1	0.33	0.15	−	−	−	9	4	7	3	4	4	C	04
179	04 55 15	−66 24		C	8.3	0.22	0.22	−	−	1	55	35	7	9	12	18	M	04
180	04 55 16.1	−65 36 17		2.1	0.4	0.22:	0.15	−	−	−	3	2	−	2	−	4	C	04
181	04 55 18.4	−68 25 15	04553−6825	10.4:	4.1	11.21	7.07	P	15	20	28	18	4	105	14	205	M	08
182	04 55 20	−69 25	04553−6933	C	C	0.22	0.30	−	−	−	−	−	4	6	9	17	S	12
183	04 55 20.5	−69 33 53	04553−6921	C	2.1:	0.67	0.67	P	−	−	28	23	4	10	5	23	M	12
184	04 55 21.5	−69 21 36		20.8:	7.0	1.33	0.37	P:	25	35	62	45	7	19	10	20	C	12
185	04 55 25	−66 57		−	−	−	0.19	−	−	−	−	−	−	−	−	5	S	04
186	04 55 25	−67 00	04556−7225:	2.1:	0.8	0.67	0.37	−	4	5	6	4	−	9	20	20	C	04
187	04 55 30	−68 28	04555−7032	C	C	−	−	−	−	1.5	−	−	3	−	10	−	M	08
188	04 55 30	−72 27	04555−6632	16.0	3.8	−	−	10×10	2	3	3	−	−	2	−	2	C	16
189	04 55 30.4	−70 32 07		2.1	1.2	0.22	0.07:	−	−	−	−	−	−	−	−	−	C	12
190	04 55 33.2	−66 32 23		62.4	31.0	1.89	1.22	P:	55	85	125	50	13	30	12	45	C	04
191	04 55 33.3	−68 41 39	04555−6841	20.8	4.6	0.33	0.33	−	18	28	29	18	4	7	7	16	C	08
192	04 55 35	−66 39	04557−6639:	C	18.6	1.78	0.44	P	−	−	90	45	12	28	−	24	M	04
193	04 55 35	−69 11	04555−6829	10.4	0.8	0.22	0.15	P:	23	28	20	18	2	4	−	11	C	12
194	04 55 35.3	−68 29 59	04556−6630	43.7	13.2	2.55	0.41	P:	25	46	54	22	4	27	8	19	C	08
195	04 55 37.9	−66 30 24		C	16.6:	0.89	0.70	P:	−	−	100	60	14	22	16	35	M	04
196	04 55 38	−70 53	04557−7052:	6.2	1.2	0.22	0.07:	−	8	11	10	7	−	2	−	2	C	12
197	04 55 40	−68 37	04557−6753	C	C	0.11	0.33	−	−	−	−	−	3	4	6	15	S	08
198	04 55 42.1	−67 53 25	04557−6920	C	1.7:	0.22	0.26	P	−	−	20	16	4	2	−	7	S	08
199	04 55 42.4	−69 20 41	04557−6952	2.1:	0.4:	0.44	0.22	−	−	5	4	3	−	8	10	16	M	12
200	04 55 42.5	−69 52 01		−	−	0.44	0.41	P	4	−	−	−	−	4	−	11	C	12
201	04 55 46.6	−65 57 21	04557−6557	4.2:	1.2:	0.11:	−	−	4	6	7	4	−	1	−	−	C	04
202	04 55 50	−68 35		10.4	5.0	0.22	0.33	−	25	30	30	18	3	5	6	15	C	08
203	04 55 57.3	−69 31 22	04559−6931	C	0.8:	0.44	0.44	P	−	−	20	18	4	8	8	20	M	12
204	04 56 10	−68 49		4.2	2.1	0.22	0.04	−	14	16	19	14	2	4	2	3	C	08
205	04 56 17.0	−66 41 40	04562−6641	C	C	0.56	0.19	−	−	−	−	19	15	20	10	15	M	04
206	04 56 20	−66 20		C	4.1	0.33	0.37	−	−	−	50	40	7	7	9	19	M	04
207	04 56 20	−69 38		C	C	0.17	0.19	−	−	−	−	−	4	3	4	9	S	12
208	04 56 20.9	−67 19 29	04563−6719	8.3:	2.1:	0.33	0.22:	−	8	12	13	8	1.5	3	1	6	C	08
209	04 56 22.8	−71 25 37	04563−7125	2.1	0.8	−	−	−	1	2	3	1	−	−	−	−	C	16
210	04 56 24.3	−66 29 48	04564−6629	C	C	10.55:	1.48:	−	−	−	−	−	30	125	20	60	M	04

Infrared Sources in the LMC (continued 4).

(1)	(2)		(3)		(4)		(5)		(6)		(7)	(8)	(9)	(10)	(11)	(12)	(13)	(14)
Number LI–LMC	RA(1950) h m s	DEC(1950) o ' "	12 μm Peak	Bg	25 μm Peak	Bg	60 μm Peak	Bg	100 μm Peak	Bg	Size arcmin	F 12μm Jy	F 25μm Jy	F 60μm Jy	F 100μm Jy	IRAS-Id	DPM field	Spectrum
							1.0E-8 Watt / (m*m sr)											
211	04 56 24.9	−70 56 48	−	−	−	−	2	−	−	−	−	−	−	0.8:	C	04564−7056	12	M
212	04 56 25	−69 11	12	8	4	3	20	18	−	−	−	0.15	0.11	0.8:	C		12	M
213	04 56 26.9	−69 35 47	9	5	4	2	22	19	21	18	−	0.15:	0.22	1.2	6.2	04564−6935	12	C
214	04 56 35.4	−66 37 21	43	17	50	14	150	60	105	60	−	0.96	4.00	37.3	93.6	04565−6637	04	C
215	04 56 40	−67 55	4	−	2	−	11	7	11	8	−	0.15	0.22	1.7	6.2		08	C
216	04 56 40.0	−69 28 56	15	4	10	4	40	25	28	21	p	0.41	0.67	6.2	14.6	04566−6928	12	C
217	04 56 41.2	−66 29 03	135	20	310	15	650	60	320	70	p	4.25	32.74	244.3	520.0	04566−6629	04	C
218	04 56 43.5	−68 57 19	9	4	3	2	−	−	−	−	−	0.19	0.11	C	C	04567−6857	08	S
219	04 56 48.1	−66 35 34	−	−	25	15	−	−	−	−	p:	−	1.11:	C	C	04568−6635	04	M
220	04 56 50	−66 35 50	5	−	1	−	−	−	−	−	−	0.19	0.11:	−	−		04	S
221	04 56 50	−70 19	5	−	1	−	9	3	9	2	−	0.19	0.11:	2.5	14.6		12	C
222	04 57 00	−66 39	21	14	−	−	−	−	−	−	−	0.26	C	C	C		04	S
223	04 57 01.1	−66 47 01	4	−	5	−	13	6	9	6	p	0.15	0.56	2.9	6.2:	04570−6647	04	C
224	04 57 06	−71 14	−	−	−	−	3	1	4	1	p:	−	−	0.8	6.2		16	C
225	04 57 08.5	−69 54 58	22	−	11	−	2	1	−	−	p	0.81	1.22	0.4:	−	04571−6954	12	M
226	04 57 09.2	−66 27 45	60	33	75	30	−	−	−	−	−	1.00:	4.99:	C	C	04571−6627	04	M
227	04 57 15	−68 08	2	−	1	−	10	7	10	8	−	0.07:	0.11:	1.2	4.2:		08	C
228	04 57 20	−68 56	14	6	4	3	38	20	33	20	−	0.30	0.11:	7.5	27.0		08	C
229	04 57 20.6	−66 23 52	35	25	15	10	−	−	−	−	−	0.37	0.56	C	C	04573−6623	04	M
230	04 57 22.5	−69 16 13	22	7	12	3	59	24	47	22	p:	0.56	1.00	14.5	52.0	04573−6916 / X04457−692	12	C
231	04 57 23.2	−70 31 24	3	−	1	−	8	2	7	3	−	0.11:	0.11:	2.5	8.3	04573−7031	12	C
232	04 57 23.3	−68 49 12	40	6	47	2	135	20	67	20	p	1.26	4.99	47.6	97.8	04573−6849 / X04457−688	08	C
233	04 57 23.8	−71 00 02	12	−	1	−	−	−	−	−	p	0.44	−	−	−	04573−7100	12	S
234	04 57 25.4	−67 25 23	3	−	1	−	5	3	5	3	−	0.11:	0.11:	0.8	4.2:	04574−6725	08	C
235	04 57 25.9	−68 29 36	80	4	120	4	305	20	135	25	p	2.81	12.88	118.0	228.8	04574−6829 / X04457−685	08	C
236	04 57 30	−68 22	4	−	3	2	17	10	15	10	−	0.15:	0.11	2.9	10.4		08	C
237	04 57 30	−69 13	15	7	3	2	−	−	−	−	−	0.30	0.11:	C	C		12	S
238	04 57 30	−71 04	−	−	1	−	5	2	4	2	−	−	0.11:	1.2	4.2		12/16	C
239	04 57 30.4	−67 07 44	2	−	1	−	7	4	8	4	−	0.07:	0.11:	1.2	8.3	04575−6707	08	C
240	04 57 32.9	−67 41 45	2	−	2	−	9	6	10	8	−	0.07:	0.22	1.2	4.2:	04575−6741	08	C

No.	RA	Dec									P					Ident	F	T
241	04 57 35	-67 17	3	-	2	-	10	5	10	7	-	0.11:	0.22	2.1	6.2		08	C
242	04 57 35	-69 35	8	3	6	3	33	22	25	20	p	0.19	0.33	4.6	10.4		12	C
243	04 57 36.1	-66 31 53	50	30	50	30	250	100	125	100	-	0.74:	2.22:	62.1	52.0	04576-6631	04	M
244	04 57 36.2	-66 19 53	20	12	12	6	55	25	25	-	-	0.30	0.67	12.4	C	04576-6619	04	M
245	04 57 37.9	-69 00 44	15	7	7	3	26	20	25	20	p:	0.30	0.44	2.5:	10.4:	04576-6900	08	C
246	04 57 40	-68 27	17	6	11	3	52	20	-	-	-	0.41	0.89	13.2	C		08	M
247	04 57 40	-69 52	4	-	-	-	3	1	2	1	-	0.15	-	0.8:	2.1:		12	C
248	04 57 40.5	-66 33 19	85	20	130	17	270	80	-	-	p:	2.40	12.54	78.7	C	04576-6633	04	M
249	04 57 56.3	-69 24 48	12	6	6	3	30	20	28	18	-	0.22	0.33	4.1	20.8	04579-6924	12	C
250	04 57 59.2	-69 04 35	14	6	4	2	20	18	20	18	p	0.30	0.22	0.8:	4.2:	04579-6904	08/12	C
251	04 58 00	-66 26	42	20	45	10	150	70	100	70	p:	0.81	3.88	33.1	62.4	04580-6626:	04	C
252	04 58 04.4	-68 11 52	4	-	1.5	-	13	8	12	10	-	0.15:	0.17	2.1	4.2:	04580-6811	08	C
253	04 58 08.7	-70 13 27	16	-	4	-	-	-	-	-	p	0.59	0.44	-	-	04581-7013	12	S
254	04 58 08.8	-67 45 32	2	-	2	-	12	7	10	8	-	0.07:	0.22	2.1	4.2:	04581-6745	08	C
255	04 58 10	-68 04	4	-	1	-	11	8	13	9	-	0.15	0.11:	1.2	8.3		08	C
256	04 58 10	-69 09	9	5	3	2	19	17	-	-	-	0.15	0.11	0.8:	C		12	M
257	04 58 20.5	-66 17	17	10	10	5	52	20	45	25	-	0.26	0.56	13.2	41.6		04	C
258	04 58 25	-70 51 44	-	-	-	-	5	2	4	2	-	-	-	1.2	4.2	04583-7051	12	M
259	04 58 25.5	-66 35 19	18	10	8	4	40	30	-	-	-	0.30	0.44	4.1	C		04	C
260	04 58 25.5	-65 53 19	2	-	1	-	5	-	3	-	p	0.07:	0.11:	2.1	6.2	04584-6553	04	C
261	04 58 27.8	-67 25 34	-	-	-	-	5	3	5	3	-	0.11:	C	0.8:	4.2:	04584-6725	08	C
262	04 58 29.5	-68 28 37	6	3	-	-	-	-	-	-	-	0.26	0.22	C	C	04584-6828	08	S
263	04 58 30	-68 57	7	-	4	2	24	16	22	20	-	-	-	3.3	4.2		08	C
264	04 58 33.0	-67 35 24	-	-	-	-	7	5	6	4	-	0.22	0.33	0.8	4.2	04585-6735	08	C
265	04 58 36.5	-70 27 28	6	-	3	-	12	5	10	4	p	0.22	0.33	2.9	12.5	04586-7027	12	C
266	04 58 39.1	-66 14 17	18	8	12	4	-	-	-	-	-	0.37	0.89	C	C	04586-6614	04	M
267	04 58 40	-69 36	5	-	4	2	25	20	20	17	-	0.19	0.22	2.1	6.2		12	C
268	04 58 45	-66 20	23	14	9	6	40	20	55	35	p:	0.33	0.33:	8.3:	41.6:	04587-6618:	04	C
269	04 58 45	-66 22	23	14	13	6	45	25	-	-	p:	0.33	0.78	8.3	C	04590-6621:	04	C
270	04 58 45	-69 58	2	-	1	-	2	1	3	2	-	0.07:	0.11:	0.4	2.1:	04586-6955:	12	C
271	04 58 46.2	-66 11 37	13	5	5	2	-	-	-	-	p:	0.30	0.33	C	C	04587-6611	04	M
272	04 58 46.6	-69 11 59	12	6	5	3	30	20	28	20	-	0.22	0.22	4.1	18.7	04587-6911	12	C
273	04 58 48.1	-68 11 39	8	-	3	1.5	-	-	-	-	p:	0.30	0.33	-	-	04588-6811	08	M
274	04 58 52.8	-69 01 52	8	4	4	1.5	23	18	20	15	-	0.15	0.28	2.1	10.4:	04588-6901	08	C
275	04 58 54.7	-68 25 07	7	3	5	-	20	12	16	12	-	0.15	0.39	3.3	8.3	04589-6825	08	C
276	04 58 56.7	-65 47 38	2	-	4	-	-	-	-	-	p	0.07:	0.44	-	-	04589-6547	04	M
277	04 58 59.7	-66 30 54	10	8	5	3	30	15	30	20	-	0.07:	0.22	6.2:	20.8:	04589-6630	04	C
278	04 59 00	-66 40	4	-	2	-	16	8	15	8	-	0.15	0.22	3.3	14.6		04	C
279	04 59 00	-70 35	-	-	-	-	3	-	2	-	-	-	-	1.2	4.2		12	C
280	04 59 00	-71 46	-	-	-	-	2	-	2	-	-	-	-	0.8	4.2		16/15	C

Infrared Sources in the LMC (continued 5).

(1) Number LI-LMC	(2) RA(1950) h m s	(2) DEC(1950) o ' "	(3) 12μm Peak	(3) Bg	(4) 25μm Peak	(4) Bg	(5) 60μm Peak	(5) Bg	(6) 100μm Peak	(6) Bg	(7) Size arcmin	(8) F 12μm Jy	(9) F 25μm Jy	(10) F 60μm Jy	(11) F 100μm Jy	(12) IRAS-Id	(13) DPM field	(14) Spectrum
							1.0E-8 Watt / (m*m sr)											
281	04 59 02.4	-69 21 48	9	5	5	2	-	-	-	-	-	0.15	0.33	C	C	04590-6921	12	M
282	04 59 05.3	-68 29 37	4	-	-	-	-	-	-	-	-	0.15:	-	C	C	04590-6829	08	S
283	04 59 15	-66 17	9	6	3	2	22	18	-	-	-	0.11	0.11	1.7	C		04	M
284	04 59 15	-66 36	6	2	3	1	10	8	-	-	-	0.15	0.22:	0.8:	C		04	M
285	04 59 19.9	-69 16 02	4	-	4	2	21	17	18	16	-	0.15:	0.22	1.7	4.2:	04593-6916	12	C
286	04 59 26.4	-69 26 40	6	3	3	2	19	17	18	16	p:	0.11:	0.11	0.8:	4.2:	04594-6926	12	C
287	04 59 40.2	-67 48 17	5	-	3	-	12	7	10	7	-	0.19	0.33	2.1	6.2	04596-6748	08	C
288	04 59 43.8	-70 54 34	-	-	-	-	3	-	2	-	-	-	-	1.2	4.2:	04597-7054	12/11	C
289	04 59 45	-66 12	9	4	4	2	17	11	22	14	-	0.19	0.22	2.5	16.6		04	C
290	04 59 50	-66 21	14	5	17	2	33	13	25	15	p:	0.33	1.66	8.3	20.8		04	C
291	04 59 52.5	-70 36 19	-	-	4	-	2	-	2	1	p	-	0.44	0.8	2.1:	04598-7036	12/11	C
292	05 00 00	-68 04	3	-	2	-	12	7	12	10	-	0.11:	0.22	2.1	4.2		08	C
293	05 00 02.0	-69 21 45	7	3	2	1	-	-	-	-	-	0.15	0.11	C	C	05000-6921	12	S
294	05 00 03	-68 39	4	-	-	-	-	-	-	-	-	0.15:	-	C	-		08	S
295	05 00 03.2	-70 13 22	18	5	10	2	56	4	35	4	p:	0.48	0.89	21.5	64.5	05000-7013	12/11	C
296	05 00 07.9	-68 46 31	4	-	2	-	9	6	11	7	-	0.15	0.22:	1.2	8.3	05001-6846	08	C
297	05 00 18.6	-67 12 21	12	4	4	-	5	3	4	3	p	0.44	0.44	0.8:	2.1:	05003-6712	08	C
298	05 00 20	-66 28	8	4	5	2	14	10	-	-	-	0.15	0.33	1.7:	C		04	M
299	05 00 20	-69 32	2	-	1.5	-	18	16	15	13	-	0.07:	0.17:	0.8	4.2		12	C
300	05 00 20	-70 45	-	-	-	-	4	1	3	1	p:	-	-	1.2	4.2		12/11	C
301	05 00 25	-68 29	11	4	1.5	-	16	10	13	10	-	0.26	0.17	2.5	6.2		08	C
302	05 00 26.4	-70 07 49	5	-	4	-	17	7	10	6	p	0.19	0.44	4.1	8.3	05004-7007	12/11	C
303	05 00 30	-70 32	6	-	-	-	-	-	-	-	-	0.22	-	-	-		12/11	S
304	05 00 31.0	-69 36 11	3	-	2	-	15	11	12	9	-	0.11:	0.22	1.7	6.2:	05005-6936	12	C
305	05 00 33.9	-65 59 02	2	-	2	-	9	4	-	-	-	0.07:	0.22	2.1	C	05005-6559	04	M
306	05 00 40	-68 10	12	7	5	2	9	7	14	11	-	0.19	0.33	0.8	6.2		08	C
307	05 00 45.2	-66 28 12	14	3	10	1	33	8	24	12	p	0.41	1.00	10.3	25.0	05007-6628	04	C
308	05 00 49.9	-67 06 53	2	-	1	-	7	3	9	5	-	0.07:	0.11:	1.7	8.3	05008-6706	08/04	W
309	05 00 50	-66 00	2	-	4	2	7	-	-	-	-	0.07:	0.22	-	-		04	M
310	05 00 57.2	-66 16 58	7	-	2	-	-	-	-	-	p	0.26	0.22	-	-	05009-6616	04	S

No.	RA	Dec	Class	Field	Cat. ID	v1	v2	v3	v4	P	n1	n2	n3	n4	s1	s2	s3	s4
311	05 01 00	-71 28 20	C	15/16	05010-6759	21.7	7.7	-	-	10x10:	1	3	-	2	-	1.5	-	-
312	05 01 01.7	-67 39 03	S	08		-	-	0.17:	0.22	-	-	-	-	-	-	-	-	6
313	05 01 05	-66 03	C	04		12.5	2.1	0.22	0.15	-	14	20	12	17	2	4	4	8
314	05 01 10	-69 02	C	08		16.6	2.5	0.22	0.19	-	18	26	17	23	2	4	5	10
315	05 01 10.9	-68 15 01	C	08	05011-6815	14.6	7.5	1.05	0.19	P	15	22	12	30	1.5	11 1.5	4	9
316	05 01 21.5	-65 58 20	C	04		6.2:	4.1	0.33	0.15:	-	12	15	12	22	1	4	2	6
317	05 01 25	-70 21	C	11/12	05013-6558	2.1:	0.8	-	-	-	1	2	3	5	-	-	-	-
318	05 01 30	-68 17	C	08/07		12.5	2.1	0.39	0.19	-	15	21	11	16	1.5	5	5	10
319	05 01 30	-70 47	C	11/12		6.2	1.2	-	-	-	3	6	3	6	-	1	-	-
320	05 01 32.1	-65 44 18	M	04	05015-6544	-	0.8:	0.11:	-	-	-	-	-	2	-	-	-	-
321	05 01 39.1	-68 05 54	S	08		-	-	0.78	2.92	P	-	-	-	-	-	7	6	85
322	05 01 40	-68 30	C	07/08	05016-6805	4.2:	1.2	0.17:	-	-	14	16	10	13	-	1.5	-	-
323	05 01 40	-69 55	C	12/11		6.2	2.5	-	-	-	4	7	4	10	-	-	-	-
324	05 01 41.4	-68 10 03	S	08	05016-6810	-	-	2.66	4.40	P	-	-	-	-	2	26	6	125
325	05 01 50	-71 06	C	15/16		4.2	1.2	-	-	-	-	2	-	3	-	-	-	-
326	05 01 54.0	-67 51 59	C	08	05019-6751	2.1:	0.8	0.44	0.26	-	8	9	6	8	-	4	-	7
327	05 01 55	-69 34	S	12/11		-	-	0.11:	0.19	-	-	-	-	-	-	1	-	5
328	05 02 00	-68 40	C	07		10.4	4.1	0.17	0.15	-	12	17	10	20	1.5	3	-	4
329	05 02 00	-68 47	C	07		8.3	2.1	0.17	0.15	-	9	13	8	13	-	1.5	-	4
330	05 02 00	-70 05	C	11		2.1:	0.4	-	-	-	2	3	4	5	-	-	-	-
331	05 02 00	-70 33	C	11/12	05020-6903	2.1	0.8	-	-	-	2	3	-	2	-	-	-	22
332	05 02 00.5	-69 03 22	M	11	05023-6822:	C	1.2:	0.33	0.41	P:	-	-	20	23	3	6	11	4
333	05 02 10	-68 23	C	07	05021-6644	8.3	1.7	0.17	0.07	-	12	16	11	15	1.5	3	2	-
334	05 02 10.3	-66 44 02	C	04		4.2	0.8	-	-	-	2	4	2	4	-	-	-	3
335	05 02 12	-71 26	M	15		-	-	0.22:	0.11:	-	-	-	5	-	-	2	-	-
336	05 02 15	-67 55	C	07/08	05023-6912	4.2:	0.8	0.17:	0.11:	-	9	11	8	10	-	1.5	-	3
337	05 02 15	-70 10	C	11/12		4.2	0.8	0.11:	0.19	-	4	6	4	6	-	1	-	5
338	05 02 19.9	-69 12 21	M	12/11		C	1.7	0.17:	-	-	-	-	14	18	1.5	3	-	-
339	05 02 20	-67 45	C	08/07	05023-6937	4.2:	1.2	0.22:	0.07:	-	8	10	6	9	-	2	-	2
340	05 02 22.7	-69 37 55	C	11/12		8.3	2.1	0.17	0.11:	-	6	10	7	12	-	1.5	-	3
341	05 02 27.1	-68 13 56	C	07	05024-6813	4.2	0.8	0.22	0.11:	1x1:	10	12	9	11	1	3	3	6
342	05 02 31.2	-69 06 24	C	11/12	05025-6906	41.3	10.3	0.78	0.72	-	31	50	32	54	5	10	11	24
343	05 02 33.9	-70 46 53	C	11	X0502-691	2.1	1.7	0.17	0.07:	P	6	7	6	10	-	1.5	-	2
344	05 02 37.4	-68 09 39	C	07	05025-7046	2.1	0.8	0.67	0.33	-	8	9	7	9	1	7	4	13
345	05 02 40	-67 04	C	04/08	05026-6809	8.3	1.2	0.22	0.22	-	9	13	7	10	-	-	-	6
346	05 02 44.2	-71 24 15	M	15	05027-7124	2.1	5.0	8.88	0.81	P	2	3	1	13	-	80	-	22
347	05 02 45.2	-69 09 00	M	11/12	05027-6908	C	2.9:	0.33	0.33	P:	-	-	33	40	4	7	10	19
348	05 02 49.5	-68 31 08	C	07	05028-6831	10.4:	4.6	0.33	0.19	P:	15	20	13	24	2	5	3	8
349	05 03 00	-66 56	C	04/03		4.2	1.7	0.22	0.15	-	10	12	8	12	1	3	-	4
350	05 03 00	-71 37	C	15		4.2	0.8	-	-	-	-	2	-	2	-	-	-	-

Infrared Sources in the LMC (continued 6).

(1)	(2) RA(1950) h m s	(2) DEC(1950) o ' "	(3) 12 µm Peak Bg		(4) 25 µm Peak Bg		(5) 60 µm Peak Bg		(6) 100 µm Peak Bg		(7) Size arcmin	(8) F 12µm Jy	(9) F 25µm Jy	(10) F 60µm Jy	(11) F 100µm Jy	(12) IRAS-Id	(13) DPM field	(14) Spectrum
Number LI-LMC			1.0E-8 Watt / (m²m sr)															
351	05 03 00.2	-65 56 50	6	-	3	-	15	9	18	12	-	0.22:	0.33	2.5	12.5	05030-6556	04	C
352	05 03 06.3	-71 54 35	4	-	-	-	-	-	-	-	-	0.15	-	-	-	05031-7154	15	S
353	05 03 10	-70 13	-	-	-	-	6	4	6	5	-	-	-	0.8	2.1:		11	C
354	05 03 14.8	-67 38 08	3	-	2	-	9	7	9	8	-	0.11:	0.22	0.8	2.1:	05032-6738	07	C
355	05 03 15	-65 53	6	-	2	1	-	-	-	-	-	0.22	0.11	-	-		04	S
356	05 03 15	-67 16	14	6	5	2	25	16	28	20	-	0.30	0.33	3.7	16.6	05032-7019	07	C
357	05 03 15.8	-70 19 18	-	-	-	-	5	3	5	3	p	-	-	0.8	4.2	05032-7041	11	C
358	05 03 17.5	-70 41 20	2	-	-	-	10	7	9	6	-	0.07:	-	1.2	6.2	05033-7122	11	C
359	05 03 21.0	-71 22 58	62	-	6	-	-	-	-	-	p	2.29	0.67	-	-	X0503-662	15	S
360	05 03 25	-66 16	2	-	2	-	5	2	5	2	-	0.07:	0.22	1.2	6.2		04/03	C
361	05 03 30	-67 50	2	-	1.5	-	12	10	13	11	-	0.07:	0.17	0.8	4.2		07	C
362	05 03 30	-68 17	-	-	1	-	8	6	8	7	-	-	0.11:	0.8	2.1		07	C
363	05 03 30.7	-65 43 54	2	-	1	-	4	2	6	2	-	0.07:	0.11:	0.8	8.3	05035-6543	04	C
364	05 03 35	-67 15	13	7	4	3	23	18	23	20	-	0.22	0.11	2.1	6.2:		07	C
365	05 03 35	-68 32	13	8	6	3	23	16	27	20	-	0.19:	0.33	2.9	14.6	05036-6832:	07	C
366	05 03 36.9	-68 59 40	-	-	-	-	17	12	16	14	-	-	-	2.1	4.2:	05036-6859	11/07	C
367	05 03 39.6	-66 49 00	9	3	7	1	12	9	14	10	p:	0.22	0.67	1.2	8.3	05036-6649	04/03	C
368	05 03 40	-68 35	13	8	6	3	-	-	2	1	-	0.19	0.33	C	C		07	M
369	05 03 40	-71 00	-	-	-	-	3	-	2	1	-	-	-	1.2	2.1		15	S
370	05 03 45	-70 46	2	-	2	-	10	8	8	7	-	0.07:	0.22	0.8	2.1:		11	C
371	05 03 51.6	-67 22 39	35	10	37	4	90	13	48	11	p	0.93	3.66	31.9	77.0	05038-6722 X0503-673	07	C
372	05 03 52.5	-68 57 15	13	3	3	-	-	-	-	-	p	0.37	0.33	C	C	05038-6857	11/07	S
373	05 03 54.3	-69 06 11	12	8	5	3	38	33	35	32	p	0.15	0.22	2.1	6.2:	05039-6906	11	C
374	05 03 55.0	-70 02 06	4	-	5	-	4	2	4	3	p	0.15	0.56	0.8:	2.1:	05039-7002	11	C
375	05 03 57.1	-67 24 37	22	12	15	4	-	-	-	-	p:	0.37	1.22	C	C	05039-6724	07	M
376	05 04 00	-71 29	4	-	-	-	-	-	-	-	-	0.15	-	-	-		15	S
377	05 04 05	-66 34	4	-	2	1	9	3	9	3	2xp:	0.26	0.35	3.1	13.6		03/04	C
378	05 04 05.1	-68 02 14	6	3	3	-	15	12	16	14	-	0.11:	0.22	1.2	4.2:	05040-6802	07	C
379	05 04 07	-66 31	5	-	-	-	-	-	-	-	-	0.19	-	-	-		03/04	S
380	05 04 10	-68 37	15	7	6	3	29	20	30	28	-	0.30	0.33	3.7	4.2:		07	C

No.	RA	Dec									flag					ID	Ref	Type
381	05 04 13.0	−65 58 25	−	−	3	−	9	4	9	6	−	−	0.33	2.1	6.2	05042-6558	04/03	C
382	05 04 13.7	−68 29 45	8	4	3	2	20	16	−	−	−	0.15:	0.11:	1.7	C	05042-6829	07	M
383	05 04 15.9	−67 20 27	20	5	8	3	−	−	−	−	P	0.56	0.56	−	−	05042-6720	07	M
384	05 04 16.6	−68 27 55	7	3	5	2	16	14	−	−	−	0.15:	0.33	0.8:	C	05042-6827	07	M
385	05 04 16.8	−71 11 08	2	−	3	−	−	−	−	−	P	0.07:	0.33	C	C	05042-7111	15	M
386	05 04 19.7	−67 15 09	12	9	6	4	22	15	20	15	−	0.11:	0.22	2.9	10.4:	05043-6715	07	C
387	05 04 25	−67 09	9	6	2	1	15	10	17	15	−	0.11	0.11:	2.1	4.2		07	C
388	05 04 30	−68 56	4	−	1.5	−	25	15	23	19	−	0.15:	0.17:	4.1	8.3		07/11	C
389	05 04 30	−69 12	10	7	5	3	−	−	−	−	−	0.11	0.22:	C	C		11	M
390	05 04 35	−69 03	9	5	5	3	34	30	32	30	−	0.15	0.22	1.7:	4.2:		11	C
391	05 04 40	−68 08 08	12	7	5	3	−	−	−	−	−	0.19	0.22	C	C		07	M
392	05 04 41.9	−65 43 45	−	−	−	−	4	2	5	3	−	−	−	0.8	4.2	05046-6543	04/03	C
393	05 04 42.8	−67 54 00	4	−	3	1.5	19	13	19	13	−	0.15	0.17	2.5	12.5	05047-6753	07	C
394	05 04 43.2	−66 44 22	16	2	26	2	47	7	23	9	P:	0.52	2.66	16.6	29.1	05047-6644	03/04	C
395	05 04 45.4	−71 10 53	2	−	5	−	16	2	11	3	P:	0.07:	0.56	5.8	16.6	05047-7110	15	C
396	05 04 47.3	−66 42 02	10	1	6	1	13	7	−	−	−	0.33	0.56	2.5	C	05047-6642	03	M
397	05 04 47.9	−66 53 28	11	4	7	1	21	8	17	12	1x1	0.39	0.93	6.1	10.9	05047-6653	03/04	C
398	05 04 50	−70 14	14	10	9	5	30	26	−	−	P:	0.15	0.44	1.7:	C	05049-7047:	11	M
399	05 04 50	−70 50	20	3	23	4	65	10	35	10	P	0.63	2.11	22.8	52.0	X0504-708	11	C
400	05 05 00	−68 33	13	8	5	2	33	24	34	31	−	0.19	0.33	3.7	6.2		07	C
401	05 05 00	−69 08	15	9	7	4	42	38	40	36	−	0.22	0.33	1.7	8.3		11	C
402	05 05 00	−69 49	−	−	1	−	6	4	7	6	−	0.11	0.11	0.8	2.1		11	C
403	05 05 00	−71 28	4	−	−	−	4	2	4	2	−	0.15:	−	0.8	4.2		15	C
404	05 05 04.5	−67 37 53	4	−	6	1.5	22	8	15	10	−	0.15:	0.50	5.8	10.4:	05050-6737	07	C
405	05 05 08.0	−68 07 31	18	7	23	4	81	20	44	18	P	0.41	2.11	25.3	54.1	05051-6807 / X0505-681	07	C
406	05 05 09.0	−68 58 11	10	6	1	1	−	−	−	−	P:	0.15	0.11:	−	−	05051-6858	11/07	S
407	05 05 10	−70 31	−	−	−	−	4	2	2	1	−	−	−	0.8	2.1		11	C
408	05 05 10.1	−67 58 44	4	−	−	−	−	−	−	−	−	0.15:	−	−	−	05051-6758	07	S
409	05 05 11.5	−70 58 30	42	−	109	3	161	9	48	7	P	1.55	11.77	62.9	85.3	05051-7058 / X0505-709	15/11	C
410	05 05 15	−66 57	3	−	5	−	16	10	−	−	p:	0.11:	0.56:	2.5:	C		03/07	M
411	05 05 15	−68 06	12	7	4	3	−	−	−	−	−	0.19	0.11	C	C		07	S
412	05 05 17.4	−70 11 29	26	10	30	7	105	45	60	35	P	0.59	2.55	24.8	52.0	05052-7011 / X0505-702	11	C
413	05 05 19.1	−69 01 37	9	6	3	3	18	16	−	−	−	0.11	0.17	0.8:	0.8:	05053-6901	11	M
414	05 05 19.3	−66 59 03	22	6	70	1.5	87	10	33	12	P	0.59	7.66	31.9	43.7	05053-6659	07/03	C
415	05 05 20	−69 21	4	−	3	1	11	8	11	9	−	0.15:	0.22	1.2	4.2		11	C
416	05 05 27.0	−67 39 19	−	−	3	−	−	−	−	−	p:	−	0.33	−	−	05054-6739	07	M
417	05 05 30	−70 09	9	8	4	3	30	26	34	30	−	0.04:	0.11	1.7	8.3		11	C
418	05 05 30	−71 05	3	−	2	4	6	4	8	5	−	0.11:	0.22	0.8	6.2		15	C
419	05 05 35	−68 11	13	7	6	1	44	35	37	33	−	0.22	0.22	3.7:	8.3:		07	C
420	05 05 38.9	−69 52 38	−	−	1	−	9	7	8	6	−	−	0.11:	0.8	4.2	05056-6952	11	C

Infrared Sources in the LMC (continued 7).

(1) Number LI-LMC	(2) RA(1950) h m s	(2) DEC(1950) o ' "	(3) 12 μm Peak	(3) 12 μm Bg	(4) 25 μm Peak	(4) 25 μm Bg	(5) 60 μm Peak	(5) 60 μm Bg	(6) 100 μm Peak	(6) 100 μm Bg	(7) Size arcmin	(8) F 12μm Jy	(9) F 25μm Jy	(10) F 60μm Jy	(11) F 100μm Jy	(12) IRAS-Id	(13) DPM field	(14) Spectrum
						1.0E-8 Watt / (m²m sr)												
421	05 05 45	-67 06	3	-	-	-	10	7	12	10	-	0.11:	-	1.2	4.2		07	C
422	05 05 45	-68 32	11	7	3	2	30	25	23	21	-	0.15	0.11:	2.1	4.2:		07	C
423	05 05 46.4	-67 56 44	-	7	1.5	-	10	9	-	-	-	-	0.17:	0.4:	-	05057-6756	07	M
424	05 05 48	-72 29	10	3	-	-	-	-	-	-	P	0.26	-	-	-		15	S
425	05 05 53.8	-68 43 04	7	-	3	1.5	-	-	-	-	P	0.26	0.17:	-	-	05058-6843	07	S
426	05 05 57.4	-66 46 38	-	-	2	-	9	6	9	7	-	-	0.22:	1.2	4.2:	05059-6646	03	C
427	05 06 00	-69 14	12	4	4	3	30	28	26	24	-	0.30	0.11:	0.8	4.2		11	C
428	05 06 00	-71 37	-	5	1	-	4	2	-	-	-	-	0.11:	0.8	-		15	M
429	05 06 00.6	-68 14 57	8	5	9	4	44	25	-	-	-	0.11:	0.56	7.9	C	05060-6814	07	M
430	05 06 05.1	-70 37 40	5	-	2	-	2	-	-	-	-	0.19	0.22	0.8	-	05060-7037	11	M
431	05 06 09.9	-65 47 01	-	-	1	-	4	2	5	4	-	-	0.11:	0.8	2.1	05061-6547	03	C
432	05 06 10	-66 47	5	-	2	-	9	6	9	7	-	0.19:	0.22	1.2	4.2:		03	C
433	05 06 10	-67 24	-	-	-	-	10	7	11	9	-	-	-	1.2	4.2		07	C
434	05 06 10.7	-68 41 38	-	-	-	-	18	12	16	15	-	-	-	2.5	2.1:	05061-6841	07	M
435	05 06 15	-68 09	14	8	7	4	-	-	-	-	-	0.22	0.33	C	C		07	M
436	05 06 20.6	-69 08 08	10	7	5	2	-	-	-	-	-	0.11:	0.33	-	-	05063-6908	11	M
437	05 06 26.7	-65 26 26	8	-	1	-	-	-	-	-	P	0.30:	0.11:	-	-	05064-6526	03	S
438	05 06 39.0	-69 03 12	2	8	4	3	23	19	20	18	-	0.07:	0.11:	1.7	4.2	05066-6903	11	C
439	05 06 39.2	-70 02 46	5	3	-	-	17	14	-	-	-	0.07:	-	1.2	-	05066-7002	11	M
440	05 06 39.3	-70 14 02	14	6	5	3	35	25	30	23	P	0.30	0.22:	4.1	14.6	05066-7014	11	C
441	05 06 40	-65 39	-	-	1	-	5	2	5	2	-	-	-	1.2:	6.2		03	C
442	05 06 40	-68 28	-	-	4	3	32	26	23	21	-	-	0.11	2.5	4.2:		07	C
443	05 06 40	-68 36	17	8	6	3	40	30	35	30	-	0.33	0.33	4.1	10.4		07	C
444	05 06 40	-69 41	6	-	1.5	-	-	-	-	-	-	0.22	0.17:	-	-		11	S
445	05 06 40	-71 02	4	-	1	-	9	5	8	4	-	0.15	0.11:	1.7	8.3		15	C
446	05 06 45	-69 35	4	-	1	-	5	3	4	3	p:	0.15:	0.11:	0.8	2.1	05068-6813	11	C
447	05 06 49.2	-68 13 15	14	7	9	4	42	24	35	23	1x1:	0.26	0.56	7.5	25.0	05068-7032	07	C
448	05 06 51.8	-70 32 10	13	3	8	1.5	34	11	25	9	-	0.55	1.00	10.8	34.8	X0506-705	11	C
449	05 06 53.8	-67 10 19	-	-	-	-	8	6	8	7	-	-	-	0.8	2.1	05068-6710	07	C
450	05 06 56.7	-70 47 46	4	-	2	-	14	8	11	9	-	0.15:	0.22:	2.5	4.2:	05069-7047	11	C

No.	Code	#	IRAS	A	B	C	D	p	n1	n2	n3	n4	n5	n6	n7	n8	Dec	RA
451	M	11		C	2.5	0.22	0.15	–	–	–	26	32	3	5	6	10	−69 17	05 07 00
452	C	11	05070−6921	6.2:	1.7:	0.11:	0.15:	–	18	21	16	20	2	3	4	8	−69 21 58	05 07 00.6
453	C	03	05070−6534	6.2	0.4	–	–	–	3	6	2	3	–	–	–	–	−65 34 20	05 07 01.4
454	C	07	05070−6757	6.2	1.2	0.11:	0.15	–	7	10	6	9	–	1	–	4	−67 57 41	05 07 03.3
455	C	11	05070−7020	2.1:	0.8	0.33	0.19	–	3	4	–	2	–	3	–	5	−70 20 43	05 07 03.3
456	C	03		4.2	1.2	–	–	–	6	8	7	10	–	–	–	–	−66 53	05 07 10
457	C	07	05072−6844	8.3	3.3	0.33	0.15	–	25	29	26	34	3	6	6	10	−68 26	05 07 15
458	M	07	05073−6850	–	1.2	0.11:	0.07:	–	–	–	20	23	2	3	–	2	−68 44 59	05 07 17.2
459	C	07		8.3:	1.2:	0.06:	0.11	–	13	17	20	23	–	2	4	7	−68 50 31	05 07 19.0
460	C	03		4.2	0.8	–	–	–	3	5	2	4	–	1.5	–	–	−66 30	05 07 20
461	C	07	05073−6752	10.4:	7.0	0.33	0.15	–	35	40	35	52	4	7	7	11	−68 36	05 07 20
462	C	11	05074−6647	4.2	0.8	0.17	–	–	10	12	8	10	–	1.5	–	–	−69 56	05 07 20
463	M	07		–	0.8:	0.44	0.22	P	–	–	4	6	–	4	–	6	−67 52 43	05 07 20.0
464	C	03		4.2	0.8	0.11:	–	–	5	7	7	9	–	1	4	–	−66 47 15	05 07 27.9
465	C	11		4.2	2.1	0.22	0.30	–	30	32	28	33	3	5	–	14	−69 06	05 07 30
466	C	11	05076−7047	2.1:	0.8	0.22	0.11	–	21	22	22	24	3	5	6	9	−69 12	05 07 30
467	C	07	05077−7119	2.1	0.8	–	–	–	8	9	7	9	–	–	–	–	−67 16	05 07 35
468	M	11		2.1	0.8	0.11:	–	P	19	20	19	21	1	2	–	–	−70 08	05 07 40
469	C	11		10.4	2.5	0.33	0.15:	–	12	17	12	18	–	3	4	8	−70 47 04	05 07 40.4
470	S	15		–	–	0.22:	0.59	–	–	–	–	–	–	2	–	16	−71 19 53	05 07 44.2
471	C	11		4.2	0.8	–	–	–	22	24	22	24	–	–	–	14	−70 12	05 07 45
472	M	07		C	3.3	0.33	0.15	–	–	–	39	47	3	6	10	12	−68 33	05 07 50
473	C	07	05080−6859	10.4:	9.5	0.22	0.15	–	35	40	36	59	5	7	8	6	−68 36	05 07 50
474	C	11	05081−6741	4.2	1.2	0.11:	0.07:	–	24	26	31	34	1	4	6	–	−69 17	05 07 50
475	C	07		4.2	1.2	0.17	–	P	18	20	24	27	1.5	3	4	–	−68 51	05 07 55
476	S	15		4.2	–	–	0.11:	–	2	4	2	4	–	–	–	3	−71 04	05 08 00
477	C	15		C	0.8	0.33	0.41	–	35	38	30	42	4	7	10	21	−71 36	05 08 00.6
478	C	11/07	05081−6855	6.2:	5.0	0.22	0.15:	P	12	13	11	13	1	3	–	4	−68 59 56	05 08 03.9
479	C	11	05082−6844	2.1:	0.8	0.11:	0.07:	–	7	9	6	11	–	1	–	–	−70 44	05 08 05
480	C	07	05083−7055	4.2	2.1	–	–	–	–	–	–	–	–	–	–	2	−67 41 38	05 08 11.2
481	M	07/11	05081−6855	C	2.1:	0.44	–	–	28	30	30	35	3	7	–	11	−68 55 33	05 08 11.9
482	C	07		4.2	3.7	0.11	0.15:	–	–	–	30	39	3	4	7	6	−68 29	05 08 12
483	M	07	05082−6844	C	2.1:	0.11:	0.07:	–	–	–	30	35	1	4	4	3	−68 44 21	05 08 15.5
484	C	11	05083−7055	C	0.8	0.11:	0.11:	–	5	6	7	9	–	1	–	–	−68 55 43	05 08 19.6
485	C	03		2.1	0.8:	0.22:	0.07:	–	–	–	–	–	–	2	–	2	−66 52	05 08 22
486	C	03		2.1:	0.8	0.33	–	–	4	5	4	6	3	3	–	–	−66 43	05 08 30
487	C	11	05079−7115:	4.2	0.8	0.11:	0.11:	–	22	24	18	20	1	2	–	–	−70 15	05 08 30
488	C	11	05085−7009	2.1	2.1	0.11:	0.11:	–	4	12	11	13	1	1	–	3	−70 32	05 08 30
489	C	15		8.3	8.3	0.11:	0.07:	–	–	8	5	10	–	1	–	3	−71 17	05 08 30
490	M	11		C	0.8:	–	–	P	–	–	18	20	1	2	–	2	−70 09 38	05 08 33.5

Infrared Sources in the LMC (continued 8).

(1) Number LI-LMC	(2) RA(1950) h m s	(2) DEC(1950) o ' "	(3) 12 μm Peak	(3) 12 μm Bg	(4) 25 μm Peak	(4) 25 μm Bg	(5) 60 μm Peak	(5) 60 μm Bg	(6) 100 μm Peak	(6) 100 μm Bg	(7) Size arcmin	(8) F 12μm Jy	(9) F 25μm Jy	(10) F 60μm Jy	(11) F 100μm Jy	(12) IRAS-Id	(13) DPM field	(14) Spectrum
491	05 08 34.2	-70 39 31	4	2	3	1.5	14	10	16	12	-	0.07:	0.17	1.7	8.3	05085-7039	11	C
492	05 08 40	-68 23	7	4	4	2	31	21	28	24	-	0.11	0.22	4.1	8.3		07	C
493	05 08 40	-68 57	16	13	7	6	-	-	-	-	-	0.11:	0.11:	C	C		07/11	C
494	05 08 40	-69 03	19	14	9	6	49	41	47	42	-	0.19	0.33	3.3	10.4		11/07	C
495	05 08 40	-69 25	23	12	8	5	45	38	40	35	-	0.41	0.33	2.9	10.4:		11	C
496	05 08 44.4	-67 13 04	5	2	-	-	10	9	11	9	-	0.11:	-	0.4:	4.2		07	C
497	05 08 50	-66 22	-	-	-	-	5	3	4	2	-	-	-	0.8	4.2	05087-6713	03	C
498	05 08 50	-70 03	2	-	2	1	16	14	14	12	-	0.07:	0.11	0.8	4.2		11	C
499	05 08 50	-70 35	4	-	2	1	13	10	13	12	-	0.15	0.11	1.2	2.1		11	C
500	05 08 53	-67 14	6	-	1	-	-	-	-	-	-	0.22	0.11:	-	-		07	S
501	05 08 54.8	-68 42 11	11	9	6	4	35	30	38	35	-	0.07:	0.22	2.1	6.2:	05089-6842	07	C
502	05 08 57.0	-69 28 28	25	18	14	9	65	42	67	43	P	0.26:	0.56	9.5	49.9	05089-6928	11	C
503	05 08 58.5	-69 06 47	18	10	6	4	34	29	35	32	-	0.30	0.22	2.1	6.2:	05089-6906	11	C
504	05 09 00	-71 21	5	-	2	1	6	2	7	5	P	0.19	0.22	1.7	4.2	05092-7121:	15	C
505	05 09 02.2	-70 50 58	-	-	1	-	8	6	-	-	-	-	0.11:	0.8:	C	05090-7050	11	M
506	05 09 05	-67 15	4	-	1.5	-	13	10	13	12	-	0.15	0.17	1.2	2.1		07	C
507	05 09 10	-68 32	18	11	6	4	42	38	-	-	-	0.26	0.22	1.7:	C	X0509-685*	07	M
508	05 09 10.3	-69 04 33	19	12	7	5	-	-	-	-	-	0.26	0.22	C	C	05091-6904	11	S
509	05 09 15	-71 53	-	-	-	-	2	-	3	2	P	-	-	0.8	2.1		15	C
510	05 09 16.1	-68 48 15	30	16	20	11	95	60	72	50	P	0.52	1.00	14.5	45.8	05092-6848	07	C
511	05 09 24	-71 35	5	-	-	-	4	2	4	3	-	0.19	-	0.8:	2.1		15	C
512	05 09 25	-70 10	5	-	1.5	-	-	-	-	-	-	0.19	0.17:	-	-		11	S
513	05 09 25.7	-67 51 03	4	-	5	1	-	-	-	-	P	0.15:	0.44	-	-	05094-6751	07	M
514	05 09 26.3	-68 33 53	20	11	23	3	68	30	52	34	P	0.33	2.22	15.7	37.4	05094-6833	07	C
515	05 09 30	-69 41	-	-	-	-	9	8	9	8	-	-	-	0.4	2.1	X0509-685*	11	C
516	05 09 30.3	-70 59 31	-	-	1	-	7	5	8	6	-	-	0.11:	0.8:	4.2:	05095-7059	11/15	C
517	05 09 31.8	-65 25 35	4	-	2	1	-	-	-	-	-	0.15:	0.22:	-	-	05095-6525	03	M
518	05 09 34.9	-68 51 44	30	18	17	14	100	85	80	70	p:	0.44	0.33	6.2	20.8:	05095-6851	07	C
519	05 09 36	-71 07	-	-	1	-	8	3	8	5	-	-	0.11:	2.1	6.2		15	C
520	05 09 38.6	-68 49 30	29	18	24	10	-	-	-	-	p:	0.41	1.55	C	C	05096-6849	07	M

No.	Type	Ref	Name	(1)	(2)	(3)	(4)	Fl	(5)	(6)	(7)	(8)	(9)	(10)	(11)	(12)	RA	Dec
521	W	11		C	0.8:	0.22	0.19	-	-	-	21	23	4	6	9	14	05 09 40	-69 21
522	S	11		-	-	-	0.15:	-	-	-	-	-	-	-	8	12	05 09 45	-69 13
523	C	11		8.3	1.2	0.22:	0.30:	-	19	23	17	20	1	3	-	8	05 09 45	-70 22
524	C	07	05098-6842	18.7	9.1	1.00	0.33	P	38	47	36	58	5	14	14	23	05 09 49.3	-68 42 23
525	C	07		14.6	3.3	0.22:	0.15	-	15	22	14	22	-	2	3	7	05 09 50	-67 58
526	S	11		-	-	-	0.11:	-	-	-	-	-	-	-	-	3	05 09 50	-69 47
527	C	11		6.2	0.8	0.11	0.19	-	9	12	8	10	1	2	-	5	05 09 50	-70 55
528	C	07	05099-6740	4.2	0.8	0.44	0.19	P	5	7	4	6	1	4	-	5	05 09 59.6	-67 40 19
529	W	07		C	C	0.33	0.19	-	-	-	-	-	7	10	15	20	05 10 00	-68 46
530	W	11		-	-	0.22	0.22	-	-	-	-	-	7	9	15	21	05 10 00	-69 28
531	W	07	05100-6850	C	C	0.67:	0.33	P:	-	-	-	-	8	14	18	27	05 10 00.0	-68 50 04
532	S	03	05100-6629	-	-	0.22	0.78	-	-	-	-	-	-	2	-	21	05 10 00.2	-66 29 03
533	C	07	05100-6812	6.2	2.1	-	-	-	14	17	13	18	-	-	-	-	05 10 03.0	-68 12 10
534	C	07/11	05101-6855A:	447.2	314.6	52.17	7.14	P	65	280	70	830	20	490	27	220	05 10 05	-68 57
535	W	11	05101-6905	C	2.1:	0.11:	0.11	-	-	-	35	40	4	5	9	12	05 10 08.8	-69 05 58
536	C	11	05101-6917	20.8:	8.3	0.22	0.41	-	35	45	35	55	7	9	14	25	05 10 09.9	-69 17 38
537	C	11	05101-6929	27.0	5.8	0.44:	0.52	P:	40	53	38	52	6	10	11	25	05 10 22.0	-69 29 34
538	C	15	05103-7131	8.3	5.0	0.56	0.11:	-	3	7	3	15	-	5	-	3	05 10 22.0	-71 31 20
539	C	11	05103-6959	2.1	0.4:	0.11:	0.19	P	7	8	8	9	-	1	-	5	05 10 22.4	-69 59 25
540	C	15		2.1	0.8	-	-	-	3	4	2	4	-	1	-	-	05 10 24	-71 39
541	W	11		C	0.8:	0.22	0.26	-	-	-	50	52	6	8	15	22	05 10 25	-69 16
542	C	11	05104-6923:	4.2:	3.3	0.33	0.19	P:	28	30	17	25	3	6	8	13	05 10 25	-69 25
543	C	07		6.2	5.4	0.22	0.11	-	20	23	11	24	2	4	6	9	05 10 30	-67 12
544	C	07	05105-6827	6.2	2.5	0.22:	0.11	-	25	28	22	28	2	4	7	10	05 10 32.4	-68 27 49
545	C	07	05105-6758	2.1:	0.8	0.11	-	-	15	16	15	17	1	2	-	-	05 10 33.3	-67 58 28
546	C	07		6.2	2.9	0.22	0.19	-	27	30	25	32	2	4	7	12	05 10 35	-68 32
547	C	07	05106-6806	2.1:	1.7	0.22	-	P:	12	13	14	18	1	3	-	-	05 10 38.2	-68 06 18
548	C	07	05106-6849	6.2:	4.1	0.33	0.15	-	37	40	35	45	6	9	12	16	05 10 38.5	-68 49 44
549	C	03	05106-6636	4.2	0.8	0.11:	-	-	3	5	4	6	-	1	-	-	05 10 39.0	-66 36 51
550	C	11	05106-6909	6.2:	2.1:	0.22:	0.11:	-	54	57	55	60	5	7	16	19	05 10 39.3	-69 09 16
551	W	11	05107-6930	C	C	0.67	0.30:	-	-	-	-	-	6	12	12	20	05 10 44.2	-69 30 07
552	C	11	05107-6953	2.1	0.8	0.22:	0.30	P	8	9	9	11	-	2	-	8	05 10 45.6	-69 53 43
553	C	07	05107-6708	16.6	8.3	0.67	0.26	P	18	26	11	31	2	8	6	13	05 10 46.1	-67 08 38
554	S	11		-	-	-	0.11:	-	-	-	-	-	1	-	7	10	05 10 50	-69 23
555	C	11		2.1	0.8	0.11:	0.19	-	6	7	5	7	-	1	-	5	05 10 50	-70 35
556	C	11	05108-7017	33.3:	4.1	1.22	0.19	P	16	17	14	24	2	13	4	9	05 10 52.3	-70 17 41
557	W	07	05108-6839	C	3.3:	0.22	0.33	P	-	-	20	28	3	5	8	17	05 10 52.9	-68 39 34
558	C	15		2.1	1.7	0.11:	0.07:	-	6	7	4	8	-	1	-	2	05 10 54	-71 05
559	C	03/07	05109-6656	8.3	1.2	-	-	-	8	12	7	10	-	-	-	-	05 10 55.3	-66 56 55
560	C	07/03	05109-6702	6.2:	1.7	0.17:	0.15:	-	12	15	10	14	-	1.5	4	8	05 10 55.9	-67 02 53

Infrared Sources in the LMC (continued 9).

(1)	(2)		(3)		(4)		(5)		(6)		(7)	(8)	(9)	(10)	(11)	(12)	(13)	(14)
Number LI-LMC	RA(1950) h m s	DEC(1950) o ' "	12 μm Peak	Bg	25 μm Peak	Bg	60 μm Peak	Bg	100 μm Peak	Bg	Size arcmin	F 12μm Jy	F 25μm Jy	F 60μm Jy	F 100μm Jy	IRAS-Id	DPM field	Spectrum
					1.0E-8 Watt / (m*m sr)													
561	05 10 56.7	-65 57 35	-	-	-	-	3	3	2	1	p:	0.11	-	1.2	2.1:	05109-6557	03	C
562	05 11 00	-68 21	9	6	4	2	22	20	24	22	-	0.15	0.22	0.8	4.2		07	C
563	05 11 00	-70 25	4	-	3	1	17	15	24	22	-	0.07:	0.22	0.8	4.2		11	C
564	05 11 00.5	-67 11 25	6	4	4	1.5	-	-	-	-	-	0.11	0.28	C	C	05110-6711	07	M
565	05 11 01.4	-72 08 13	4	1	-	-	-	-	-	-	p	-	-	-	-	05110-7208	15	S
566	05 11 05	-69 03	23	18	11	8	-	-	-	-	-	0.19	0.33	C	C	05110-6616	11	M
567	05 11 05.5	-66 16 35	5	-	3	-	-	-	-	-	p	0.19	0.33	-	-		03	M
568	05 11 10	-68 45	16	10	6	4	40	28	33	29	-	0.22	0.22	5.0	8.3	05112-6843:	07	C
569	05 11 15	-69 41	9	2	1.5	-	-	-	-	-	p:	0.26	0.17:	-	-		11	S
570	05 11 17.3	-67 55 49	16	5	5	2	15	12	16	15	p	0.41	0.33	1.2	2.1:	05112-6755	07	C
571	05 11 18.4	-67 39 57	9	-	3	1.5	17	14	16	15	p	0.33	0.17	-	-	05113-6739	07	S
572	05 11 20	-67 47	4	-	1.5	-	-	-	-	-	-	0.15	0.17:	1.2	2.1		07	C
573	05 11 20	-68 57	22	15	13	9	14	11	-	-	-	0.26	0.44	C	C		07/11	M
574	05 11 20	-70 08	-	-	1	-	16	12	12	10	-	-	0.11:	1.2	4.2		11	C
575	05 11 20.1	-69 39 07	8	-	3	1	-	-	-	-	p:	0.30	0.22	1.7:	C	05113-6939	11	M
576	05 11 24	-71 13	6	-	-	-	4	2	4	3	-	0.22	-	0.8:	2.1		15	S
577	05 11 30	-66 30	-	-	-	-	-	-	-	-	-	-	-	C	C		03	C
578	05 11 30	-66 56	5	-	1	-	13	9	13	11	-	0.19	0.11:	C	C		03	S
579	05 11 30	-67 27	3	-	1.5	-	32	26	27	24	-	0.11:	0.17:	1.7	4.2		07	C
580	05 11 30	-68 39	12	8	4	3	-	-	-	-	p	0.15	0.11	2.5	6.2		07	C
581	05 11 38.4	-71 45 04	2	-	-	-	4	2	4	3	-	0.07:	-	0.8	2.1	05116-7145	15	C
582	05 11 40	-69 12	30	18	13	8	80	70	75	70	-	0.44	0.56	4.1:	10.4:		11	C
583	05 11 48	-71 07	-	-	1.5	-	8	5	8	6	-	-	0.17:	1.2	4.2		15	C
584	05 11 48.4	-70 18 37	9	3	3	1	13	9	15	14	-	0.22	0.22	1.7:	2.1:	05118-7018	11	M
585	05 11 49.7	-69 36 16	35	9	7	4	33	30	-	-	p	0.96	0.33	1.2	-	05118-6936	11	C
586	05 11 50	-69 06	33	24	13	10	70	65	75	70	-	0.33	0.33	2.1:	10.4:		11	C
587	05 11 50	-69 20	-	-	-	-	30	28	26	25	-	-	-	0.8	2.1		11	C
588	05 11 51.7	-68 47 17	-	-	4	3	-	-	-	-	-	-	0.11:	-	-	05118-6847	07	M
589	05 11 55.6	-68 52 28	8	6	4	3	36	30	27	24	-	0.07:	0.11:	2.5	6.2:	05119-6852	07	C
590	05 12 00	-69 46	8	5	4	2	25	21	28	25	-	0.11:	0.22	1.7	6.2		11	C

Cl	Cd	Name	f1	f2	f3	f4	p	a	b	c	d	e	f	g	h	Dec	RA	No.
C	11	05120-6931	10.4	5.8	0.44	0.30	p:	37	42	33	47	6	10	14	22	-69 31 19	05 12 04.3	591
C	07	05120-6717	14.6:	3.3	0.22	0.07:	-	18	25	16	24	2	4	4	6	-67 17 46	05 12 05.0	592
M	11	05121-7042	2.1	0.8	0.11:	0.07:	p:	9	10	8	10	-	1	-	2	-70 42 18	05 12 08.1	593
S	07		C	C	0.44	0.26	-	-	-	-	-	2	6	6	13	-67 15	05 12 10	594
S	07		-	-	0.11:	0.15	-	-	-	-	-	1	2	5	9	-67 53	05 12 10	595
C	03		2.1	0.8	-	-	p:	1	2	-	2	-	3	-	-	-66 04	05 12 15	596
C	07	05122-6831	8.3:	1.2	0.17	-	-	16	20	18	21	1.5	3	-	-	-68 31 22	05 12 15.4	597
C	07		4.2:	2.1	0.17	0.15	-	20	22	23	28	1.5	7	6	10	-68 35	05 12 20	598
C	07	05124-6712	29.1	6.2	0.56	0.33	p:	18	32	19	34	2	6	6	15	-67 12 22	05 12 29.5	599
M	11		C	2.1:	0.22:	0.19	-	-	-	45	50	4	6	10	15	-69 23	05 12 30	600
C	11		4.2:	0.8:	0.22:	0.22	-	33	35	29	31	3	5	10	16	-69 43	05 12 30	601
C	11		2.1	0.4:	0.11:	-	-	7	8	8	9	-	1	-	-	-70 14	05 12 30	602
M	11	05125-7035	C	0.8:	0.44	0.41	p	-	-	11	13	2	6	6	17	-70 35 52	05 12 32.4	603
C	07	05125-6802	4.2:	1.7	0.17	-	-	10	12	10	14	-	1	-	-	-68 02 59	05 12 32.7	604
C	11	05125-7053	2.1:	0.8:	-	-	-	7	8	6	8	-	1.5	-	-	-70 53 27	05 12 33.4	605
C	15		4.2	0.8	-	0.07:	-	3	5	2	4	-	-	-	2	-71 31	05 12 36	606
C	07		4.2	0.8	0.17	0.11:	-	17	19	17	19	3	6	-	3	-68 22	05 12 40	607
M	11	05126-6937	C	1.7	0.22	-	-	-	-	38	42	1.5	4	-	-	-69 37 32	05 12 40.0	608
C	15	05127-7113	6.2	0.8	0.11:	0.11:	-	5	8	7	9	6	1	-	3	-71 13 46	05 12 42.0	609
C	11	05127-7030: X0512-705	10.4:	8.3	0.78	0.44	p:	30	35	25	45	4	11	7	19	-70 32	05 12 45	610
M	11	05127-6911	C	C	0.33	-	p:	-	-	-	-	8	11	-	-	-69 11 23	05 12 45.8	611
C	07	05128-6723	-	-	0.33	0.26	p	-	-	-	-	2	5	6	13	-67 23 08	05 12 49.6	612
C	07		4.2	1.2	0.17	0.11	-	21	23	19	22	1.5	3	5	8	-67 38	05 12 50	613
C	11		10.4	1.7	0.33	0.44	-	60	65	54	58	8	11	20	32	-69 07	05 12 50	614
S	11	05128-6919	-	-	-	0.22	-	-	-	-	-	-	-	13	19	-69 19 27	05 12 52.7	615
C	15		16.6	2.5	0.33	0.30	-	5	13	6	12	3	3	1	9	-71 12	05 12 57	616
C	07		6.2	1.2	0.11:	0.11	-	13	16	12	15	1	2	4	7	-68 13 55	05 12 57.2	617
C	03		4.2	0.8	-	-	-	4	6	4	6	-	-	-	-	-66 43	05 13 00	618
M	11	05129-6813	C	1.2	0.22:	0.11:	-	-	-	20	23	3	5	6	9	-70 26	05 13 00	619
C	11		25.0	19.0	2.00	0.74	p	27	39	27	73	4	22	8	28	-70 27 59	05 13 00.2	620
C	15	05130-7027	6.2	0.8	0.17	0.11:	-	5	8	7	9	-	1.5	-	3	-71 06 11	05 13 02.6	621
M	11		C	0.8	0.22	0.22	-	-	-	12	14	3	5	5	11	-70 36	05 13 10	622
S	11	05132-6941	C	C	0.22:	0.26:	-	-	-	-	-	7	9	14	21	-69 41 08	05 13 12.0	623
C	07		4.2	2.1	-	-	-	18	20	15	20	-	-	-	-	-67 41	05 13 15	624
C	03	05132-6654	4.2:	1.2:	-	-	-	8	10	7	10	-	-	-	-	-66 54 28	05 13 17.0	625
C	15		2.1	0.8	-	-	-	3	4	2	4	-	-	-	-	-71 28	05 13 18	626
M	11		C	3.3	0.22	0.15	-	-	-	62	70	9	11	15	19	-69 11	05 13 20	627
C	11	05133-6948	6.2	1.2	-	-	-	14	17	16	19	-	-	-	-	-69 48 21	05 13 20.3	628
C	11	05133-6931	6.2:	2.1:	0.22:	0.19	-	37	40	41	46	6	8	10	15	-69 31 53	05 13 21.1	629
C	11	05133-7001	4.2	0.8	-	-	-	6	8	7	9	-	-	-	-	-70 01 28	05 13 21.1	630

Infrared Sources in the LMC (continued 10).

(1) Number LI-LMC	(2) RA(1950) h m s	(2) DEC(1950) o ' "	(3) 12 μm Peak	(3) Bg	(4) 25 μm Peak	(4) Bg	(5) 60 μm Peak	(5) Bg	(6) 100 μm Peak	(6) Bg	(7) Size arcmin	(8) F 12μm Jy	(9) F 25μm Jy	(10) F 60μm Jy	(11) F 100μm Jy	(12) IRAS-Id	(13) DPM field	(14) Spectrum
							1.0E-8 Watt / m*m sr											
631	05 13 22.6	-69 37 07	20	14	9	6	60	50	50	45	-	0.22:	0.33:	4.1:	10.4:	05133-6937	11	C
632	05 13 25.8	-67 31 57	18	7	9	4	45	28	-	-	p:	0.41	0.56	7.0	C	05134-6731	07	W
633	05 13 30	-69 44	19	13	5	3	43	40	46	45	-	0.22	0.22	1.2	2.1:		11	C
634	05 13 35	-69 39	33	20	17	13	70	65	-	-	-	0.48	0.44	2.1:	C		11	W
635	05 13 40.2	-69 25 37	190	27	400	30	770	150	260	115	p	6.03	41.07	256.7	301.6	05136-6925	11	C
636	05 13 43.8	-69 14 17	33	13	30	7	100	60	68	60	p	0.74	2.55	16.6	16.6:	05137-6914	11	C
637	05 13 45	-69 01	-	-	3	2	31	28	27	26	-	-	0.11:	1.2	2.1		11/07	C
638	05 13 45.4	-67 02 10	5	2	2	1	14	10	14	11	-	0.11:	0.11	1.7	6.2:	05137-6702	07/03	C
639	05 13 47.3	-67 14 30	180	5	17	1.5	-	-	-	-	-	6.47	1.72	12.4	-	05137-6714	07	S
640	05 13 50	-69 21	35	22	30	18	160	130	110	70	p:	0.48	1.33	12.4	83.2		11	C
641	05 13 53.7	-67 10 26	17	7	7	2	36	26	28	17	-	0.37	0.56	4.1	22.9	05138-6710	07	C
642	05 13 55.9	-67 30 39	50	14	90	5	175	28	76	22	p	1.33	9.43	60.9	112.3	05139-6730	07	C
643	05 14 00	-67 24	11	6	6	2	20	16	-	-	-	0.19	0.44	1.7:	C	X0513-674	07	W
644	05 14 00	-69 09	15	11	6	4	-	-	-	-	-	0.15	0.22:	-	-		11	W
645	05 14 00	-70 00	-	-	-	-	10	8	-	-	-	-	-	0.8	-		11	W
646	05 14 02.1	-67 26 12	18	7	18	3	59	27	-	-	p	0.41	1.66	13.2	C	05140-6726	07	W
647	05 14 06	-71 11	4	-	4	-	10	6	10	5	p	0.15	0.44	1.7	10.4		15	C
648	05 14 07.0	-69 38 57	40	20	32	13	135	65	85	50	pxl	0.92	2.52	30.9	74.5	05141-6938	11	C
649	05 14 07.3	-66 27 41	3	-	1	-	7	4	7	4	p:	0.11:	0.11:	1.2	6.2	05141-6627	03	C
650	05 14 12	-71 42	-	-	-	-	3	2	4	2	-	-	-	0.4	4.2		15	C
651	05 14 12	-71 48	-	-	1	-	4	2	4	2	-	-	0.11:	0.8	4.2		15	C
652	05 14 15	-66 19	-	-	-	-	5	2	4	2	-	-	-	1.2	4.2		03	C
653	05 14 15	-68 50	9	6	5	3	32	24	28	25	-	0.11	0.22	3.3	6.2		07	C
654	05 14 15	-69 17	20	16	9	7	-	-	-	-	-	0.15	0.22	C	C		11	W
655	05 14 15	-70 18	13	6	3	1	21	17	20	18	p	0.26	0.22	1.7	4.2:	05146-7018:	11	C
656	05 14 20	-67 34	15	8	7	4	-	-	-	-	-	0.26	0.33	C	C		07	W
657	05 14 30	-67 38	10	5	3	1.5	22	20	-	-	-	0.19	0.17	0.8:	C		07	W
658	05 14 30	-70 47	-	-	-	-	7	6	8	7	-	-	-	0.4	2.1		11	C
659	05 14 40	-69 13	16	11	6	5	48	44	45	40	-	0.19	0.11:	1.7:	10.4:		11	C
660	05 14 40	-70 14	16	7	4	-	23	17	26	11	-	0.33	0.44	2.5	31.2	05145-7012:	11	C

No.	RA (1950)	Dec (1950)	1	2	3	4	fl	Name	Ref	Type
661	05 14 45	−68 25	4.2	1.2	0.17:	−	−	05148−6715	07	C
662	05 14 48.5	−67 15 22	14.6	5.4	0.56	0.22	p	05148−6730	07	C
663	05 14 53.4	−67 30 36	C	0.8:	0.44	0.26	−		07	M
664	05 14 55.4	−72 05 57	−	−	0.17:	0.22:	p:	05149−7205	15	S
665	05 14 58	−69 33	25.0	18.6	1.55	0.11:	−	05148−6933:	11	C
666	05 15 00	−66 00	2.1:	0.8	−	−	−		03	C
667	05 15 00	−66 29	4.2	1.7	−	−	−	X0514−664	03	C
668	05 15 00	−69 28	20.8	6.2	0.22	0.19	−		11	M
669	05 15 00	−69 30	C	4.1	0.22	0.19	−		11	M
670	05 15 00	−71 33	−	−	−	0.15:	−		15	S
671	05 15 03.5	−69 42 36	−	−	0.44	0.30	p	05150−6942	11	W
672	05 15 06.6	−68 58 01	6.2	1.2	0.22	0.15	−	05151−6858	07/11	C
673	05 15 11.7	−69 05 13	4.2:	1.2	0.11:	0.11	−	05151−6905	11	C
674	05 15 17.5	−67 59 34	6.2	2.1	0.17	0.07:	p:	05152−6759	07	C
675	05 15 19.0	−66 22 24	4.2:	0.8	0.11:	−	−	05153−6622	03	C
676	05 15 20	−71 06	6.2	1.2	−	−	p:		15	C
677	05 15 24	−71 41	2.1	1.2	−	−	−		15	C
678	05 15 24.2	−65 35 48	C	1.2:	2.00	4.07	p	05154−6535	03	S
679	05 15 25.8	−69 22 02	−	C	0.22	0.15	−	05154−6922	11	M
680	05 15 26.8	−67 34 44	4.2:	1.7	0.11:	0.11:	−	05154−6734	07	C
681	05 15 30	−69 04	4.2	1.2	0.22	0.19	−	05155−6914	11/07	C
682	05 15 31.3	−69 14 25	4.2:	1.7	−	0.07:	−	05155−7004	11	C
683	05 15 31.6	−70 04 41	6.2	0.8:	0.22:	0.19	−	05155−7036	11	C
684	05 15 33.9	−70 36 53	10.4	1.7	0.44	0.22	−	X0515−706	11	C
685	05 15 34.4	−72 15 34	2.1:	0.4:	−	−	−	05155−7215	15	C
686	05 15 36	−71 04	−	−	−	0.26	p		15	S
687	05 15 37.6	−68 52 24	−	−	0.11:	0.11:	−	05156−6852	07	S
688	05 15 38.7	−69 55 30	2.1	0.8	0.17:	0.11	−	05156−6955	11	C
689	05 15 40	−68 17	2.1	2.1	0.17	0.11	−		07	C
690	05 15 40	−69 01	4.2:	1.2:	0.22	0.11	−		11/07	C
691	05 15 44.0	−66 45 01	14.6	2.5	0.22	0.19	p:	05157−6645	03	C
692	05 15 49.3	−68 02 16	10.4	6.2	0.67	0.37	p	05158−6802	07	C
693	05 15 49.5	−68 08 01	6.2:	0.4:	0.28	0.93	−	05158−6808	07	C
694	05 15 50	−68 27	−	−	0.11	0.11	−		07	C
695	05 15 50	−69 27	C	C	0.22:	0.26	−		11	S
696	05 15 50	−70 31	−	−	0.11:	0.19	−		11	S
697	05 15 55	−70 01	4.2	1.7	0.11:	0.11	−		11	C
698	05 15 59.5	−70 37 39	C	0.8:	0.22	0.19	−	05159−7057	11	M
699	05 16 00	−68 06	−	−	0.11:	0.26	−		07	S
700	05 16 00	−68 11	−	−	0.11:	0.11	−		07	S

Infrared Sources in the LMC (continued 11).

(1)	(2)		(3)		(4)		(5)		(6)		(7)	(8)	(9)	(10)	(11)	(12)	(13)	(14)
Number LI-LMC	Position RA(1950) h m s	DEC(1950) o ' "	12 μm Peak	Bg	25 μm Peak	Bg	60 μm Peak	Bg	100 μm Peak	Bg	Size arcmin	F 12μm Jy	F 25μm Jy	F 60μm Jy	F 100μm Jy	IRAS-Id	DPM field	Spec-trum
					1.0E-8 Watt / (m²m sr)													
701	05 16 00	-69 48	8	6	4	3	26	23	-	-	-	0.07	0.11	1.2	C		11	M
702	05 16 00	-71 03	2	-	1	-	4	2	4	2	-	0.07:	0.11:	0.8	4.2		15	C
703	05 16 00	-71 22	-	-	-	-	4	2	6	4	-	-	-	0.8	4.2		15	C
704	05 16 05	-66 55	7	-	1	-	5	4	6	5	p:	0.26:	0.11:	0.4	2.1		03	C
705	05 16 09.9	-66 12 10	-	-	-	-	3	-	4	2	-	-	-	1.2:	4.2	05161-6612	03	C
706	05 16 10	-68 21	11	7	4	3	28	22	27	24	-	0.15	0.11	2.5	6.2		07	C
707	05 16 10	-69 23	18	14	9	7	75	71	52	50	-	0.15	0.22	1.7:	4.2:		11	C
708	05 16 10.9	-69 40 27	-	-	2	1	16	13	15	13	-	-	0.11:	1.2	4.2:	05161-6940	11	C
709	05 16 30	-68 46	12	8	7	4	40	30	-	-	-	0.15	0.33	4.1:	C		07	M
710	05 16 30	-68 49	13	8	7	5	50	40	45	38	-	0.19	0.22	4.1	14.6		07	C
711	05 16 30	-69 20	20	16	12	9	-	-	-	-	-	0.15:	0.33:	C	C		11	M
712	05 16 30	-69 48	11	7	4	2	38	32	33	30	-	0.15	0.22	2.5	6.2		11	C
713	05 16 30	-69 50	11	6	4	3	-	-	-	-	-	0.19	0.11:	C	C		11	S
714	05 16 33	-70 31	3	-	2	-	-	-	-	-	-	0.11:	-	-	-		11	S
715	05 16 34.6	-71 50 47	3	-	2	-	8	3	5	2	-	0.11:	0.22	2.1	6.2	05165-7150	15	C
716	05 16 35	-69 12	13	8	8	6	58	41	42	37	-	0.19	0.22	7.0	10.4	05164-6911:	11	C
717	05 16 40	-68 14	9	5	4	2	25	18	25	23	-	0.15	0.22	2.9	4.2		07	C
718	05 16 40	-68 18	10	7	4	2	-	-	-	-	-	0.11	0.22	C	C		07	M
719	05 16 44.1	-68 25 17	14	6	4	2	20	18	23	18	p	0.30	0.22	0.8	10.4:	05167-6825	07	C
720	05 16 44.9	-67 29 16	5	-	2	-	14	11	16	13	-	0.19	0.22:	1.2	6.2	05167-6729	07	C
721	05 16 50	-68 03	12	6	4	2	23	19	23	22	p:	0.22	0.22	1.7:	2.1:		07	C
722	05 16 50	-69 57	10	3	-	-	-	-	-	-	p:	0.26	-	-	-		11	S
723	05 16 55.3	-67 22 55	15	3	10	1	33	10	23	11	p	0.44	1.00	9.5	25.0	05169-6722 X0516-674	07	C
724	05 16 55.4	-68 52 57	13	10	6	4	46	43	-	-	-	0.11	0.22:	1.2:	C	05169-6852	07	M
725	05 17 00.0	-69 30 40	12	7	-	-	-	-	-	-	-	0.19:	-	-	-	05169-6930	11	S
726	05 17 01.1	-71 37 11	-	-	4	-	-	-	-	-	p	-	0.44	-	-	05170-7137	15	M
727	05 17 02.6	-71 56 45	5	-	5	2	5	2	4	2	-	0.19	0.56	1.2	4.2	05170-7156	15	C
728	05 17 03.4	-69 27 09	33	20	12	9	100	80	-	-	-	0.48	0.33:	8.3:	C	05171-6927	11	M
729	05 17 06.2	-66 03 20	2	-	3	-	9	3	-	-	p:	0.07:	0.33	2.5	C	05171-6603	03	M
730	05 17 10	-68 22	11	6	4	2	-	-	-	-	-	0.19:	0.22:	-	-		07	M

No.	RA	Dec	q	p	Ident	S1	S2	S3	S4	str	a	b	c	d	e	f	g	h
731	05 17 10	−69 19	C	11		10.4:	4.1:	0.33:	0.30:	−	80	75	130	120	15	12	25	33
732	05 17 11.8	−70 48 04	W	11	05171−7048	−	0.4:	0.17	0.26	P	−	−	4	3	−	1.5	−	7
733	05 17 15	−68 04	C	07		6.2	3.3	−	−	−	25	22	26	18	5	6	10	14
734	05 17 15	−68 56	W	07		C	2.1:	0.11	0.15	−	17	15	45	40	6	5	10	7
735	05 17 20	−67 55	C	07		4.2	1.2	0.17	0.15	−	−	15	14	11	3	1.5	3	7
736	05 17 20	−69 11	S	11		−	−	0.11	0.15	−	−	−	−	−	7	6	9	13
737	05 17 20	−71 17	C	15		6.2:	1.2	0.22	0.15	P:	8	5	7	4	3	1	−	4
738	05 17 27.4	−69 36 47	C	11	05174−6936	16.6:	16.6	2.22	0.52	P	38	30	70	30	26	6	8	22
739	05 17 27.7	−66 34 01	C	03	05174−6634	4.2	0.8	−	0.07:	P:	8	6	6	4	7	6	−	2
740	05 17 30	−69 42	W	11		C	C	0.22	0.15	−	−	−	−	−	7	5	7	11
741	05 17 32.6	−66 45 53	C	03	05175−6645 / X0517−667	67.0	43.3	10.39	1.33	1x2	34	5	85	5	52	−	18	−
742	05 17 38.5	−69 22 43	S	11	05176−6922	C	C	0.22:	0.37	−	−	−	−	−	17	15	28	38
743	05 17 38.5	−69 58 31	C	11	05176−6958	10.4	3.3:	0.44:	0.19:	−	42	37	44	36	10	6	−	15
744	05 17 40	−67 37	C	07		4.2	1.7	0.17:	−	−	11	9	12	8	1.5	−	−	−
745	05 17 40	−68 59	C	07/11		10.4	2.5	0.22	0.30	−	45	40	50	44	7	5	11	19
746	05 17 40	−71 25	C	15		4.2	0.8	−	−	P:	6	4	4	2	−	−	8	−
747	05 17 42.9	−68 23 40	C	07	05177−6823	4.2:	2.1:	0.56	0.19:	−	17	15	20	15	7	2	3	8
748	05 17 45	−66 06	C	03		17.7	3.2	0.50	0.56:	3x1:	12	5	11	6	2	−	−	6
749	05 17 45	−69 02	W	11/07		C	1.2:	0.22	0.11	−	−	−	45	42	6	4	9	12
750	05 17 49.1	−68 38 40	S	07	05178−6838	−	−	0.06:	0.26	P:	−	−	−	−	2	1.5	−	7
751	05 17 50	−68 52	S	07		C	C	0.11	0.15	−	−	−	38	−	6	5	8	12
752	05 17 50	−70 01	W	11		C	0.8	0.11	0.19	−	−	−	16	36	4	3	9	14
753	05 17 55	−67 57	W	07	05179−6758:	C	0.8	0.17:	−	−	−	−	14	−	3	1.5	−	−
754	05 17 56.1	−69 29 42	C	11	05179−6929	10.4:	10.3	0.44	0.37	−	72	67	95	70	13	9	14	24
755	05 17 57.8	−69 49 24	C	11	05179−6949	12.5	6.2	0.33	0.30	−	47	41	60	45	7	4	8	16
756	05 17 59.3	−69 18 37	S	11	05179−6918	C	C	C	0.41	−	−	−	−	53	−	−	−	47
757	05 18 00	−68 50	W	07		8.3	9.5	0.22	0.19	−	41	37	30	−	7	5	−	14
758	05 18 00	−69 05	C	11/07		10.4	2.5	0.33	0.22	−	45	40	58	64	9	6	12	18
759	05 18 00	−69 09	C	11		27.0	6.6	0.44	0.33	−	63	50	68	84	12	8	13	22
760	05 18 04.2	−66 23 49	W	03	05180−6623	C	0.8	0.22	−	−	−	−	4	6	2	−	−	−
761	05 18 05	−65 35 01	S	03		−	−	−	0.19	−	−	−	−	−	−	−	5	−
762	05 18 08.9	−71 35 01	C	15	05181−7135	4.2	−	0.11:	0.48	P	4	2	−	−	1	−	−	13
763	05 18 12.4	−72 44 56	S	15	05182−7244	−	−	−	0.26:	P	−	−	−	−	−	−	3	10
764	05 18 13.3	−69 18 59	C	11	05182−6918	41.6	26.9	1.78	0.56	−	130	110	220	150	50	34	30	45
765	05 18 13.8	−69 24 42	C	11	05182−6924	20.8:	8.3	0.44	0.41	−	90	80	155	130	20	16	28	39
766	05 18 14.2	−71 18 00	C	15	05182−7117 / X0518−713	95.8	49.4	5.46	2.23	2x2	44	5	85	5	23	1	27	−
767	05 18 15	−69 48	W	11		C	0.8:	0.11	0.15	−	−	−	60	55	7	6	7	11
768	05 18 15	−69 55	C	11	05181−6953:	10.4	4.1	0.22	0.22	−	45	40	55	45	7	5	12	18
769	05 18 15	−70 19	C	11		6.2	2.5	0.11	0.07	−	19	16	23	17	3	2	5	5
770	05 18 20	−69 33	W	11	05181−6932:	−	−	0.33	0.26	−	−	−	−	−	9	6	15	22

Infrared Sources in the LMC (continued 12).

(1)	(2)			(3)		(4)		(5)		(6)		(7)	(8)	(9)	(10)	(11)	(12)	(13)	(14)
Number LI-LMC	RA(1950) h m s	DEC(1950) o ' "		12 μm Peak	Bg	25 μm Peak	Bg	60 μm Peak	Bg	100 μm Peak	Bg	Size arcmin	F 12μm Jy	F 25μm Jy	F 60μm Jy	F 100μm Jy	IRAS-Id	DPH field	Spectrum
	Position			1.0E-8 Watt / (m×m sr)															
771	05 18 24.6	-66 40 35		-	-	2	-	9	6	10	8	p:	-	0.22	1.2	4.2	05184-6640	03	C
772	05 18 28.7	-69 35 42		26	10	11	6	70	60	-	-	p:	0.59	0.56	4.1:	c	05184-6935	11	W
773	05 18 29.7	-70 40 43		-	-	-	-	10	8	10	9	-	-	-	0.8	2.1	05184-7040	11	C
774	05 18 30	-65 58		4	-	1.5	-	7	5	9	6	-	0.15:	0.17:	0.8	6.2	-	03	C
775	05 18 30	-67 36		6	-	1	-	-	-	-	-	-	0.22	0.11:	-	-	-	07	S
776	05 18 32.3	-67 29 37		-	-	1	-	8	7	9	7	-	-	0.11:	0.4:	4.2:	05185-6729	07	C
777	05 18 33.1	-68 06 29		9	4	5	1.5	16	14	15	14	-	0.19:	0.39	0.8	2.1:	05185-6806	07	C
778	05 18 36.2	-68 56 58		8	5	4	3	27	24	23	21	-	0.11:	0.11:	1.2	4.2:	05186-6856	07	C
779	05 18 40	-67 04		2	-	1	-	10	6	9	7	-	0.07:	0.11:	1.7	4.2	-	07/03	C
780	05 18 41.1	-68 11 56		-	-	1.5	-	17	13	15	13	-	-	0.17:	1.7	4.2:	05186-6811	07	C
781	05 18 43.6	-70 04 42		16	8	6	4	38	26	36	25	-	0.30:	0.22	5.0	22.9	05187-7004	11	C
782	05 18 44.1	-70 33 32		5	-	7	-	7	6	7	6	-	0.19	0.78	0.4:	2.1:	05187-7033	11	C
783	05 18 48.5	-67 07 48		2	-	2	-	11	8	9	8	-	0.07:	0.22	1.2	2.1:	05188-6707	07	C
784	05 18 50	-69 10		18	13	11	8	90	85	72	68	-	0.19	0.33	2.1:	8.3:	-	11	C
785	05 18 50.2	-69 43 01		48	28	35	20	210	150	100	60	p	0.74	1.66	24.8	83.2:	05188-6943	11	C
786	05 18 55.2	-70 08 39		13	7	8	4	28	24	-	-	p	0.22	0.44	1.7	c	05189-7008	11	W
787	05 19 00	-66 18		5	-	2	-	7	4	10	7	-	0.19	0.22:	1.2	6.2	-	03	C
788	05 19 00	-66 31		5	-	2	-	10	6	12	8	-	0.19	0.22:	1.7	8.3	-	03	C
789	05 19 00	-69 18		70	30	100	60	320	180	170	130	p	1.48	4.44	58.0	83.2	05192-6917:	11	C
790	05 19 00	-69 54		17	13	8	6	53	50	45	42	-	0.15	0.22	1.2	6.2	-	11	C
791	05 19 00	-71 30		3	-	-	-	-	-	-	-	p:	0.11:	-	-	-	15	S
792	05 19 00.2	-69 28 11		39	23	20	14	140	100	85	70	p:	0.59	0.67	16.6:	31.2	05190-6928	11	C
793	05 19 03.5	-67 48 23		10	2	3	1	-	-	-	-	p	0.30	0.22	-	c	05190-6748	07	S
794	05 19 10	-69 37		-	-	16	12	-	-	-	-	-	-	0.44	c	c	05191-6936	11	W
795	05 19 10	-70 09		13	9	7	4	20	18	20	18	-	0.15	0.33	0.8	4.2:	05193-7009:	11	C
796	05 19 14.2	-68 33 49		4	-	2	1	20	17	20	16	-	0.15:	0.11:	1.2	8.3	05192-6833	07	C
797	05 19 15	-67 59		9	4	4	3	20	18	20	18	-	0.19	0.11	0.8:	4.2:	-	07	C
798	05 19 16.5	-68 24 22		13	6	14	3	52	23	33	24	p	0.26	1.22	12.0	18.7	05192-6824	07	C
799	05 19 20	-70 22		4	-	2	1	18	15	15	14	-	0.15	0.11:	1.2	2.1	-	11	C
800	05 19 23.6	-67 54 39		14	6	3	2	-	-	-	-	p	0.30	0.11:	-	-	05193-6754	07	S

No.	RA	Dec									p					Name	Code	Type
801	05 19 25	-67 44	5	-	3	1.5	11	15	14	16	-	0.19	0.17:	1.7	4.2	-	07	C
802	05 19 30	-67 05	5	-	1	-	6	8	-	-	-	0.19	0.11:	0.8	-	-	07/03	M
803	05 19 30	-67 15	2	-	1	-	8	10	8	9	-	0.07:	0.11:	0.8	2.1	-	07	C
804	05 19 30	-69 41	110	30	130	30	200	500	100	220	p	2.96	11.10	124.2	249.6	05196-6941:	11	C
805	05 19 30	-69 53	24	16	8	6	-	-	-	-	-	0.30	0.22	C	C	-	11	S
806	05 19 30.3	-69 33 09	28	24	16	13	120	125	80	85	-	0.15:	0.33	2.1:	10.4	05195-6933	11	C
807	05 19 30.7	-69 12 00	75	30	85	23	180	260	100	130	p	1.66	6.88	33.1	62.4:	05195-6911	11	C
808	05 19 33.1	-69 21 47	20	13	12	9	80	90	75	80	p:	0.26	0.33:	4.1:	10.4:	05195-6921	11	C
809	05 19 36	-71 18	5	-	2	-	4	8	6	8	-	0.19	0.22	1.7	4.2	-	15	C
810	05 19 36.4	-69 23 21	27	13	14	9	-	-	-	-	-	0.52	0.56	C	C	05196-6923	11	M
811	05 19 39.4	-69 15 28	52	29	40	22	140	165	100	140	p	0.85	2.00	10.3:	83.2:	05196-6915	11	C
812	05 19 40	-67 57	11	7	4	3	-	-	-	-	-	0.15	0.11	C	C	-	07	S
813	05 19 44.3	-69 50 20	25	12	23	7	48	76	40	50	-	0.48	1.78	11.6	20.8:	05197-6950	11	C
814	05 19 47	-66 30	9	2	-	-	-	2	-	-	p:	0.26	-	-	-	-	03	S
815	05 19 48	-71 49	-	-	1	-	-	2	2	3	-	-	0.11:	0.8	2.1	-	15	C
816	05 19 48.4	-69 41 40	80	30	100	60	-	-	-	-	p:	1.85	4.44	C	C	05198-6941	11	M
817	05 19 51.7	-65 49 08	4	-	1	-	3	7	4	9	p:	0.15:	0.11:	1.7	10.4	05198-6549	03	C
818	05 20 00	-66 37	2	-	2	2	8	13	9	17	-	0.07:	0.22	2.1	16.6	-	03	C
819	05 20 09.9	-70 13 06	11	8	4	2	24	28	19	23	-	0.11:	0.22:	1.7	8.3	05201-7013	11	C
820	05 20 10	-68 26	8	4	3	2	22	26	24	26	-	0.15	0.11:	1.7	4.2	-	07	C
821	05 20 10	-68 50	11	9	6	4	35	42	30	32	-	0.07:	0.22:	2.9	4.2:	05202-6933	07	C
822	05 20 12.2	-69 33 33	30	18	17	12	80	105	50	60	-	0.44	0.56	10.3	20.8:	05202-6655	11	C
823	05 20 16.4	-66 55 49	13	3	25	2	7	73	13	32	p	0.37	2.55	27.3	39.5	X0520-668	03	C
824	05 20 19.8	-71 16 48	3	-	2	-	3	8	5	10	p:	0.11:	0.22	2.1	10.4	05203-7116	15	M
825	05 20 20	-69 13	17	8	8	5	50	60	-	-	-	0.33	0.33	4.1:	C	05205-6913:	11	C
826	05 20 20	-70 48	5	-	2	-	11	13	12	15	-	0.19	0.22	0.8	6.2	05204-6833	11	C
827	05 20 25.1	-68 33 15	8	6	4	3	28	32	32	35	-	0.07:	0.11:	1.7	6.2:	05204-6840	07	C
828	05 20 27.4	-68 40 58	11	6	8	5	32	43	32	34	-	0.19	0.33	4.6	4.2:	05204-6904	07	M
829	05 20 29.0	-69 04 02	4	2	5	3	25	29	22	25	-	0.07:	0.22:	1.7	6.2:	-	11/07	C
830	05 20 30	-66 10	6	-	1	-	7	11	6	12	p:	0.22	0.11:	1.7	12.5	-	03	C
831	05 20 33.4	-66 49 33	15	4	13	2	16	60	13	34	p	0.41	1.22	18.2	43.7	05205-6649	03	C
832	05 20 40	-69 40	15	11	10	7	-	-	-	-	-	0.15	0.33	C	C	-	11	M
833	05 20 42	-66 36	6	-	-	-	-	-	-	-	-	0.22	-	-	-	-	03	S
834	05 20 45	-68 14	11	6	5	2	25	30	25	30	-	0.19	0.33	2.1	10.4	-	07	S
835	05 20 45	-68 51	24	14	7	5	-	-	-	-	p:	0.37	0.22	C	C	-	07	C
836	05 20 45	-69 58	8	6	5	2	-	-	-	10	-	0.07:	0.33:	-	-	05207-6956:	11	M
837	05 20 50	-67 13	-	-	1	-	10	13	9	10	-	-	0.11:	1.2	2.1	-	07	C
838	05 20 50	-71 01	5	-	-	-	-	-	-	-	p:	0.19	-	-	-	05209-7101:	11/15	S
839	05 20 52.3	-67 55 46	17	10	12	8	50	60	40	50	-	0.26	0.44	4.1	20.8:	05208-6755	07	S
840	05 20 52.3	-68 06 45	13	10	6	4	45	50	42	45	-	0.11	0.22	2.1	6.2:	05208-6806	07	C

Infrared Sources in the LMC (continued 13).

(1)	(2)			(3)		(4)		(5)		(6)		(7)	(8)	(9)	(10)	(11)	(12)	(13)	(14)
Number LI-LMC	Position RA(1950) h m s	DEC(1950) o ' "		12 μm Peak Bg		25 μm Peak Bg		60 μm Peak Bg		100 μm Peak Bg		Size arcmin	F 12μm Jy	F 25μm Jy	F 60μm Jy	F 100μm Jy	IRAS-Id	DPM field	Spectrum
							1.0E-8 Watt / (m×m sr)												
841	05 21 00	-68 02		16	12	10	7	-	-	-	-	-	0.15	0.33:	C	C		07	M
842	05 21 14.7	-68 30	48	8	6	4	3	33	29	37	32	-	0.07:	0.11:	1.7:	10.4	05212-6830	07	C
843	05 21 15	-68 35		10	8	4	3	45	35	35	33	-	0.07:	0.11:	4.1	4.2:		07	C
844	05 21 15	-70 19		-	-	2	1	19	16	20	18	-	-	0.11:	1.2	4.2		11	C
845	05 21 15	-70 46		5	-	2	-	13	11	12	10	-	0.19	0.22	0.8	4.2		11	C
846	05 21 19.7	-66 07	04	2	-	7	-	15	5	8	4	p	0.07:	0.78	4.1	8.3	05213-6607	03	C
847	05 21 20	-68 51		26	12	12	8	68	40	60	45	-	0.52	0.44	11.6	31.2		07	C
848	05 21 20	-69 36		16	8	8	4	69	50	39	30	-	0.30	0.44	7.9	18.7		11	S
849	05 21 20	-70 07		20	15	7	6	-	-	-	-	-	0.19	0.11:	C	C		11	C
850	05 21 21.4	-70 12	31	15	11	8	4	26	23	27	25	-	0.15	0.44	1.2:	4.2:	05213-7012	11	C
851	05 21 24.3	-65 32	04	8	-	23	-	42	4	18	5	p	0.30:	2.55	15.7	27.0	05214-6532	03	C
852	05 21 25	-65 56		5	-	2	-	9	3	8	5	Lxp:	0.24	0.26	2.7	6.3		03	C
853	05 21 25	-68 45		15	8	7	5	45	38	43	40	-	0.26	0.22	2.9	6.2		07	S
854	05 21 25	-69 03		8	5	5	4	34	32	31	29	-	0.11:	0.11	0.8	4.2		11/07	C
855	05 21 29.1	-67 49	45	18	7	28	4	50	35	50	45	p	0.41	2.66	6.2	10.4:	05214-6749	07	C
856	05 21 30	-69 54		20	15	6	5	40	37	42	40	-	0.19:	0.11:	1.2	4.2:		11	C
857	05 21 30	-70 10		16	11	6	4	32	29	32	30	-	0.19	0.22	1.2	4.2		11	C
858	05 21 30	-71 17		4	-	-	-	-	-	-	-	-	0.15	-	-	-		15	S
859	05 21 30.7	-67 08	00	3	-	2	1	13	10	13	12	-	0.11:	0.11:	1.2	2.1:	05215-6707	07	C
860	05 21 36.7	-67 27	33	-	-	2	1	10	8	10	8	-	-	0.11	0.8	4.2	05216-6727	07	C
861	05 21 37.4	-67 53	55	105	18	130	9	125	62	75	65	p	3.22	13.43	26.1	20.8:	05216-6753	07	M
862	05 21 40	-66 45		4	-	4	1	-	-	-	-	-	0.15:	0.33	C	C		03	M
863	05 21 40	-70 16		12	8	5	3	-	-	-	-	-	0.15	0.22	C	C		11	M
864	05 21 40.7	-71 45	58	5	-	4	-	13	2	12	4	p	0.19	0.44	4.6	16.6	05216-7145	15	C
865	05 21 45	-70 14		16	12	6	4	-	-	-	-	-	0.15	0.22:	C	C		11	M
866	05 21 45.7	-70 01	54	40	25	17	10	72	50	64	55	p:	0.56	0.78	9.1	18.7	05217-7001	11	C
867	05 21 48.3	-69 15	20	-	-	4	2	26	22	25	21	-	-	0.22:	1.7	8.3	05218-6915	11	C
868	05 21 50	-68 41		13	7	6	4	35	31	-	-	-	0.22	0.22	1.7	C		07	M
869	05 21 50	-69 33		8	4	2	1	-	-	-	-	-	0.15:	0.11:	-	-		11	S
870	05 21 55.1	-67 44	02	10	6	4	2	17	14	23	17	-	0.15	0.22	1.2	12.5	05219-6744	07	C

No.	RA	Dec	T	n	Ident.	f1	f2	f3	f4	p	d1	d2	d3	d4	d5	d6	d7	d8
871	05 21 55.1	−72 08 27	S	15	05219−7208	−	−	0.17:	0.30	p	−	−	−	−	1.5	1.5	−	8
872	05 21 59.3	−69 43 06	C	11	05219−6943	72.8	66.2	6.66	1.85	p	75	110	260	100	15	75	30	80
873	05 22 00	−68 31	C	07		2.1	1.7	0.22	0.19	−	35	36	35	31	5	7	9	14
874	05 22 00	−68 37	C	07		2.1	0.8:	0.22	0.30	−	35	36	37	35	5	7	8	16
875	05 22 00	−71 19	C	15		4.2	0.8	−	0.07:	−	6	8	6	4	−	−	−	2
876	05 22 03.5	−67 58 16	M	07	05220−6758	C	20.7:	8.88	0.56:	−	−	−	250	200	20	100	30	45
877	05 22 08.4	−69 23 49	C	11	05221−6923	16.6	5.0	0.17:	0.30	p	15	23	30	18	1.5	3	6	14
878	05 22 10.0	−65 46 06	C	03	05221−6546 / X0522−657	72.4	11.3	1.07	1.27	2xp:	3	35	27	5	8	8	2	22
879	05 22 10.5	−67 49 55	M	07	05221−6749	10.4:	18.6	2.66	0.63	p	45	50	90	45	4	28	7	24
880	05 22 10.6	−68 00 32	C	07	05221−6800	312.0	246.3	32.19	3.96	p	140	290	770	175	50	340	33	140
881	05 22 13.8	−69 58 28	C	11	05222−6958	4.2:	1.2	0.22	0.41	p:	50	52	48	16	8	10	25	36
882	05 22 15	−66 46 45	M	03		C	2.9	0.22	0.19	−	−	−	23	−	1	3	4	9
883	05 22 15	−69 12	C	11		4.2	1.2	0.11	−	−	23	25	31	28	4	5	−	−
884	05 22 15	−70 51	C	11		2.1	0.4	−	−	−	7	8	9	8	−	−	−	−
885	05 22 15.7	−67 37 42	C	07	05222−6737	10.4	5.0	0.67	0.30	p	15	20	24	12	2	8	2	10
886	05 22 20	−70 13	M	11		C	0.8:	0.11:	0.11:	−	−	−	25	23	3	4	12	15
887	05 22 23.7	−68 01 28	M	07	05223−6801	C	C	22.20	1.11:	−	−	−	−	−	50	250	50	80
888	05 22 29.0	−68 07 18	M	07	05224−6807	C	37.3:	6.22:	0.74:	−	−	−	170	80	14	70	30	50
889	05 22 30	−66 33	M	03		C	C	0.22:	0.19:	−	−	−	−	−	−	2	−	5
890	05 22 30	−70 09	M	11		C	C	0.11:	0.11:	−	−	−	−	−	6	7	12	15
891	05 22 34.3	−68 42 34	C	07	05225−6842	6.2	2.1	0.11	0.07:	−	25	28	33	28	4	5	6	8
892	05 22 35	−68 13	M	07		C	4.1	0.89	0.22	−	−	−	60	50	7	15	12	18
893	05 22 40	−67 24	S	07		−	−	0.11:	0.19:	−	−	−	−	−	−	1	−	5
894	05 22 41.1	−67 58 22	M	07	05226−6758	52.0:	70.4	4.55	1.11:	p	125	150	330	160	9	50	30	60
895	05 22 41.5	−65 44 35	M	03	05226−6544	C	9.7	0.80	0.54	1xp:	−	−	−	−	2	8	3	15
896	05 22 45.3	−67 30 32	M	07	05227−6730	−	1.2	0.17	−	−	60	75	12	9	−	1.5	−	−
897	05 22 46.0	−69 52 44	C	11	05227−6952	31.2	8.3	0.22	0.37:	−	−	−	80	60	8	10	30	40
898	05 22 47.6	−67 10 09	C	07	05227−6710	33.3	9.9	0.33	0.11	p:	18	34	38	14	4	7	9	12
899	05 22 49.1	−69 45 12	M	11	05228−6945	C	16.6:	3.33	0.74:	−	−	−	120	80	20	50	40	60
900	05 22 49.9	−66 43 51	C	03	05228−6643 / X0522−667	62.4	17.8	2.00	0.70	p:	14	44	63	20	2	20	5	24
901	05 22 52.0	−68 25 05	C	07	05228−6825	12.5:	3.7	0.33	0.22	−	42	48	44	35	6	9	12	18
902	05 22 52.4	−67 46 37	C	07	05228−6746	10.4	2.1	0.11:	0.15	−	13	18	23	18	2	3	−	4
903	05 22 54.8	−69 31 23	C	11	05229−6931	2.1:	1.2	−	−	−	12	13	17	14	−	1	−	−
904	05 22 55	−67 21	S	07		−	−	0.11:	0.19:	−	−	−	−	−	−	−	−	5
905	05 23 00	−67 39	C	07		2.1	1.7	0.17	0.15	−	14	15	18	14	−	1.5	−	4
906	05 23 00	−68 51	C	07		4.2:	1.2:	0.33:	0.26	−	34	36	40	37	4	7	7	14
907	05 23 00.2	−68 47 56	C	07	05230−6847	10.4:	2.1	0.22:	0.07:	−	25	30	35	30	4	6	6	8
908	05 23 01.3	−68 58 19	M	07	05230−6858	−	−	0.33:	0.22	−	−	−	−	−	4	6	9	15
909	05 23 02.2	−68 35 08	C	07	05230−6835	10.4	2.1	0.22	0.15	−	26	31	38	33	4	6	7	11
910	05 23 02.3	−71 37 53	C	15/14	05230−7137 / X0522−716	35.4	9.1	1.00	0.26	p	6	23	27	5	1	10	4	11

Infrared Sources in the LMC (continued 14).

(1)	(2)		(3)		(4)		(5)		(6)		(7)	(8)	(9)	(10)	(11)	(12)	(13)	(14)
Number	Position		12 μm		25 μm		60 μm		100 μm		Size	F 12μm	F 25μm	F 60μm	F 100μm	IRAS-Id	DPM	Spec-trum
LI-LMC	RA(1950) h m s	DEC(1950) o ' "	Peak	Bg	Peak	Bg	Peak	Bg	Peak	Bg	arcmin	Jy	Jy	Jy	Jy		field	
							1.0E-8 Watt / (m*m sr)											
911	05 23 03.2	-68 07 11	105	20	170	20	340	125	150	100	p	3.14	16.65	89.0	104.0	05230-6807	07	C
912	05 23 07.7	-70 30 29	5	-	2	-	13	10	12	8	-	0.19:	0.22	1.2	8.3	05231-7030	11/10	C
913	05 23 10	-66 48	9	5	4	2	-	-	-	-	-	0.15	0.22	C	C		03	W
914	05 23 10	-67 10	17	7	6	3	-	-	-	-	-	0.37	0.33	C	C		07	S
915	05 23 10	-69 13	7	4	6	4	34	30	-	-	-	0.11:	0.22	1.7	C		11/10	W
916	05 23 10	-70 12	13	5	5	3	21	18	19	17	p:	0.30	0.22	1.2	4.2	05230-7011:	11/10	C
917	05 23 12.4	-69 41 48	20	15	20	18	-	-	-	-	-	0.19:	0.22:	C	C	05232-6941	11/10	W
918	05 23 13.8	-71 11 25	-	-	2	1	3	2	-	-	p:	-	0.22	0.4	-	05232-7111	15/14	W
919	05 23 14.5	-66 26 20	11	3	8	1	27	10	22	15	p:	0.30	0.78	7.0	14.6	05232-6626	03	C
920	05 23 16.1	-71 42 23	6	1	3	1	15	8	18	7	-	0.19	0.22:	2.9	22.9	05232-7142	15/14	C
921	05 23 17.5	-69 53 48	40	28	24	10	105	60	-	-	p:	0.44:	1.55	18.6	C	05232-6953	11/10	W
922	05 23 20	-66 47	15	5	6	2	30	14	-	-	p:	0.37	0.44	6.6	C		03	W
923	05 23 20	-69 27	5	2	-	-	17	15	18	15	-	0.11:	-	0.8	6.2		11/10	C
924	05 23 20	-71 23	3	1	2	-	9	5	9	7	p:	0.11	0.22:	1.7	4.2	05231-7120:	15/14	C
925	05 23 23.7	-68 02 50	85	30	70	15	250	170	130	110	p	2.03	6.10	33.1	41.6	05233-6802	07	C
926	05 23 25	-67 12	9	7	4	3	-	-	-	-	-	0.07	0.11	C	C		07	W
927	05 23 25	-69 02	15	11	9	6	72	60	-	-	-	0.15	0.33	5.0:	C		11/07	W
928	05 23 30	-68 35	11	7	4	3	40	38	34	33	-	0.15	0.11:	0.8:	2.1:		07	C
929	05 23 30	-71 38	9	2	5	2	14	10	-	-	p:	0.26:	0.33:	1.7:	C		15/14	C
930	05 23 34.3	-70 04 17	-	-	-	-	26	23	24	21	-	-	-	1.2	6.2	05235-7004	11/10	C
931	05 23 35	-68 21	16	10	8	6	47	42	46	43	-	0.22	0.22	2.1	6.2		07	C
932	05 23 35.7	-65 44 35	11	-	2	1	4	2	8	4	p	0.41	0.22	0.8:	8.3:	05235-6544	03/02	C
933	05 23 37.0	-67 26 48	11	2	8	1	33	9	22	15	p	0.33	0.78	9.9	14.6	05236-6726	07/06	C
934	05 23 38.1	-71 18 49	-	-	3	1	13	5	11	7	p:	-	0.22:	3.3	8.3	05236-7118	15/14	C
935	05 23 40	-69 58	17	9	6	4	38	36	-	-	p:	0.30	0.22	0.8:	C		11/10	W
936	05 23 42	-66 57	5	-	2	-	9	6	9	7	-	0.19	0.22:	1.2:	4.2:	05237-6655:	03/02	C
937	05 23 42.8	-70 00 45	9	4	6	2	34	32	-	-	-	0.19:	0.44	0.8:	C	05237-7000	11/10	W
938	05 23 43.8	-67 55 15	12	8	20	5	60	50	-	-	-	0.15:	1.66	4.1:	C	05237-6755	07/06	W
939	05 23 45	-68 50	8	4	4	3	30	24	-	-	-	0.15:	0.11:	2.5:	C		07/06	C
940	05 23 50	-68 17	14	8	7	4	48	43	42	40	-	0.22	0.33	2.1	4.2		07/06	C

No.	RA	Dec									p					ID	Plates	Type
941	05 23 50	−69 35	8	5	5	4	34	30	26	23	−	0.11:	0.11	1.7	6.2		11/10	C
942	05 23 50	−69 51	23	14	7	5	42	40	−	−	−	0.33	0.22:	0.8:	C		11/10	W
943	05 23 52.4	−68 02 42	46	25	64	22	150	75	100	80	p:	0.78	4.66	31.0	41.6:		07/06	C
944	05 23 55	−69 13	22	14	12	8	83	77	−	−	−	0.30	0.44	2.5	C		11/10	W
945	05 23 55	−69 27	8	4	4	3	32	30	25	23	−	0.15:	0.11:	0.8	4.2	05238−6802	11/10	C
946	05 23 58.3	−67 59 54	26	13	44	15	100	55	−	−	p:	0.48	3.22	18.6	C	05239−6759	07/06	W
947	05 24 00	−69 05	16	12	10	7	75	66	62	56	−	0.15	0.33	3.7	12.5		11/10	C
948	05 24 00.8	−68 09 48	12	9	20	8	60	50	−	−	p:	0.11:	1.33	4.1	C	05240−6809	06/07	W
949	05 24 01.8	−68 44 40	9	9	4	2	34	19	23	17	p:	0.19	0.22	6.2:	12.5:	05240−6844	06/07	C
950	05 24 05	−70 11	−	−	3	2	23	20	21	18	−	−	0.11:	1.2	6.2		10/11	C
951	05 24 06	−71 15	4	−	1.5	−	−	−	30	−	−	0.15	0.17:	C	C	05242−6623:	14/15	W
952	05 24 08	−66 26	12	6	5	3	40	25	−	20	p:	0.22	0.22:	6.2:	20.8:	05241−6632	02/03	C
953	05 24 08.7	−66 32 20	9	6	5	3	22	15	−	−	p:	0.11:	0.22	2.9:	C		02/03	W
954	05 24 10	−67 33	9	6	−	−	160	−	−	−	−	0.11:	−	−	−		10/11	S
955	05 24 10	−69 42	45	35	34	14	130	130	−	−	−	0.37	2.22	12.4:	C		10/11	W
956	05 24 15	−67 29	12	5	7	3	47	20	35	24	p:	0.26	0.44	11.2	22.9	05242−6748	06/07	C
957	05 24 16.1	−67 48 18	5	−	4	2	−	−	50	40	p	0.19	0.22	−	−		06/07	W
958	05 24 20	−67 02	17	13	10	7	70	60	13	−	−	0.15	0.33	4.1:	20.8:		10/11	C
959	05 24 20	−70 40	4	4	1	−	13	10	19	8	−	0.15	0.11:	1.2	10.4		10/11	C
960	05 24 25.9	−71 22 40	5	5	3	1	20	8	−	8	p:	0.15	0.22	5.0	22.9	05244−7122	14/15	C
961	05 24 26.9	−68 32 32	32	15	23	8	82	55	−	−	p	0.63	1.66	11.2	C	05244−6832	06/07	W
962	05 24 30	−66 47	4	−	3	−	13	6	13	10	−	0.15	0.33:	2.9:	6.2:		02/03	C
963	05 24 30	−70 31	5	2	2	−	12	8	13	8	−	0.22	0.22	1.7	10.4		10/11	C
964	05 24 31.4	−67 12 03	11	4	5	2	32	19	27	22	−	0.26	0.33	5.4	10.4	05245−6712	06	C
965	05 24 35	−69 13	24	15	11	9	94	81	75	55	p:	0.33	0.22	5.4	41.6	05248−6915:	10	C
966	05 24 39.0	−71 37 22	10	2	5	1	18	11	19	11	p	0.30:	0.44	2.9	16.6	05246−7137	14/15	C
967	05 24 40	−68 14	14	9	7	5	43	34	40	35	−	0.19	0.22	3.7	10.4		06	C
968	05 24 40	−69 23	24	16	14	9	90	75	70	60	−	0.30	0.56	6.2:	20.8:		10/11	C
969	05 24 40	−69 32	11	7	6	4	43	40	32	30	−	0.15	0.22	1.2	4.2		10/11	C
970	05 24 40	−70 19	−	−	2	−	11	9	8	7	−	−	0.22:	0.8	2.1:		10/11	C
971	05 24 40.2	−70 03 49	38	5	5	2	−	−	−	−	p	1.22	0.33	−	−	05246−7003	10/11	S
972	05 24 40.7	−66 09 27	3	−	−	−	18	13	−	−	−	0.11:	−	2.1:	C		02	W
973	05 24 40.9	−69 44 05	53	45	25	20	200	150	−	−	−	0.30	0.56:	20.7:	C	05246−6944	10/11	W
974	05 24 45	−68 26	17	14	7	5	48	44	−	−	−	0.11:	0.22:	1.7:	C		06	W
975	05 24 45	−69 03	16	12	6	4	−	−	−	−	−	0.15:	0.22:	C	C		10/06	W
976	05 24 45.0	−69 41 30	65	37	50	25	−	−	−	−	p	1.04	2.77	C	C	05247−6941	10/11	W
977	05 24 49.9	−69 27 48	17	12	10	7	76	60	53	40	−	0.19	0.33	6.6	27.0	05248−6927	10/11	C
978	05 24 50	−68 32	22	16	13	8	70	60	−	−	p:	0.22	0.56	4.1:	C		06	W
979	05 24 50	−68 36	15	8	7	4	50	42	−	−	−	0.26	0.33	3.3:	C		06	W
980	05 24 50	−68 52	13	5	−	−	−	−	−	−	−	0.30:	−	−	−		06	S

Infrared Sources in the LMC (continued 15).

Brightness columns (3)–(6) are in units of 1.0E-8 Watt / (m²m sr). Flux columns (8)–(11) are in Jy.

(1) Number LI–LMC	(2) RA(1950) h m s	(2) DEC(1950) o ′ ″	(3) 12 µm Peak	(3) Bg	(4) 25 µm Peak	(4) Bg	(5) 60 µm Peak	(5) Bg	(6) 100 µm Peak	(6) Bg	(7) Size arcmin	(8) F 12µm	(9) F 25µm	(10) F 60µm	(11) F 100µm	(12) IRAS-Id	(13) DPM field	(14) Spectrum
981	05 24 50	-72 01	5	-	-	-	-	-	-	-	-	0.19	-	-	-		14/15	S
982	05 24 50.6	-70 07 41	9	5	5	3	-	-	-	-	-	0.15	0.22	-	-	05248-7007	10/11	M
983	05 24 51.0	-66 29 14	18	6	10	4	53	20	50	20	p	0.44	0.67	13.7	62.4	05248-6629	02	C
984	05 24 51.2	-69 15 02	24	14	10	7	90	70	70	60	-	0.37	0.33:	8.3:	20.8:	05248-6915	10/11	C
985	05 24 54	-71 37	10	3	4	1	-	-	-	-	p:	0.26:	0.33:	C	C		14	M
986	05 24 56.6	-69 56 28	17	4	8	5	40	30	-	-	-	0.48	0.33	4.1:	C	05249-6956	10/11	M
987	05 25 00	-69 18	24	12	9	7	-	-	-	-	-	0.44	0.22:	C	C	05249-6916:	10/11	S
988	05 25 03.3	-71 34 34	9	2	5	1	21	9	18	12	p	0.26:	0.44	5.0	12.5	05250-7134	14	C
989	05 25 05.9	-71 41 27	-	-	2	-	11	9	-	-	-	-	0.22	0.8	C	05250-7141	14	M
990	05 25 09.7	-68 01 52	8	3	4	2	21	18	20	18	-	0.19	0.22:	1.2:	4.2:	05251-6801	06	C
991	05 25 10	-68 19	18	14	8	6	48	44	-	-	-	0.15	0.22	1.7:	C		06	M
992	05 25 10	-70 01	11	8	3	2	25	21	20	18	-	0.11	0.11:	1.7	4.2		10	C
993	05 25 10.5	-69 53 01	14	10	-	-	-	-	-	-	-	0.15	-	C	C	05251-6953	10	S
994	05 25 11.9	-68 21 46	22	16	12	8	68	45	60	48	p:	0.22	0.44	9.5	25.0	05251-6821	06	S
995	05 25 12	-71 37	9	4	3	2	-	-	-	-	p:	0.19	0.11	C	C		14	S
996	05 25 12.3	-66 03 24	4	3	5	1	27	12	-	-	p:	0.04:	0.44	6.2	C	05252-6603	02	M
997	05 25 18	-71 53	5	-	1	-	-	-	-	-	p	0.19	0.11:	-	-		14	S
998	05 25 18.4	-69 08 16	12	8	-	-	-	-	-	-	-	0.15	C	C	C	05253-6908	10	S
999	05 25 18.8	-68 30 53	48	18	52	8	147	52	87	47	p	1.11	4.88	39.3	83.2	05253-6830	06	C
1000	05 25 20	-67 13	9	6	4	2	28	18	-	-	-	0.11	0.22	4.1	C		06	M
1001	05 25 20	-70 10	-	-	-	-	20	18	17	16	-	-	-	0.8	2.1		10	C
1002	05 25 23.1	-66 18 57	40	15	25	9	-	-	-	-	p	0.93	1.78:	C	C	05253-6618	02	M
1003	05 25 26.4	-67 32 13	16	6	18	7	75	40	47	35	p	0.37	1.22	14.5	25.0	05254-6732	06	C
1004	05 25 30	-66 33	4	2	3	1	28	12	-	-	-	0.07	0.22	6.6	C		02	M
1005	05 25 30	-69 14	17	11	12	8	82	75	-	-	-	0.22	0.44	2.9	C		10	M
1006	05 25 30	-69 22	25	15	14	9	80	70	72	58	p:	0.37	0.56	4.1:	29.1	05253-6921:	10	C
1007	05 25 30	-71 51	5	-	1.5	-	-	-	-	-	p:	0.19	0.17	-	C		14	S
1008	05 25 32.3	-69 43 28	35	20	35	15	160	100	-	-	p:	0.56	2.22:	24.8	C	05255-6943	10	M
1009	05 25 34.7	-66 20 21	25	20	25	8	110	60	-	-	-	0.19:	1.89	20.7:	C	05255-6620	02	M
1010	05 25 40	-66 15	42	20	20	12	-	-	-	-	p:	0.81	0.89	C	C	X0525-662	02	M

No.	RA (1950)	Dec (1950)	n1	n2	n3	n4	n5	n6	n7	n8	p	F12	F25	F60	F100	IRAS name	02/06	Cl
1011	05 25 40	-66 59	-	-	-	-	6	4	6	4	-	-	-	0.8	4.2		06	C
1012	05 25 40	-68 23	16	9	7	5	45	40	-	-	p:	0.26	0.22	2.1:	C		10	M
1013	05 25 40	-69 50	14	8	6	3	25	21	-	-	p:	0.22	0.33	1.7	C		14	M
1014	05 25 42.1	-71 35 45	7	2	5	1	-	-	-	-	p	0.19:	0.44	C	—	05257-7135	02	C
1015	05 25 46.6	-66 17 36	75	10	135	8	240	20	130	20	-	2.40	14.10	91.1	228.8	05257-6617	02	C
1016	05 25 47.9	-71 30 25	3	-	2	1	12	10	12	9	-	0.11:	0.11:	1.2	4.2	05257-7130	14	C
1017	05 25 50.0	-67 13	13	5	9	3	34	21	43	22	p	0.30	0.67	8.7	27.0	05260-6711:	06	C
1018	05 25 50.3	-69 28 57	17	8	11	5	35	25	60	30	p	0.33	0.67	12.4	20.8	05258-6928	10	C
1019	05 25 52.9	-65 47 56	22	3	3	1	12	10	10	5	p	0.70	0.22:	2.1	4.2:	05258-6547 (X0525-658)	02	C
1020	05 25 56.8	-66 11 54	36	15	18	8	60	50	78	40	-	0.78:	1.11	15.7	20.8:	05259-6611	02	C
1021	05 25 58.1	-69 52 58	47	20	16	8	95	45	80	45	p:	1.00	0.89	20.7	72.8	05259-6952	10	C
1022	05 25 59.5	-66 07 03	24	9	24	8	72	25	50	30	p:	0.56	1.78	19.5	41.6:	05259-6607	02	C
1023	05 26 00	-68 42	13	9	8	5	60	50	-	-	-	0.15	0.33	4.1:	C		06	M
1024	05 26 00	-69 22	22	12	9	6	62	50	-	-	p	0.37	0.33	5.0	C		10	M
1025	05 26 00	-69 55	33	28	13	8	80	70	-	-	-	0.19	0.56	4.1:	C		10	C
1026	05 26 00	-70 06	-	-	-	-	22	20	20	18	-	-	-	0.8	4.2:		10	C
1027	05 26 00	-70 19	3	-	2	-	12	10	9	7	-	0.11:	0.22:	0.8	4.2		10	C
1028	05 26 00	-71 06	7	-	1	-	-	-	-	-	-	0.26	0.11:	-	-		14	S
1029	05 26 02.2	-67 17 23	4	-	3	2	-	-	-	-	-	0.15:	0.11:	C	C	05260-6717	06	S
1030	05 26 03.5	-68 57 54	12	6	5	3	50	35	50	27	-	0.22:	0.22:	6.2:	47.8:	05260-6857	06	C
1031	05 26 05.1	-70 10 23	-	-	3	1.5	-	-	-	-	-	-	0.17	-	-	05260-7010	10	M
1032	05 26 06.8	-70 01 55	12	8	4	3	33	27	30	20	-	0.15	0.11:	2.5	20.8	05261-7001	10	C
1033	05 26 08.4	-66 14 34	30	15	15	8	80	60	-	-	-	0.56	0.78:	8.3:	C	05261-6614	02	M
1034	05 26 08.9	-67 29 10	11	6	15	8	65	50	45	32	-	0.19	0.78	6.2	27.0:	05261-6729	06	C
1035	05 26 09.1	-66 22 46	16	5	6	2	60	50	-	-	p:	0.41	0.44:	4.1:	C	05261-6622	02	M
1036	05 26 10	-67 51	8	5	5	4	34	28	30	25	-	0.11:	0.11	2.5	10.4		06	C
1037	05 26 11.1	-67 33 15	9	5	17	12	70	60	-	-	-	0.15:	0.56:	4.1:	C	05261-6733	06	M
1038	05 26 11.5	-66 09 27	25	15	10	5	-	-	-	-	-	0.37:	0.56:	C	C	05261-6609	02	M
1039	05 26 18.7	-68 00 30	2	-	1	-	20	16	60	45	-	0.07:	0.11:	1.7:	C	05263-6800	06	M
1040	05 26 20	-68 42	19	8	13	8	72	55	-	-	-	0.41	0.56	7.0	31.2:	05266-6842:	06	C
1041	05 26 20.4	-68 38 29	16	10	16	4	82	48	55	36	-	0.22	1.33	14.1	39.5	05263-6838	06	C
1042	05 26 21	-65 58	9	5	3	1	15	10	20	15	-	0.15	0.22:	2.1	10.4		02	C
1043	05 26 21.6	-67 39 59	14	8	16	7	73	45	45	35	-	0.22	1.00	11.6	20.8	05263-6739	06	C
1044	05 26 22.8	-67 24 39	11	5	8	5	40	30	35	22	-	0.22	0.33	4.1:	27.0:	05263-6724	06	C
1045	05 26 28.3	-67 33 20	13	7	23	10	80	60	55	45	-	0.22	1.44	8.3:	20.8:	05264-6733	06	C
1046	05 26 30	-68 48	16	9	15	12	80	50	-	-	-	0.26:	0.33:	12.4:	C		06	M
1047	05 26 30	-69 09	13	8	6	4	44	35	37	34	-	0.19	0.22	3.7	6.2:	05263-6905:	10	C
1048	05 26 30	-70 39	2	-	1	-	13	11	11	9	-	0.07:	0.11:	0.8	4.2		10	C
1049	05 26 33.8	-68 52 48	40	25	60	30	150	100	110	47	-	0.56:	3.33:	20.7:	131.0:	05265-6852	06	C
1050	05 26 34.0	-68 10 47	8	3	4	2	19	15	21	16	-	0.19	0.22	1.7:	10.4:	05265-6810	06	C

Infrared Sources in the LMC (continued 16).

(1)	(2)		(3)		(4)		(5)		(6)		(7)	(8)	(9)	(10)	(11)	(12)	(13)	(14)
Number LI-LMC	Position RA(1950) h m s	DEC(1950) o ' "	12 µm Peak	Bg	25 µm Peak	Bg	60 µm Peak	Bg	100 µm Peak	Bg	Size arcmin	F 12µm Jy	F 25µm Jy	F 60µm Jy	F 100µm Jy	IRAS-Id	DPM field	Spec-trum
					1.0E-8 Watt / (m²m sr)													
1051	05 26 35	-67 45	11	6	8	5	42	34	35	30	-	0.19	0.33	3.3	10.4:		06	C
1052	05 26 36	-67 42	13	3	15	5	55	40	41	35	p	0.37	1.11	6.2	12.5		06	C
1053	05 26 38.2	-65 41 53	6	2	2	-	7	4	9	6	p:	0.15	0.22:	1.2	6.2:	05266-6743:	02	C
1054	05 26 40	-67 18	7	-	5	3	29	18	25	18	-	0.26:	0.22:	4.6	14.6	05266-6541	06	C
1055	05 26 40	-67 26	11	6	7	6	-	-	-	-	p:	0.19	0.11:	C	C		06	S
1056	05 26 40	-67 36	14	8	11	8	90	65	60	55	-	0.22	0.33	10.3	10.4:		06	C
1057	05 26 40	-69 24	10	4	9	6	-	-	-	-	-	0.22	0.33	C	C		10	M
1058	05 26 40.5	-71 38 25	15	4	6	1	33	9	28	11	p	0.41	0.56	9.9	35.4	05266-7138	14	C
1059	05 26 42.8	-69 13 17	13	4	7	5	-	-	-	-	p	0.33	0.22:	-	-	05267-6913	10	S
1060	05 26 44.7	-69 41 08	-	-	-	-	11	8	10	8	-	-	-	1.2	4.2:	05267-6941	10	C
1061	05 26 45	-66 12	13	6	4	2	32	25	-	-	-	0.26	0.22	2.9:	C		02	M
1062	05 26 45	-68 18	4	2	4	2	28	21	23	19	-	0.07:	0.22	2.9	8.3		06	C
1063	05 26 45	-68 37	15	6	8	4	55	43	-	-	p:	0.33	0.44	5.0	C		06	M
1064	05 26 45	-68 56	15	6	15	8	82	35	45	30	-	0.33	0.78	19.5	31.2:	05268-6836:	06	C
1065	05 26 45	-69 22	12	4	15	8	60	30	42	30	p:	0.30	0.78	12.4	25.0	05268-6920:	10	M
1066	05 26 50	-68 32	-	-	2	-	28	23	25	23	-	-	0.22:	2.1	4.2		06	C
1067	05 26 50.5	-65 57 44	3	1	3	1.5	12	9	-	-	-	0.11:	0.17:	1.2:	C		02	C
1068	05 26 50.5	-67 52 58	13	6	6	4	36	25	31	22	-	0.26	0.22	4.6	18.7	05268-6557	02	C
1069	05 27 00	-66 07	9	5	3	1	23	9	20	10	-	0.15	0.22	5.8	20.8:	05268-6752	02	C
1070	05 27 00	-68 59	10	6	5	3	20	18	-	-	p:	0.15	0.22	0.8:	C		06/10	M
1071	05 27 00	-71 08	4	-	2	1	13	10	14	11	-	0.15:	0.11:	1.2	6.2	05273-7107:	14	C
1072	05 27 00	-71 41	9	1	2	1	12	10	14	12	p:	0.30:	0.11:	0.8:	4.2:	X0526-716B	14	C
1073	05 27 00.0	-66 52 10	3	-	2	-	10	4	7	3	p:	0.11:	0.22:	2.5	8.3	05269-6652	02	C
1074	05 27 01.4	-68 27 56	4	2	5	2	28	25	24	20	p:	0.07:	0.33	1.2:	8.3	05270-6827	06	C
1075	05 27 05	-72 31	5	-	-	-	-	-	-	-	-	0.19:	-	-	-		14	S
1076	05 27 05.8	-68 51 36	112	25	250	30	460	73	110	47	p	3.22	24.42	160.2	131.0:	05270-6851	06	M
1077	05 27 06.6	-70 06 10	12	4	-	-	-	-	-	-	-	0.30	-	-	-	05271-7006	10	S
1078	05 27 11.6	-69 09 31	17	8	9	6	55	35	42	30	p:	0.33	0.33	8.3	25.0	05271-6909	10	C
1079	05 27 15	-70 11	-	-	-	-	22	20	17	16	-	-	-	0.8	2.1		10	C
1080	05 27 15.4	-67 35 07	15	9	10	8	80	65	59	50	-	0.22	0.22:	6.2:	18.7:	05272-6735	06	C

No.	Type	n	Identification	v1	v2	v3	v4	P	c1	c2	c3	c4	c5	c6	c7	c8	RA (1950)	Dec (1950)
1081	M	06	05274-6841:	–	1.2:	0.22	0.44	p	–	–	18	18	2	4	2	14	05 27 16	-68 40
1082	M	02	05273-6624	–	–	0.44	0.30	p	–	–	15	–	–	4	1	9	05 27 18.1	-66 24 52
1083	C	06	05272-6731:	C	8.3:	0.78:	0.30	–	11	13	–	90	18	25	12	20	05 27 20	-67 31
1084	C	06		4.2:	1.2	–	–	–	13	15	70	17	–	–	–	–	05 27 30	-68 06
1085	C	14		4.2:	0.8:	0.22	0.19	–	13	15	14	12	1	3	4	9	05 27 36	-71 12
1086	C	14	05276-7125	35.4:	9.5:	0.33	0.22	p:	13	30	9	32	1	4	4	10	05 27 40.9	-71 25 31
1087	C	10		6.2:	3.3	0.22	0.11	–	24	27	20	28	3	5	9	12	05 27 45	-70 33
1088	C	10		4.2	0.8	0.17:	0.07:	–	14	16	13	15	2	3	2	4	05 27 45	-70 59
1089	C	14	05277-6729	6.2	1.7	0.17:	0.15:	–	10	13	6	10	–	1.5	–	4	05 27 45	-71 43
1090	C	06		62.4:	41.4	5.22	0.74	p	60	90	100	200	18	65	16	36	05 27 46.4	-67 29 31
1091	C	10	05277-7024	4.2:	1.2:	–	–	–	15	17	16	19	–	–	–	–	05 27 47.3	-70 24 24
1092	M	10	05278-6942	–	–	0.44	0.37	–	–	–	–	–	4	4	4	14	05 27 48.2	-69 42 05
1093	C	06	X0527-714	8.3	2.1	0.22	0.11:	–	18	22	18	23	2	2	–	3	05 27 50	-67 48
1094	M	14		C	C	0.78	0.33	–	–	–	–	–	9	9	5	14	05 27 50	-71 27
1095	S	06		–	–	–	0.11	–	–	–	–	–	1	1	5	8	05 27 53	-68 07
1096	C	02	05278-6651	2.1:	0.8	–	–	–	2	3	3	5	–	–	–	–	05 27 53.0	-66 51 24
1097	C	10	05279-7036	16.6	3.3	0.33	0.22	–	35	43	40	48	5	8	12	18	05 27 55	-70 03
1098	C	10	X0527-706	25.0	3.7	0.44	0.30	p:	22	34	28	37	6	10	10	18	05 27 58.1	-70 36 12
1099	C	10		6.2	2.5	–	0.15:	–	12	15	12	18	–	–	–	4	05 28 00.3	-69 29
1100	M	10	05280-6910	4.2	10.3	23.31	3.88	p	30	32	35	60	10	220	7	112	05 28 00.3	-69 10 25
1101	M	10	05280-7039	C	1.7:	0.11	0.15:	–	–	–	12	16	3	4	6	10	05 28 03.8	-70 39 10
1102	C	06	05280-6727	31.2:	8.3:	3.44	0.85	p	65	80	80	100	13	44	15	38	05 28 05.9	-67 27 49
1103	S	10	05281-6915	–	–	0.22	0.30	p	–	–	–	–	2	4	5	13	05 28 07.4	-69 15 45
1104	C	14		6.2	1.2	0.22	0.26	–	15	18	11	14	2	4	6	13	05 28 09	-71 16
1105	C	10		6.2:	1.7:	0.11:	0.15:	–	30	33	30	34	3	4	10	14	05 28 10	-70 14
1106	C	14	05281-7126	35.4:	9.1:	0.44	0.22	p	13	30	9	31	3	7	4	10	05 28 10.0	-71 26 40
1107	C	06/02		2.1:	0.4:	0.11:	0.19	p	3	4	3	4	–	1	–	5	05 28 15	-67 02
1108	C	10		4.2	0.8:	–	0.11:	–	18	20	18	20	–	–	7	10	05 28 15	-70 27
1109	C	06		14.6	5.4	0.22	0.11:	–	18	25	15	28	3	5	3	6	05 28 20	-68 13
1110	M	10/06		C	12.4:	0.44	0.19	–	–	–	40	70	5	9	9	14	05 28 20	-69 04
1111	C	10	05284-6923	25.0	6.2	0.33	0.22	p:	15	27	20	35	3	6	8	14	05 28 26.4	-69 23 39
1112	C	06		4.2	2.5	0.11:	0.11:	–	14	16	16	22	–	1	–	3	05 28 30	-67 43
1113	C	10		10.4	2.5	0.11	0.11:	–	20	25	22	28	2	3	6	9	05 28 30	-70 16
1114	C	10		4.2	1.2	0.17	0.11:	–	13	15	11	14	–	1.5	–	–	05 28 30	-70 48
1115	C	10	05285-6955	27.0	3.7	0.33	0.48	p	20	33	25	34	4	7	9	22	05 28 33.4	-69 55 36
1116	M	06	05285-6529	C	C	0.22	0.15	–	–	–	–	–	3	5	5	9	05 28 35	-67 30
1117	S	02	X0528-690	–	–	0.17:	0.59:	p	–	–	–	–	–	1.5	–	16	05 28 35.2	-65 29 12
1118	C	10		58.2	10.3	0.33	–	–	30	58	50	75	9	12	–	–	05 28 40	-69 06
1119	C	10		–	–	0.11:	0.26	–	–	–	–	–	2	3	6	13	05 28 40	-70 00
1120	C	10		8.3	2.5	0.11	0.11	–	30	34	30	36	3	4	9	12	05 28 40	-70 05

Infrared Sources in the LMC (continued 17).

(1)	(2)		(3)		(4)		(5)		(6)		(7)	(8)	(9)	(10)	(11)	(12)	(13)	(14)
Number	Position		12 μm		25 μm		60 μm		100 μm		Size	F 12μm	F 25μm	F 60μm	F 100μm	IRAS-Id	DPM	Spec-
LI-LMC	RA(1950)	DEC(1950)	Peak	Bg	Peak	Bg	Peak	Bg	Peak	Bg		Jy	Jy	Jy	Jy		field	trum
	h m s	o ′ ″			1.0E-8 Watt / (m²·m sr)						arcmin							
1121	05 28 40	-70 14	15	10	4	3	40	30	33	30	-	0.19:	0.11	4.1	6.2:		10	C
1122	05 28 40	-70 54	10	6	3	2	18	14	17	15	-	0.15	0.11	1.7:	4.2:		10	C
1123	05 28 40	-70 57	11	8	3	2	20	18	20	15	-	0.11	0.11:	0.8:	C		10	M
1124	05 28 40	-71 23	8	5	2	1	12	10	-	-	P:	0.11	0.11	0.8:	10.4:		14	C
1125	05 28 40.6	-68 09 32	10	5	5	3	-	-	20	15	-	0.19	0.22	-	-	05286-6809	06	M
1126	05 28 42.1	-66 16 26	5	-	-	-	-	-	-	-	-	0.19	-	C	-	05287-6616	02	S
1127	05 28 43.1	-69 10 59	45	15	23	10	-	-	-	-	P	1.11	1.44	C	C	05287-6910	10	M
1128	05 28 50	-65 57	9	5	8	3	7	4	7	5	-	-	-	1.2	4.2		02	C
1129	05 28 50	-68 27	9	5	8	3	37	20	30	24	-	0.15:	0.56	7.0	12.5:		06	C
1130	05 28 58.9	-66 17 41	5	2	4	-	-	-	-	-	P	0.11	0.44	-	-	05289-6617	02	M
1131	05 29 00	-67 20	8	4	5	3	22	16	-	11	-	0.15	0.22	2.5	C		06	M
1132	05 29 00	-69 37	9	6	4	2	20	16	13	11	-	0.11:	0.22	1.7:	4.2:		10	C
1133	05 29 00	-71 16	8	5	3	2	16	12	20	16	P:	0.11	0.11:	1.7:	8.3:		14	C
1134	05 29 03.8	-67 56 25	-	-	-	-	16	14	13	11	-	-	-	0.8	4.2:	05290-6756	06	C
1135	05 29 06.4	-66 43 31	8	-	1.5	-	-	-	-	-	P	0.30	0.17:	-	-	05291-6643	02	S
1136	05 29 07	-67 23	9	5	5	3	28	14	23	14	-	0.15	0.22	5.8	18.7	05291-6700	06	C
1137	05 29 08.1	-67 00 03	2	-	2	-	11	8	8	6	-	0.07:	0.22	1.2	4.2:		06/02	C
1138	05 29 10	-66 51	-	-	1	-	4	2	2	1	-	-	0.11:	0.8:	2.1:		02	C
1139	05 29 10	-69 04	12	8	9	5	-	-	-	-	-	0.15:	0.44	C	C		10/06	M
1140	05 29 10	-70 34	2	-	-	-	11	9	11	10	-	0.07:	-	0.8	2.1:		10	C
1141	05 29 15	-67 03	5	-	1	-	13	6	8	5	-	0.19:	0.11:	2.9	6.2	X0529-670	06/02	C
1142	05 29 15	-70 07	22	15	4	3	-	-	-	-	P	0.26	0.11:	-	-		10	S
1143	05 29 15	-70 10	-	-	6	4	30	28	27	24	-	-	0.22	0.8	6.2		10	C
1144	05 29 15	-70 59	15	6	6	4	32	26	28	24	P:	0.33	0.22	2.5	8.3:		10	C
1145	05 29 20	-69 09	13	9	8	6	-	-	-	-	-	0.15	0.22:	C	C		10	M
1146	05 29 20	-69 45	15	8	-	-	-	-	-	-	-	0.26	-	-	-		10	S
1147	05 29 20	-70 16	10	7	-	-	-	-	-	-	-	0.11	-	-	-		10	S
1148	05 29 20	-70 23	12	7	-	-	26	22	20	18	P:	0.19	-	1.7	4.2		10	C
1149	05 29 20.7	-67 15 44	11	-	3	-	-	-	-	-	P	0.41	0.33	-	-	05293-6715	06	S
1150	05 29 21.4	-69 11 57	14	10	7	5	-	-	-	-	-	0.15:	0.22:	C	C	05293-6911	10	M

No.	RA	Dec	(1)	(2)	(3)	(4)	(5)	(6)	(7)	(8)						IRAS		
1151	05 29 21.4	-70 13 08	-	-	-	-	32	28	25	20	-	-	-	1.7:	10.4	05293-7013	10	C
1152	05 29 22.3	-69 06 30	22	12	10	6	65	40	-	-	-	0.37:	0.44:	10.3	C	05293-6906	10	M
1153	05 29 27.1	-71 04 44	25	5	10	3	-	-	-	-	p	0.74	0.78	C	C	05294-7104	14	M
1154	05 29 27.8	-67 33 07	4	-	3	-	17	13	16	13	-	0.15:	0.33	1.7	6.2	05294-6733	06	C
1155	05 29 30	-66 58	5	-	2	-	-	-	-	-	-	0.19	0.22	-	-		02/06	M
1156	05 29 30	-71 14	10	5	4	2	27	15	30	20	p:	0.19	0.22	5.0	20.8:		14	C
1157	05 29 31.6	-71 21 41	8	3	2.5	1	-	-	-	-	p	0.19	0.17	1.7	-	05295-7121	14	S
1158	05 29 35	-70 43	2	-	1	-	12	9	10	8	-	0.07:	0.11:	1.2	4.2		10	C
1159	05 29 42.9	-65 17 14	5	-	-	-	-	-	-	-	-	0.19:	-	-	-	05297-6517	02	S
1160	05 29 45	-70 09	18	12	7	4	30	28	-	-	p:	0.22	0.33	0.8:	C		10	M
1161	05 29 47.3	-68 28 56	18	14	9	6	52	35	45	28	-	0.15:	0.33	7.0	35.4	05297-6828	06	C
1162	05 29 50	-67 48	-	-	1	6	19	15	15	12	-	-	0.11:	1.7	6.2		06	C
1163	05 29 50.4	-69 11 25	17	5	7	5	26	24	-	-	-	0.44	0.22:	0.8:	-	05298-6911	10	M
1164	05 29 52.2	-69 57 27	31	8	15	3	30	27	-	-	p	0.85	1.33	1.2:	C	05298-6957	10	M
1165	05 29 53.3	-67 17 02	-	-	2	-	10	7	7	6	-	-	0.22:	1.2	2.1:	05298-6717	06	C
1166	05 29 53.8	-68 42 38	11	7	6	4	40	30	-	-	-	0.15	0.22	4.1:	C	05298-6842	06	M
1167	05 29 55	-68 32	22	13	13	7	60	40	45	40	-	0.33	0.67	8.3	10.4:		06	C
1168	05 29 55	-69 51	13	8	6	3	39	40	18	16	-	0.19	0.33	3.7	4.2:		10	C
1169	05 29 57.1	-71 04 01	14	6	10	6	61	25	45	20	p:	0.30	0.44	14.9	52.0	05299-7104	14	C
1170	05 29 59.9	-67 20 44	5	-	2	-	-	-	-	-	-	0.19	0.22	-	-	05299-6720	06	M
1171	05 30 00	-69 54	11	8	5	3	36	30	28	24	-	0.11	0.22	2.5	8.3		10	C
1172	05 30 01.4	-68 59 32	11	4	3	2	10	8	-	-	p:	0.26	0.11	-	-	05300-6859	06/10	S
1173	05 30 04.6	-70 49 04	3	-	1	-	-	-	-	-	-	0.11:	0.11:	0.8:	C	05300-7049	10	M
1174	05 30 05	-70 18	12	7	-	-	32	-	-	-	p:	0.19	-	-	-		10	S
1175	05 30 05.1	-70 14 53	8	4	-	-	32	27	24	21	-	0.15:	-	2.1	6.2:	05300-7014	10	C
1176	05 30 05.3	-66 59 46	4	-	4	1	20	6	13	4	p	0.15	0.33	5.8	18.7	05300-6659	02/06	C
1177	05 30 05.6	-66 51 15	7	-	10	-	-	-	-	-	-	0.26:	0.22	-	-	05300-6651	02	S
1178	05 30 10	-71 13	15	7	12	5	50	25	50	30	-	0.30	0.11:	10.3:	41.6:		14	C
1179	05 30 12.2	-70 56 53	6	-	2	2	35	30	-	-	-	0.22:	0.22:	2.1	C	05302-7056	10	M
1180	05 30 15	-70 08	18	12	13	3	32	28	30	24	-	0.22	0.22	1.7	12.5		10	C
1181	05 30 15.1	-69 34 14	-	-	2	3	26	20	15	13	-	-	0.11	2.5	4.2:	05302-6934	10	C
1182	05 30 15.7	-71 02 32	-	-	8	7	60	40	50	30	-	-	0.33	8.3:	41.6:	05302-7102	14/10	C
1183	05 30 16.5	-71 05 33	22	8	-	6	50	30	50	40	p	0.52	0.67	8.3:	20.8:	05302-7105	14	C
1184	05 30 20	-66 04	-	-	1	-	9	4	10	5	-	-	0.22:	2.1	10.4	05301-6602:	02	C
1185	05 30 20	-68 38	19	14	4	10	70	45	-	-	-	0.19	0.33	10.3:	C		06	M
1186	05 30 20.1	-66 55 04	6	-	2	1	55	52	-	-	-	0.22	0.22	-	-	05303-6655	02	M
1187	05 30 23.6	-68 32 54	15	12	8	6	-	-	50	44	p	0.11:	0.22:	1.2:	12.5:	05303-6832	06	C
1188	05 30 24.5	-70 00 23	13	5	-	-	10	5	-	-	-	0.30	-	-	-	05304-7000	10	S
1189	05 30 24.6	-71 36 30	7	-	1	1	-	-	13	9	p	0.26	0.11:	2.1	8.3	05304-7136	14	C
1190	05 30 25.8	-67 22 23	13	-	4	2	-	-	-	-	p	0.48	0.22	-	-	05304-6722	06	S

Infrared Sources in the LMC (continued 18).

(1)	(2) RA(1950) h m s	(2) DEC(1950) ° ' "	(3) 12 μm Peak	Bg	(4) 25 μm Peak	Bg	(5) 60 μm Peak	Bg	(6) 100 μm Peak	Bg	(7) Size arcmin	(8) F 12μm Jy	(9) F 25μm Jy	(10) F 60μm Jy	(11) F 100μm Jy	(12) IRAS-Id	(13) DPM field	(14) Spectrum
Number LI-LMC							1.0E-8 Watt / (m²m sr)											
1191	05 30 28.2	-68 28 18	12	8	7	5	40	30	-	-	-	0.15:	0.22	4.1:	C	05304-6828	06	M
1192	05 30 28.7	-69 24 56	3	-	5	3	27	20	18	15	-	0.11:	0.22	2.9	6.2	05304-6924	10	C
1193	05 30 30	-67 40	4	2	3	1	21	15	16	11	-	0.07	0.22	2.5	10.4		06	C
1194	05 30 30	-69 12	14	9	7	5	34	30	30	28	-	0.19	0.22	1.7	4.2		10	C
1195	05 30 30	-69 44	-	-	5	3	29	22	21	18	-	-	0.22	2.9	6.2		10	C
1196	05 30 35	-69 42	-	-	4	2	-	-	-	-	-	-	0.22	-	-		10	M
1197	05 30 40	-66 45	3	-	2	1	-	-	-	-	-	0.11:	0.22	-	-		02	M
1198	05 30 40.3	-70 32 52	8	-	3	-	-	-	-	-	-	0.30	0.33	-	-	05306-7032	10	M
1199	05 30 42.4	-71 07 15	22	8	20	8	100	40	70	45	p	0.52	1.33	24.8	52.0	05307-7107	14	C
1200	05 30 50	-71 31	4	-	2	-	10	8	14	13	3x3	0.57	0.73	1.6	2.9		14	C
1201	05 30 55.6	-70 56 53	14	3	6	3	32	28	30	26	-	0.41:	0.33	1.7	8.3:	05309-7056	10	C
1202	05 30 59.5	-68 08 53	4	-	3	2	-	-	-	-	-	0.15:	0.11:	C	C	05309-6808	06	S
1203	05 31 00	-67 58	9	6	5	3	30	20	35	29	-	0.11	0.22	4.1	12.5:		06	C
1204	05 31 00	-68 41	20	15	15	10	93	80	-	-	-	0.19:	0.56:	5.4:	C		06	M
1205	05 31 00	-69 33	4	2	4	3	20	18	-	-	-	0.07:	0.11	0.8:	C		10	M
1206	05 31 00	-71 14	10	5	4	2	22	12	30	20	-	0.19	0.22	4.1	20.8:	05314-7115:	14	C
1207	05 31 00.2	-67 22 18	8	2	8	2	20	11	15	9	p	0.22	0.67	3.7	12.5	05310-6722	06	C
1208	05 31 02.6	-71 10 00	45	10	42	6	150	100	100	70	p	1.29	4.00	20.7	62.4:	05310-7110	14	C
1209	05 31 03.1	-69 13 47	12	9	9	6	-	-	-	-	-	0.11:	0.33:	C	C	05310-6913	10	M
1210	05 31 04.0	-68 14 05	8	3	4	2	30	26	28	22	p:	0.19	0.22	1.7	12.5	05310-6814	06	C
1211	05 31 05	-68 45	12	8	7	5	-	-	-	-	-	0.15	0.22:	C	C		06	M
1212	05 31 06.4	-69 18 02	11	5	5	4	-	-	-	-	-	0.22	0.11:	C	C	05311-6918	10	S
1213	05 31 09.2	-68 36 38	44	20	50	13	140	90	110	75	p	0.89	4.11	20.7	72.8	05311-6836	06	C
1214	05 31 09.3	-67 23 58	3	-	4	2	16	12	17	12	-	0.11:	0.22	1.7:	10.4:	05311-6723	06	C
1215	05 31 10	-69 08	11	8	11	8	-	-	-	-	-	0.11	0.33	C	C		10	M
1216	05 31 15	-67 52	12	6	6	4	42	35	35	30	-	0.22	0.22	2.9	10.4:		06	C
1217	05 31 15	-69 46	12	7	5	3	-	-	-	-	p:	0.19	0.22:	-	-		10	M
1218	05 31 16.4	-69 04 45	13	7	10	7	60	50	45	40	p:	0.22	0.33	4.1	10.4:	05312-6904	10	C
1219	05 31 20	-67 46	9	6	6	4	30	25	-	-	-	0.11	0.22	2.1:	C		06	M
1220	05 31 20	-69 36	-	-	-	-	27	23	21	19	-	-	-	1.7	4.2:		10	C

No.	RA	Dec	Type	n	Ident					P								
1221	05 31 20	−70 12 03	C	10	05313-6913	4.2	1.7	0.22	−	−	22	24	24	28	2	4	−	−
1222	05 31 21.9	−69 13 03	M	10	05313-6920	C	4.1:	0.22:	0.44	−	−	−	40	50	9	11	10	22
1223	05 31 23.5	−69 20 51	C	10	05314-7101	6.2	2.1	0.22	0.37	P	21	24	27	32	5	7	5	15
1224	05 31 25.3	−71 01 51	M	14/10	05314-6910	C	4.1:	0.33	−	−	−	−	30	40	4	7	−	−
1225	05 31 26.6	−69 10 21	C	10		31.2	8.3	0.33:	0.26	P:	40	55	45	65	8	11	12	19
1226	05 31 27	−66 08	C	02		16.6	3.3	0.22	0.19	−	6	14	5	13	−	2	−	5
1227	05 31 30	−67 59	M	06		C	4.1	0.11:	0.15	−	−	−	30	40	3	4	6	10
1228	05 31 30	−68 03	C	06		10.4:	2.1	0.33	0.22	−	30	35	30	35	3	6	6	12
1229	05 31 30	−68 22	C	06		20.8:	2.9	0.22	0.11	−	35	45	35	42	5	7	10	13
1230	05 31 30	−70 16	C	10		4.2	1.7	0.22	−	−	21	23	30	34	2	4	−	−
1231	05 31 30	−71 10	C	14	05313-7109:	197.6	76.6	7.10	2.44	P	30	125	40	225	6	70	10	76
1232	05 31 30.8	−71 45 07	C	14	05315-7145	2.1:	0.8	0.78	0.15:	P	4	5	2	4	−	7	4	4
1233	05 31 33.6	−68 33 33	C	06	05315-6833	20.8:	10.3:	1.22	0.56	−	90	100	100	125	13	24	20	35
1234	05 31 35.4	−66 31 52	S	02	05315-6631	−	−	0.22:	0.33	P	−	−	−	−	−	2	−	9
1235	05 31 39.0	−66 16 02	C	02	05316-6616	6.2:	2.5	0.22	−	−	6	9	4	10	−	2	−	−
1236	05 31 40	−67 01	C	06/02	05316-7124	4.2:	0.8	0.11:	0.33	−	−	4	3	5	−	1	−	11
1237	05 31 40.9	−71 24 46	C	14	05316-6604	20.8	3.3	0.22	0.41	P	2	20	8	16	1.5	3.5	2	14
1238	05 31 41.5	−66 04 53	M	02	05317-6907	−	−	0.44	−	P	10	−	−	−	1	5	3	−
1239	05 31 45.3	−69 07 39	M	10	05318-6956	C	C	0.78	−	−	−	−	−	−	7	14	−	−
1240	05 31 48.6	−69 56 22	C	10		14.6	5.8	0.22	−	−	25	32	30	44	4	6	−	−
1241	05 31 51	−66 43	M	02	05318-6824	−	C	0.17:	0.11:	−	−	−	−	−	−	1.5	−	3
1242	05 31 51.7	−68 24 36	M	06	05318-7247	C	C	0.33:	0.22:	P:	−	−	−	−	6	9	12	18
1243	05 31 52.3	−72 47 56	S	14	05319-6723	−	−	−	0.44:	P:	12	15	11	19	−	−	8	12
1244	05 31 54.4	−67 23 52	C	06		6.2:	3.3	0.33	0.22	−	28	26	5	5	2	5	2	8
1245	05 31 55	−68 06	C	06		4.2:	2.5	0.11:	0.22	−	−	−	−	10	3	4	6	12
1246	05 31 55	−71 04	M	14	05319-6836	C	C	0.44	0.37	−	−	−	18	−	7	11	8	18
1247	05 31 55.9	−68 36 46	C	06	05320-6832	20.8:	10.3:	2.33	0.33	P:	90	100	125	100	14	35	23	32
1248	05 32 00.5	−70 03	C	10	05320-7020	4.2:	1.7	0.17	−	−	11	13	22	18	1.5	3	−	−
1249	05 32 00.5	−68 32 03	C	06		31.2:	20.7	1.00	0.67	P:	100	115	155	105	17	26	30	48
1250	05 32 01.6	−70 20 18	M	10		−	0.8:	0.56	0.11	−	−	−	24	22	2	7	4	7
1251	05 32 01.8	−71 06 12	C	14	05320-7106	243.4	105.6	12.32	2.22	P	137	285	30	115	4	115	10	70
1252	05 32 05	−69 49	C	10	05320-6626	10.4:	4.1	0.11:	0.11:	P:	20	45	50	60	3	4	9	12
1253	05 32 05.0	−66 26 08	C	02	X0552-664	104.0:	19.9:	3.44	0.78	−	40	60	60	12	1	32	4	25
1254	05 32 07.8	−69 41 35	C	10	05321-6941	4.2:	1.2	0.11	0.30	−	10	12	31	28	5	6	8	16
1255	05 32 09.0	−68 28 47	M	06	05321-6828	C	12.4	2.33	0.37	P	−	−	130	100	17	38	25	35
1256	05 32 10	−66 23	M	02		C	0.8:	0.22:	0.19:	−	−	−	10	12	−	2	−	5
1257	05 32 10	−69 00	M	06/10		C	C	0.33:	0.15	−	−	−	−	−	6	9	10	14
1258	05 32 10	−70 32	C	10		4.2:	0.8	−	−	−	9	10	10	8	23	−	−	−
1259	05 32 10.8	−67 44 30	C	06	05321-6744	C	C	5.99	1.48	P	7	−	60	−	−	77	20	60
1260	05 32 15	−71 24	M	14		C	0.8:	0.22:	0.07	−	−	−	12	12	−	2	3	5

Infrared Sources in the LMC (continued 19).

(1)	(2)			(3)		(4)		(5)		(6)		(7)	(8)	(9)	(10)	(11)	(12)	(13)	(14)
Number LI-LMC	Position RA(1950) h m s	DEC(1950) o ' "		12 μm Peak Bg		25 μm Peak Bg		60 μm Peak Bg		100 μm Peak Bg		Size arcmin	F 12μm Jy	F 25μm Jy	F 60μm Jy	F 100μm Jy	IRAS-Id	DPM field	Spectrum
						1.0E-8 Watt / (m*m sr)													
1261	05 32 15.2	-67 48 28		20	15	16	8	100	80	-	-	-	0.19	0.89:	8.3:	C	05322-6748	06	M
1262	05 32 20	-66 05		-	-	-	-	12	7	12	8	-	-	-	2.1	8.3		02	C
1263	05 32 20	-68 19		12	10	7	5	42	35	-	-	-	0.07	0.22	2.9:	C		06	M
1264	05 32 20	-70 24		7	4	3	2	29	24	23	20	-	0.11:	0.11:	2.1	6.2		10	C
1265	05 32 24.0	-69 24 07		-	-	-	-	28	25	-	-	-	-	-	1.2	-	05324-6924	10	M
1266	05 32 25.5	-65 51 34		115	-	14	-	-	-	-	-	P	4.25	1.55	-	-	05324-6551	02	S
1267	05 32 28.1	-68 12 32		-	-	5	3	30	27	28	24	-	-	0.22:	1.2	8.3	05324-6812	06	C
1268	05 32 29.3	-66 19 17		2	-	1.5	-	11	6	10	8	-	0.07:	0.17	2.1	4.2	05324-6619	02	C
1269	05 32 30	-69 39		10	7	6	5	43	40	30	28	-	0.11:	0.11:	1.2	4.2:		10	C
1270	05 32 30	-69 49		15	10	7	5	70	60	48	40	-	0.19	0.22	4.1	16.6:		10	M
1271	05 32 30	-69 56		12	9	6	4	52	48	37	35	-	0.11:	0.22	1.7	4.2		10	C
1272	05 32 30	-71 18		18	6	4	2	20	10	25	18	P	0.44	0.22	4.1:	14.6:		14	C
1273	05 32 30.0	-66 29 21		48	2	62	-	205	3	110	2	P	1.70	6.88	83.6	224.6	05325-6629	02	C
1274	05 32 34.9	-67 43 41		78	30	105	25	350	75	180	45	P	1.78	8.88	113.8	280.8	05325-6743	06	C
1275	05 32 35.2	-69 11 07		10	7	7	5	34	30	-	-	-	0.11:	0.22	1.7:	-	05325-6911	10	M
1276	05 32 35.7	-71 06 17		16	10	20	6	100	40	-	-	-	0.22:	1.55	24.8:	C	05325-7106	14	M
1277	05 32 37.0	-68 58 46		25	12	13	8	92	60	62	45	P	0.48	0.56	13.2	35.4:	05326-6858	06/10	C
1278	05 32 38.5	-70 04 19		-	-	5	2	-	-	-	-	-	-	0.33	-	-	05326-7004 X0532-687	10	M
1279	05 32 39.4	-68 42 11		38	10	24	13	132	63	90	65	-	1.04	1.22	28.6	52.0	05326-6842	06	C
1280	05 32 40	-67 10		14	-	6	2	-	-	-	-	-	0.52	0.44	-	-		06	S
1281	05 32 45.1	-67 57 08		14	3	7	2	-	-	-	-	P	0.41	0.56	-	-	05327-6757	06	M
1282	05 32 50	-67 32		24	15	14	10	80	50	60	40	-	0.33	0.44	12.4	41.6		06	C
1283	05 32 51	-71 13		12	5	-	-	-	-	-	-	p:	0.26	-	C	C		14	S
1284	05 32 52.5	-68 27 08		40	22	23	18	-	-	-	-	p:	0.67	0.56:	C	C	05328-6827	06	S
1285	05 32 54.5	-71 15 18		30	4	17	2	62	10	48	13	p:	0.96	1.66	21.5	72.8	05329-7115	14	C
1286	05 32 54.7	-67 08 54		23	-	18	1.5	3	2	-	-	p	0.85	1.83	0.4:	-	05329-6708	06	M
1287	05 32 55.3	-69 40 26		10	8	6	4	42	36	33	30	-	0.07:	0.22	2.5:	6.2:	05329-6940	10	C
1288	05 33 00	-67 36		12	9	9	6	70	54	-	-	-	0.11	0.33	6.6:	C		06	M
1289	05 33 00	-67 38		15	10	10	7	70	60	45	40	-	0.19	0.33	4.1	10.4:		06	C
1290	05 33 00	-69 36		14	10	7	4	50	40	35	30	-	0.15	0.33	4.1	10.4:	05331-6935:	10	C

No.	Sp	R	IRAS	(4)	(5)	(6)	(7)	P	a	b	c	d	e	f	g	h	Dec (1950)	RA (1950)
1291	C	10	–	14.6	2.9	0.22	0.15	–	25	32	30	37	3	5	7	11	−70 13 18	05 33 00
1292	M	06	05330−6743	C	C	4.22	1.18	P	–	–	–	–	23	61	20	52	−67 43 18	05 33 00.9
1293	C	06	05330−6826	62.4	31.9	3.88	0.93	P	70	100	90	167	18	53	18	43	−68 26 03	05 33 02.1
1294	S	02	05331−6650	–	–	0.11:	0.26	P	–	–	–	–	–	1	–	7	−66 50 05	05 33 08.3
1295	C	10	05332−7025	–	–	–	0.15:	–	–	–	–	–	–	–	4	8	−70 25 29	05 33 14.8
1296	S	06	–	10.4:	8.3:	0.22	0.19	–	40	45	50	70	7	9	10	15	−67 30	05 33 15
1297	C	10	–	–	–	–	0.15	–	–	–	–	–	–	–	7	11	−70 11	05 33 15
1298	C	10	05333−6948	87.4	70.4	8.88	1.33	P	50	92	70	240	10	90	12	48	−69 48 24	05 33 19.9
1299	C	10	–	6.2:	2.1	0.33	0.11	–	45	48	50	55	–	9	10	13	−69 06	05 33 19.8
1300	C	06/10	05334−6858:	52.0	18.6	0.78	0.41	P:	45	70	65	110	9	16	12	23	−69 00	05 33 22
1301	C	10	05334−6935	10.4:	4.1:	0.22:	0.07:	–	35	40	40	50	7	9	8	10	−69 35 54	05 33 26.7
1302	C	14	–	2.1	0.4	0.17:	0.19	–	2	3	1	2	–	1.5	–	5	−71 53	05 33 27
1303	S	02	05334−6604	12.5	5.0	0.33	0.15	P	10	16	6	18	–	3	2	6	−66 04 19	05 33 29.7
1304	M	06	05334−6706	–	–	0.22	0.37	P	–	–	–	–	–	2	–	10	−67 06 17	05 33 29.8
1305	M	06	–	C	6.2	0.33	0.19	–	–	–	45	60	6	9	7	12	−67 28	05 33 30
1306	S	10	–	–	–	–	0.19:	–	–	–	–	–	–	–	10	15	−69 09 45	05 33 30
1307	M	06	05335−6803	C	1.2	0.22:	0.07:	–	–	–	15	18	–	2	–	2	−68 03	05 33 34.0
1308	M	10	–	C	2.1:	0.22	0.11:	–	–	–	40	45	5	7	7	10	−69 24	05 33 35
1309	M	10	05335−6927	–	2.1	0.22	–	–	–	–	30	35	4	6	–	–	−69 27 40	05 33 35.9
1310	C	06	–	4.2:	1.2	–	0.11:	–	13	15	14	17	–	1	–	3	−67 54	05 33 40
1311	M	06	–	C	2.1	0.11:	–	–	–	–	17	22	2	3	–	–	−68 07	05 33 40
1312	M	10	–	C	2.1	0.22	0.11:	–	–	–	48	53	5	7	7	10	−69 42	05 33 40
1313	S	10	–	–	–	0.11:	0.26	–	–	–	–	–	–	1	–	7	−70 33	05 33 40
1314	C	10	05337−6923	14.6	10.3:	0.22:	0.15	P:	28	35	45	70	5	7	6	10	−69 23 05	05 33 42.2
1315	M	06	05338−6725:	–	C	0.33	0.19	–	–	–	–	–	7	10	6	11	−67 27	05 33 45
1316	M	10	–	–	–	0.22	0.19	P:	–	–	–	–	5	7	10	15	−69 44	05 33 45
1317	M	02	05338−6617	31.2	6.2	0.11:	–	–	–	–	–	–	1	1	–	–	−66 17 29	05 33 48.7
1318	S	06	05338−6822	–	–	0.33	0.41	P:	37	52	38	53	6	9	12	23	−68 22 21	05 33 51.9
1319	C	14	05338−7159	12.5:	4.6:	0.22	0.70	P	–	–	–	–	–	2	8	19	−71 59 42	05 33 51.9
1320	M	06	05338−6735	2.1:	0.8	0.11:	0.15:	–	34	40	38	49	6	7	8	12	−67 35 24	05 33 52.7
1321	C	02	05338−6645	C	C	0.11:	0.19:	P:	1	2	1	3	–	1	–	–	−66 45 12	05 33 52.8
1322	M	10	–	106.1	64.2	0.22	1.41	–	–	–	65	220	6	8	6	11	−69 26	05 33 55
1323	C	06	05339−6847	25.0:	5.0	6.88	0.44	P	42	93	80	92	8	70	10	48	−68 47 52	05 33 58.6
1324	C	10	05339−6954	4.2	1.7	0.67	–	–	50	62	–	5	9	15	13	25	−69 54 27	05 33 58.9
1325	C	02	05339−6631	12.5	–	–	–	P:	3	5	1	–	–	–	–	–	−66 31 01	05 33 59.2
1326	C	02	–	C	1.7:	0.22:	0.11:	–	9	15	10	14	–	2	–	3	−66 07	05 34 00
1327	M	06	–	10.4:	1.7:	0.33	0.19	P:	–	–	30	34	3	6	7	12	−68 39	05 34 00
1328	C	10	–	C	4.1	0.22	0.11:	–	30	35	50	60	6	8	8	11	−69 42	05 34 00
1329	C	06	–	–	2.1	0.22	0.11	–	–	–	33	38	4	6	8	11	−68 18	05 34 10
1330	C	10	–	8.3:	4.1	0.22	0.19	P:	58	62	60	70	8	10	10	15	−69 13	05 34 15

Infrared Sources in the LMC (continued 20).

(1)	(2)		(3)		(4)		(5)		(6)		(7)	(8)	(9)	(10)	(11)	(12)	(13)	(14)
Number LI-LMC	RA(1950) h m s	DEC(1950) o ' "	12 μm Peak	Bg	25 μm Peak	Bg	60 μm Peak	Bg	100 μm Peak	Bg	Size arcmin	F 12μm Jy	F 25μm Jy	F 60μm Jy	F 100μm Jy	IRAS-Id	DPM field	Spectrum
							1.0E-8 Watt / (m²m sr)											
1331	05 34 17.2	-67 27 17	8	4	9	5	40	30	-	-	p:	0.15	0.44	4.1:	C	05342-6727	06	M
1332	05 34 20	-70 14	14	4	7	4	44	40	39	36	-	0.15	0.33	1.7	6.2		10	C
1333	05 34 22.4	-68 27 30	11	6	5	4	30	25	35	25	-	0.19	0.11	2.1:	20.8:	05343-6827	06	M
1334	05 34 30	-67 55	2	-	3	2	17	14	-	-	-	0.07:	0.11	1.2	C		06	C
1335	05 34 30	-69 36	17	13	14	10	80	75	65	60	-	0.15:	0.44	2.1:	10.4:		10	C
1336	05 34 35	-69 01	17	11	13	9	-	-	-	-	p:	0.22	0.44	C	C		10/06	M
1337	05 34 36	-66 14	5	2	5	1.5	20	7	17	10	2xp	0.19:	0.62	6.7	15.9		02	C
1338	05 34 36.2	-67 29 05	9	4	6	1.5	40	30	37	33	-	0.19	0.22:	4.1:	8.3:	05346-6729	06	C
1339	05 34 40	-70 01	10	6	5	4	-	-	-	-	-	0.15:	0.11:	-	-		10	S
1340	05 34 40	-70 20	-	-	4	3	25	22	24	22	-	-	0.11:	1.2	4.2		10	C
1341	05 34 41.0	-69 49 13	215	15	220	30	170	120	80	70	-	7.40	21.09	20.7	20.8:	05346-6949	10	M
1342	05 34 45	-69 12	18	12	15	12	-	-	-	-	-	0.22	0.33	C	C		10	M
1343	05 34 45	-71 05	7	-	1	-	10	8	13	10	p	0.26	0.11:	0.8	6.2		14	C
1344	05 34 47.2	-68 37 04	6	2	7	3	29	17	20	16	p:	0.15	0.44	5.0	8.3	05347-6837	06	C
1345	05 34 48.4	-70 24 48	16	3	4	1.5	-	-	-	-	-	0.48	0.28	-	-	05348-7024	10	S
1346	05 34 52.3	-68 14 12	18	7	6	4	-	-	-	-	-	0.41	0.22	-	-	05348-6814	06	S
1347	05 34 55.2	-70 42 40	3	-	2	-	13	11	11	9	p:	0.11:	0.22	0.8	4.2	05349-7042	10	C
1348	05 35 00	-67 19	5	-	2	-	11	8	9	7	-	0.19	0.22	1.2	4.2		06	C
1349	05 35 00	-68 08	11	6	6	4	27	20	34	20	-	0.19	0.22	2.9	29.1		06	C
1350	05 35 00	-68 24	13	8	6	4	-	-	-	-	-	0.19	0.22	C	C		06	M
1351	05 35 03	-66 21 40	7	2	4	1	20	5	19	10	lxp	0.24:	0.39	6.6:	19.1	05350-6637	02	C
1352	05 35 03.5	-66 37 48	-	-	2	1	9	1	6	3	-	-	0.11	3.3	6.2	05351-6943	02	C
1353	05 35 06.9	-69 43 48	17	12	17	15	-	-	-	-	-	0.19	0.22	C	-	X0535-660	10	M
1354	05 35 10	-66 02	10	5	4	1.5	40	13	-	-	-	0.19	0.28	11.2	C		02	M
1355	05 35 10	-68 00	4	-	3	1.5	19	14	18	17	-	0.15	0.17	2.1	2.1:		06	C
1356	05 35 10	-68 16	12	7	6	4	32	25	35	25	-	0.19	0.22	2.9	20.8		06	C
1357	05 35 10	-68 57	15	7	15	9	110	70	60	50	-	0.30	0.67	16.6	2.1:		06	M
1358	05 35 10	-69 48	19	13	20	16	130	120	-	-	-	0.22	0.44	4.1:	C		10	M
1359	05 35 12.8	-69 33 36	35	20	27	20	150	100	92	70	-	0.56	0.78	20.7	45.8	05352-6933	10	C
1360	05 35 20	-67 05	5	-	1	-	-	-	-	-	-	0.19	0.11:	-	-		06	S

No.	Type	Code	Ident.	F1	F2	F3	F4	P	n1	n2	n3	n4	n5	n6	n7	n8	RA (1950)	Dec (1950)
1361	C	10		20.8:	8.3:	0.56	0.48	–	110	100	140	120	28	33	30	43	05 35 20	−69 06
1362	C	10	05353−7012	10.4:	1.2:	0.22:	0.26	–	40	35	43	40	4	6	12	19	05 35 20.2	−70 12 58
1363	M	06	05353−6841	C	2.1:	0.11:	0.07:	–	–	–	25	20	3	4	–	2	05 35 21.0	−68 41 40
1364	M	06		–	–	0.11:	0.11	–	–	–	–	–	3	5	7	10	05 35 25	−67 46
1365	C	10		10.4:	2.5	0.22	0.07:	–	40	35	64	58	8	10	8	10	05 35 30	−69 25
1366	S	02/06	05354−6657	–	–	0.22	0.33	P	–	–	–	–	–	2	–	150	05 35 30.0	−66 57 53
1367	C	06	05355−6736	384.8	265.0	35.30	4.74	P	55	240	700	60	330	12	22	9	05 35 30.1	−67 36 34
1368	C	06	X0535−676	–	1.7	–	0.07:	–	–	–	17	21	–	–	–	2	05 35 35	−68 28
1369	M	06/10		4.2:	8.3:	0.67	0.15	–	15	–	100	–	15	21	13	17	05 35 35	−68 58
1370	M	10	05355−6915	C	C	1.33:	–	–	35	40	80	52	40	–	–	–	05 35 35.2	−69 15 55
1371	M	10	05356−6954	C	C	0.11:	–	–	–	–	–	–	8	9	–	–	05 35 39.7	−69 54 40
1372	C	06/02		4.2	0.8	0.11:	–	–	3	5	6	4	1	–	20	–	05 35 40	−67 01
1373	M	10		62.4:	8.3	0.78	0.22	P	80	110	160	140	20	27	20	26	05 35 40	−69 51
1374	M	10		C	4.1:	0.22	0.15:	–	–	–	50	–	5	7	8	12	05 35 40	−70 05
1375	M	10		C	C	0.22	0.15	–	–	–	–	–	4	6	12	16	05 35 40	−70 13
1376	C	02	05356−6604	97.2	43.3	6.39	1.25	2x1	54	12	93	13	1	33	5	22	05 35 40.5	−66 04 03
1377	C	06		14.6	3.3	0.22:	0.19	–	34	27	42	34	5	7	7	12	05 35 50	−68 47
1378	S	10		–	–	0.11:	0.30	P:	28	20	32	24	6	7	11	19	05 35 50	−70 01
1379	C	06		16.6	3.3	0.22	0.11	–	–	–	–	–	4	6	7	10	05 35 55	−67 46
1380	S	14	05359−7110	–	–	0.22	1.04	P	–	–	–	–	–	2	1	29	05 35 55.3	−71 10 01
1381	C	06		6.2:	3.3	0.22	0.15	–	18	–	22	14	1	3	–	4	05 36 00	−67 59
1382	M	02	05360−6648	–	0.8:	0.22	0.22	P	15	–	4	2	–	2	–	6	05 36 00.8	−66 48 26
1383	C	10	05360−6914 X0535−692	228.8	124.2	5.55	1.48	P:	280	170	500	200	80	130	60	100	05 36 02.2	−69 14 22
1384	M	10/06		C	4.1:	0.22	0.22	–	–	–	130	120	30	32	25	31	05 36 05	−69 04
1385	S	14	05361−7142	–	–	–	0.33	P	–	–	–	–	–	–	–	9	05 36 06.1	−71 42 21
1386	M	02		C	5.0	0.22	0.26	–	–	–	22	10	10	8	7	11	05 36 10	−66 37
1387	M	06		C	C	0.22:	0.15	–	16	–	–	18	9	–	16	20	05 36 10	−67 32
1388	M	10		C	41.4:	2.22	0.52	–	–	–	300	200	60	40	40	54	05 36 10	−69 12
1389	M	10		C	4.1:	0.56	0.07:	–	–	–	110	100	15	10	10	12	05 36 10	−69 23
1390	M	10		C	C	0.33	0.15	–	–	–	–	–	11	8	18	22	05 36 10	−70 03
1391	C	10		4.2:	0.8	0.22	0.11	–	18	16	20	18	4	11	8	37	05 36 10	−70 36
1392	M	06	05362−6735	C	33.1	4.22	0.44	P	–	–	150	70	54	25	–	110	05 36 12.3	−67 35 37
1393	M	10	05362−6926	C	4.1	0.22:	–	–	10	–	75	65	10	–	1	7	05 36 13.6	−69 26 44
1394	C	14	05360−7121:	6.2	1.2	0.56	0.22	P	7	–	8	5	5	5	2	45	05 36 15	−71 22
1395	S	02	05363−6619	–	–	0.44	1.59	P	–	–	–	–	6	6	2	22	05 36 19.5	−66 19 09
1396	C	06	05363−6720	6.2	1.2	0.11:	0.15	–	12	–	13	10	3	4	–	4	05 36 20.8	−67 20 56
1397	C	10	05363−6940	208.0	78.7	9.99	2.59	P	220	120	420	230	140	50	40	110	05 36 21.0	−69 40 34
1398	C	02		10.6	3.5	–	–	1xp	17	12	18	10	4	–	1	9	05 36 24	−66 16
1399	C	02/06	05364−6657	–	–	0.33	0.33	P	–	–	14	–	1	4	–	1	05 36 27.8	−66 57 25
1400	C	02	05364−6601	4.2:	1.7	0.11:	0.19	–	12	10	14	10	–	1	–	5	05 36 28.5	−66 01 27

Infrared Sources in the LMC (continued 21).

(1)	(2)			(3)		(4)		(5)		(6)		(7)	(8)	(9)	(10)	(11)	(12)	(13)	(14)
Number LI-LMC	RA(1950) h m s	DEC(1950) o ' "		12 μm Peak	Bg	25 μm Peak	Bg	60 μm Peak	Bg	100 μm Peak	Bg	Size arcmin	F 12μm Jy	F 25μm Jy	F 60μm Jy	F 100μm Jy	IRAS-Id	DPM field	Spectrum
							1.0E-8 Watt / (m² m sr)												
1401	05 36 30	-70 45 17		-	-	-	-	19	16	19	17	-	-	-	1.2	4.2:		10	C
1402	05 36 32.1	-66 27 17		16	10	9	3	45	15	30	20	p:	0.22:	0.67	12.4:	20.8:	05365-6627	02	C
1403	05 36 32.8	-69 34 05		44	28	40	27	160	90	95	70	-	0.59	1.44	29.0	52.0	05365-6934	10	C
1404	05 36 35	-70 03		23	18	11	8	93	75	70	55	-	0.19	0.33	7.5	31.2		10	C
1405	05 36 36.8	-70 50 43		-	-	2	1	8	6	17	14	-	-	0.11:	0.8:	6.2	05366-7050	10	C
1406	05 36 38.0	-69 43 00		46	30	50	40	260	200	-	-	p:	0.59	1.11	24.8	C	05366-6942	10	M
1407	05 36 42.8	-69 48 38		-	-	-	-	160	150	-	-	-	-	-	4.1:	C	05367-6948	10	M
1408	05 36 43.6	-66 26 09		14	8	8	3	-	-	-	-	p:	0.22	0.56	C	C	05367-6626	02	M
1409	05 36 45	-66 39		5	-	9	6	22	10	-	-	-	0.19	0.33	5.0	C		02	M
1410	05 36 45	-67 15		3	-	2	-	14	10	11	9	-	0.11:	0.22:	1.7	4.2		06	C
1411	05 36 45	-70 40		-	-	3	2	19	17	-	-	-	-	0.11:	0.8	C		10	M
1412	05 36 47.9	-66 45 24		4	-	2	-	8	5	-	-	-	0.15:	0.22	1.2:	-	05367-6645	02	M
1413	05 36 48.6	-69 24 43		37	10	25	18	-	-	-	-	p	1.00	0.78	C	C	05368-6924	10	S
1414	05 36 50	-66 36		-	-	4	2	22	9	22	12	1xp	-	0.26	5.8	21.3		02	C
1415	05 36 50	-67 07		5	-	4	1.5	20	10	14	9	-	0.19	0.28	4.1	10.4		06	C
1416	05 36 50	-68 24		3	-	-	-	13	11	11	9	-	0.11:	-	0.8	4.2:		06	C
1417	05 36 53.5	-69 20 12		33	22	48	33	210	180	120	100	p	0.41	1.66	12.4:	41.6:	05368-6920	10	C
1418	05 36 55	-66 35		8	-	3	1	-	-	-	-	-	0.30	0.22	-	-		02	S
1419	05 37 00	-66 32		9	3	6	2	26	17	-	-	-	0.22	0.44	3.7	C		02	M
1420	05 37 00	-67 02		4	-	1.5	-	15	8	13	8	-	0.15	0.17:	2.9	10.4		06/02	C
1421	05 37 00	-70 10		18	12	7	5	43	40	42	40	-	0.22	0.22	1.2	4.2:	05369-7011:	10	C
1422	05 37 00.8	-66 23 44		23	8	22	4	80	12	68	15	p	0.56	2.00	28.2	110.2	05370-6623	02	C
1423	05 37 02.5	-66 52 20		6	2	4	1	17	5	14	5	p:	0.15	0.33	5.0	18.7	X0536-664	02	C
1424	05 37 04.5	-70 19 23		10	4	7	3	21	18	-	-	p	0.22	0.44	1.2	C	05370-6652	10	M
1425	05 37 07.6	-69 31 27		11	9	24	12	60	55	-	-	p	0.07:	1.33	2.1:	C	05370-7019	10	M
1426	05 37 08	-66 22		22	8	13	4	40	14	-	-	-	0.52	1.00	10.8	C	05371-6931	02	M
1427	05 37 08.7	-70 45 15		12	5	-	-	-	-	-	-	p:	0.26	-	-	-		10	S
1428	05 37 10	-67 34		20	14	10	7	57	28	55	35	-	0.22	0.33	12.0	41.6	05371-7045	06	C
1429	05 37 10	-69 15		78	50	130	100	500	400	250	200	-	1.04	3.33	41.4	104.0	05373-6915:	10	C
1430	05 37 13.7	-66 28 45		22	6	12	2	52	14	40	20	p	0.59	1.11	15.7:	41.6:	05372-6628	02	C

ID	Class	No.	Name					P									Dec	RA
1431	C	06	05372-6816	8.3	2.1	0.39	0.19	p:	8	12	8	13	5	1.5	7	2	-68 16 03	05 37 15.8
1432	C	10		2.1	0.8	0.11	-	-	15	16	13	15	2	1	-	30	-70 49	05 37 20
1433	C	10		62.4:	41.4:	2.77:	1.11:	-	120	150	400	500	125	100	60	40	-69 50	05 37 30
1434	W	10	05375-6949:	C	C	0.78	0.56	-	-	-	-	-	34	27	55	40	-69 50	05 37 30
1435	W	10	05375-7002	C	4.1:	0.33	0.33	-	-	-	80	90	13	10	29	20	-70 02 21	05 37 30.8
1436	C	10		62.4	20.7	1.22	1.00	p:	140	170	200	250	36	25	62	35	-69 47	05 37 40
1437	C	10	05373-6957:	20.8	4.1:	0.33	0.19	-	75	85	100	110	13	13	25	20	-69 58	05 37 40
1438	C	14		12.5	1.7	0.22	0.22	-	13	19	9	13	3	1	8	2	-71 03	05 37 40
1439	C	06	05377-6728	16.6:	3.7:	0.22:	0.15	-	12	20	16	25	4	2	9	5	-67 28 21	05 37 43.9
1440	C	06		4.2	2.1	0.22	0.26	-	11	13	11	16	2	-	7	-	-67 09	05 37 45
1441	W	10		C	4.1:	0.78	0.26	P	-	-	150	160	32	25	33	26	-69 40	05 37 45
1442	C	02	05377-6646	4.2:	0.8:	0.22	0.11:	-	6	8	6	8	3	-	3	-	-66 46 43	05 37 45.8
1443	W	06		C	C	0.22	0.15	-	-	-	-	-	5	3	8	4	-68 38	05 37 50
1444	C	02	05379-6641	12.5	5.8	0.44	0.15	P	12	18	8	22	6	2	5	1	-66 41 52	05 37 58.9
1445	C	02		4.2	1.7	-	-	-	7	9	5	9	-	-	-	-	-66 13	05 38 00
1446	C	06		41.6:	10.3:	0.33	0.07	-	40	60	50	75	10	7	14	12	-68 51	05 38 00
1447	W	10		C	8.3:	0.89	0.22	-	-	-	200	220	43	35	32	26	-69 22	05 38 00
1448	C	10	05381-6912	312.0:	124.2:	22.20	2.96	-	400	550	800	1100	450	250	200	120	-69 12 47	05 38 09.5
1449	C	06/02	05382-6701	2.1:	0.8:	0.11:	-	-	8	9	7	9	2	1	-	-	-67 01 59	05 38 13.5
1450	C	06	05382-6836	10.4	2.1	0.22:	0.11:	-	11	16	15	20	4	2	5	2	-68 36 21	05 38 16.3
1451	W	10	05382-6936	C	C	0.33	-	-	-	-	-	-	30	27	10	-	-69 36 07	05 38 17.7
1452	C	14		10.4:	1.2	0.06	0.26	-	15	20	9	12	2	1.5	6	3	-71 03	05 38 21
1453	C	06	05383-6755	4.2:	2.1	0.22:	0.22	-	13	15	13	18	2	-	7	-	-67 55 45	05 38 22.5
1454	W	14		-	1.7	0.17	0.19	P	-	-	11	15	1.5	-	5	2	-71 16	05 38 24
1455	C	06	05384-6746	4.2:	1.2	0.11:	0.19	-	12	14	11	14	3	2	5	-	-67 46 32	05 38 29.6
1456	C	02		8.3	2.5	0.17:	-	-	8	12	6	12	1.5	-	-	-	-66 25	05 38 30
1457	C	06		4.2:	0.8:	0.17:	0.19	-	10	12	11	13	1.5	-	5	-	-67 06	05 38 30
1458	S	10		C	C	C	0.37:	-	-	-	-	-	-	-	50	40	-69 18	05 38 30
1459	W	10		C	1.2	0.22	0.19	-	-	-	50	53	9	7	25	20	-70 17	05 38 30
1460	C	10	05386-6955	8.3:	4.1:	0.22	0.22	-	93	97	100	110	14	12	26	-	-69 55 59	05 38 38.2
1461	W	10		C	C	1.11:	0.44	-	-	-	-	-	40	30	50	38	-69 33	05 38 40
1462	C	10	05386-7008:	31.2	3.3	0.56:	0.52	-	75	90	75	83	15	-	54	40	-70 10	05 38 45
1463	C	10		10.4	1.7	0.22	0.15	-	45	50	38	42	7	5	24	20	-70 24	05 38 45
1464	W	10	05387-7003	C	2.1:	0.22:	-	-	-	-	75	80	10	8	-	-	-70 03 58	05 38 47.2
1465	C	14		6.2	0.8:	0.11:	-	-	12	15	8	10	1	-	-	-	-71 18	05 38 48
1466	C	10	05389-7027:	10.4:	2.1	0.22	0.15	-	30	35	30	35	6	4	20	16	-70 28	05 38 50
1467	W	10/06		C	C	2.77:	0.93:	-	-	-	-	-	125	100	75	50	-69 01	05 38 55
1468	W	06	05389-6844	C	4.1:	0.33:	0.07:	-	-	-	40	50	7	4	12	10	-68 44 28	05 38 55.2
1469	C	10	05389-6908	3120.0	2794.5	471.75:	74.00	-	700	2200	750	7500	5000	750	2200	200	-69 08 02	05 38 57.4
1470	W	10	05389-6922	C	2.11	2.22	2.11	P	-	-	-	-	60	40	77	20	-69 22 08	05 38 57.4

Infrared Sources in the LMC (continued 22).

(1)	(2)			(3)		(4)		(5)		(6)		(7)	(8)	(9)	(10)	(11)	(12)	(13)	(14)
Number LI-LMC	Position RA(1950) h m s	DEC(1950) o ' "		12 μm Peak	Bg	25 μm Peak	Bg	60 μm Peak	Bg	100 μm Peak	Bg	Size arcmin	F 12μm Jy	F 25μm Jy	F 60μm Jy	F 100μm Jy	IRAS-Id	DPM field	Spectrum
						1.0E-8 Watt / (m²m sr)													
1471	05 38 57.6	-70 42 40		44	9	50	4	120	20	56	20	p	1.29	5.11	41.4	74.9	05389-7042	10	C
1472	05 38 57.8	-68 54 59		30	25	30	28	-	-	-	-	-	0.19:	0.22:	C	C	X0538-707	06	M
1473	05 38 58	-71 00		10	5	4	2	19	13	24	18	p:	0.19	0.22	2.5	12.5	05389-6854	14/10	C
1474	05 39 00	-67 19		7	3	3	1.5	18	14	17	14	-	0.15	0.17	1.7	6.2		06	C
1475	05 39 00	-70 20		28	24	9	7	-	-	-	-	-	0.15	0.22	C	C		10	M
1476	05 39 04	-71 41		3	-	-	-	-	-	-	-	-	0.11:	-	-	-		14	S
1477	05 39 09.7	-69 26 28		50	35	80	50	300	225	150	125	-	0.56	3.33	31.0	52.0:	05391-6926	10	C
1478	05 39 15	-67 47		5	-	4	1.5	13	10	13	11	-	0.19	0.28:	1.2:	4.2:		06	C
1479	05 39 15	-69 47		40	30	25	20	250	200	170	150	-	0.37	0.56	20.7:	41.6:		10	C
1480	05 39 18	-66 34		2	-	1	-	10	7	10	8	-	0.07:	0.11:	1.2	4.2		02	C
1481	05 39 20	-67 55		2	-	-	-	15	13	14	12	-	0.07:	-	0.8:	4.2:		06	C
1482	05 39 20	-69 15		120	75	300	200	1000	800	500	400	-	1.66	11.10:	82.8:	208.0:		10	C
1483	05 39 20	-69 30		80	60	-	-	-	-	-	-	p:	0.74	C	C	C		10	S
1484	05 39 20	-70 35		14	11	4	3	20	18	31	29	-	0.11	0.11:	0.8:	4.2:		10	C
1485	05 39 21.5	-69 36 16		60	50	1	-	-	-	-	-	-	0.37:	C	C	C	05393-6936	10	S
1486	05 39 27.2	-70 15 14		40	33	15	12	110	100	110	100	-	0.26	0.33	C	C	05394-7015	10	M
1487	05 39 30	-69 56		35	30	18	15	10	5	11	7	-	0.19:	0.33	4.1:	20.8:		10	C
1488	05 39 33.1	-68 20 13		4	-	2	-	15	12	-	-	-	0.15:	0.22:	2.1	8.3	05395-6820	06	C
1489	05 39 35.0	-71 03 31		-	-	3	2	15	12	-	-	-	-	0.11	1.2:	C	05395-7103	14	M
1490	05 39 37.4	-69 31 56		205	85	405	125	1100	500	450	300	p	4.44	31.08	248.4	312.0	05396-6931	10	C
1491	05 39 37.8	-71 08 04		8	4	3	2	17	12	23	18	-	0.15	0.11	2.1	10.4	05396-7108	14	C
1492	05 39 40	-67 02		5	-	2	-	14	9	13	7	-	0.19	0.22:	2.1	12.5		06/02	C
1493	05 39 40	-68 56		50	33	47	30	300	170	200	150	-	0.63	1.89	53.8	104.0:		06/10	C
1494	05 39 50	-69 08		200	120	600	500	-	-	-	-	-	2.96	11.10:	C	C	05399-6906:	10	M
1495	05 39 50	-69 19		50	40	54	40	300	250	-	-	1x1:	0.37:	1.55	20.7:	C		10	M
1496	05 39 51.6	-67 19 50		4	-	4	2	21	14	15	12	-	0.15:	0.22	2.9	6.2	05398-6719	06	C
1497	05 40 02.2	-70 13 49		45	33	35	15	120	75	95	80	-	0.44	2.22	18.6	31.2	05400-7013	10	C
1498	05 40 03	-66 40		-	-	-	-	9	4	6	4	-	-	-	0.8:	4.2		02	C
1499	05 40 03	-66 48		-	-	-	-	9	5	8	4	1x1:	-	-	1.9	8.7		02	C
1500	05 40 05.3	-70 01 31		36	30	15	12	84	75	-	-	-	0.22	0.33	3.7:	C	05400-7001	10	M

No.	RA (1950)	Dec (1950)	c1	c2	c3	c4	c5	c6	c7	c8	flag	S12	S25	S60	S100	Designation	Type	Code
1501	05 40 06.4	-69 47 37	320	200	600	400	1500	500	600	300	p:	4.44	22.20	414.0:	624.0:	05401-6947	C	10
1502	05 40 06.4	-70 20 06	-	-	-	-	-	-	-	-	-	-	0.22	-	-	05401-7020	M	10
1503	05 40 09.0	-69 40 13	530	125	1200	200	2100	500	720	350	p	14.98	111.00	662.4	769.6	05401-6940	C	10
1504	05 40 10	-70 30	14	11	4	4	27	25	38	36	-	0.11:	0.11:	0.8:	4.2:		C	10
1505	05 40 10	-71 10	12	6	4	2	-	-	-	-	-	0.22	0.22	C	C		C	14
1506	05 40 13.2	-69 56 46	52	35	25	17	-	-	-	-	p:	0.63	0.89	C	C	05402-6956	M	10
1507	05 40 15	-69 29	65	45	100	75	300	200	75	70	p:	0.74	2.77	41.4:	C	05402-6927:	M	10
1508	05 40 17.9	-70 09 18	33	27	12	10	65	58	9	7	-	0.22	0.22:	2.9:	10.4:	05402-7009	C	10
1509	05 40 18.9	-68 30 29	5	-	4	1.5	9	9	9	-	-	0.19:	0.28	0.8:	4.2:	05403-6830	C	06
1510	05 40 20	-67 32	2	-	-	-	9	8	8	-	-	0.07:	-	0.4	2.1:		C	06
1511	05 40 22.7	-70 46 41	16	12	-	1	28	25	35	32	-	0.15	-	1.2	6.2:	05403-7046	C	10
1512	05 40 25.7	-67 38 33	4	-	3	-	13	10	13	9	-	0.15:	0.22:	1.2	8.3	05404-6738	C	06
1513	05 40 28.1	-66 19 03	-	-	4	-	-	-	-	-	p	-	0.44	-	-	05404-6619	M	02
1514	05 40 29.1	-69 33 15	-	-	80	70	240	200	-	-	-	-	1.11	16.6:	C	05404-6933	M	10
1515	05 40 30	-68 39	10	6	3	2	-	-	-	-	p:	0.15	0.11:	C	C		S	06
1516	05 40 30	-69 08	75	50	125	100	-	-	-	-	-	0.93	2.77:	C	C		M	10
1517	05 40 30	-71 07	22	6	3	2	-	-	-	-	-	0.59	0.11	-	624.0:		S	14
1518	05 40 33.3	-69 46 10	310	200	700	400	1500	500	600	300	p	4.07	33.30	414.0:	C	05405-6946	C	10
1519	05 40 33.5	-69 00 54	30	25	35	30	-	-	-	-	-	0.19:	0.56	C	C	05405-6900	M	10/06
1520	05 40 35.2	-69 35 47	70	60	80	70	-	-	-	-	p	0.37:	1.11:	C	C	05405-6935	M	10
1521	05 40 36.3	-71 11 30	59	5	80	2	160	11	83	15	p	2.00	8.66	61.7	141.4	05406-7111	C	14
1522	05 40 36.7	-69 24 14	48	25	32	25	-	-	-	-	p	0.85	0.78	C	C	X0540-712	S	10
1523	05 40 40	-69 51	65	50	55	40	-	3	-	-	-	0.56	1.66	C	C	05406-6924	M	10
1524	05 40 41.1	-66 08 19	-	-	2	-	5	-	9	4	-	-	0.22	0.8	10.4	05406-6608	C	02
1525	05 40 45	-69 42	100	70	140	85	550	500	-	-	-	1.11:	6.10	20.7:	C	X0540-661	M	10
1526	05 40 45	-70 34	24	20	12	8	33	30	-	-	-	0.15	0.44	1.2:	C	05409-6942:	M	10
1527	05 40 46.7	-68 12 56	8	3	4	1.5	9	5	15	7	-	0.19	0.28	1.7	16.6:	05407-6812	C	06
1528	05 40 48	-66 13	5	-	1	-	5	4	6	4	-	0.19	0.11:	0.4:	4.2:	X0540-682	C	02
1529	05 40 48	-71 28	4	-	3	1	12	8	17	9	3x3	0.57	0.73	3.4	23.3		C	14
1530	05 40 52	-72 30	5	-	-	-	-	-	-	-	-	0.19:	-	-	-		S	14
1531	05 40 53.0	-71 35 13	2	-	2	1	9	5	12	8	-	0.07:	0.11	1.7	8.3	05408-7135	C	14
1532	05 40 54	-71 14	20	10	17	3	45	15	6	4	p:	0.37	1.55	12.4:	4.2	05409-6615	M	14
1533	05 40 57.5	-66 15 31	-	-	1	-	5	3	-	-	-	-	0.11:	0.8:	20.8:	05409-6838	C	02
1534	05 40 59.4	-68 38 40	5	3	3	2	30	17	25	15	-	0.07:	0.56:	5.4:	C		C	06
1535	05 41 00	-68 54	33	25	20	15	-	-	-	-	-	0.30	-	C	-		M	06
1536	05 41 00	-70 40	17	13	6	5	33	30	42	40	-	0.15:	0.11:	1.2	4.2:	05410-6520	C	10
1537	05 41 01.3	-65 20 56	4	-	1	-	-	-	9	7	-	0.15:	0.11:	-	-	05410-6726	S	02
1538	05 41 02.7	-67 26 52	3	-	-	-	10	7	7	-	-	0.11:	-	1.2	4.2	05410-6927	C	06
1539	05 41 05.0	-69 27 06	30	20	25	-	120	100	-	-	-	0.37	0.56	8.3:	C	05410-6954	M	10
1540	05 41 05.1	-69 54 16	40	30	-	-	-	-	-	-	p:	0.37	-	-	-		S	10

Infrared Sources in the LMC (continued 23).

Peak/Bg columns (3)–(6) in units of 1.0E-8 Watt / (m²m sr).

(1)	(2) Position		(3) 12 μm		(4) 25 μm		(5) 60 μm		(6) 100 μm		(7) Size	(8) F 12μm	(9) F 25μm	(10) F 60μm	(11) F 100μm	(12)	(13) DPM	(14) Spec-
Number LI-LMC	RA(1950) h m s	DEC(1950) o ' "	Peak	Bg	Peak	Bg	Peak	Bg	Peak	Bg	arcmin	Jy	Jy	Jy	Jy	IRAS-Id	field	trum
1541	05 41 08.2	-70 12 04	36	30	18	13	70	50	70	60	-	0.22	0.56	8.3	20.8	05411-7012	10	C
1542	05 41 10	-70 47	19	14	7	5	30	25	45	40	-	0.19	0.22	2.1	10.4:		10	C
1543	05 41 10.4	-69 23 41	35	28	-	-	-	-	-	-	-	0.26	-	-	-	05411-6923	10	S
1544	05 41 11.9	-70 03 34	23	18	11	8	73	60	65	55	-	0.19	0.33	5.4	20.8	05411-7003	10	C
1545	05 41 15	-67 11	-	-	1.5	-	12	9	10	8	-	-	0.17	1.2	4.2		06	C
1546	05 41 15	-67 55	-	-	1	-	15	11	14	11	-	-	0.11:	1.7	6.2		06	C
1547	05 41 15	-68 58	33	25	25	20	210	170	150	125	-	0.30	0.56	16.6	52.0		06/10	C
1548	05 41 15	-69 35	50	35	35	29	125	100	80	60	-	0.56	0.67	10.3:	41.6:	05413-6934:	10	C
1549	05 41 15	-69 47	65	45	60	50	-	-	-	-	-	0.74	1.11	C	C		10	M
1550	05 41 19.0	-70 29 34	18	14	15	7	48	35	-	-	p	0.15	0.89	5.4	C	05413-7029	10	M
1551	05 41 21.6	-70 35 30	27	21	13	8	53	38	55	45	p	0.22	0.56	6.2	20.8	05413-7035	10	C
1552	05 41 22.2	-69 38 37	30	20	35	30	-	-	-	-	-	0.37:	0.56:	C	C	05413-6938	10	M
1553	05 41 22.9	-69 19 58	30	20	17	15	-	-	-	-	p	0.37	0.22:	C	C	05413-6919	10	S
1554	05 41 23.4	-66 55 11	-	-	1	-	6	4	6	5	-	-	0.11:	0.8:	2.1:	05413-6655	02/06	C
1555	05 41 28.2	-70 44 55	21	15	10	6	34	27	-	-	p	0.22	0.44	2.9	C	05414-7044	10	M
1556	05 41 31.4	-72 04 26	24	-	2	-	-	-	-	-	p	0.89	0.22	-	-	05415-7204	14	S
1557	05 41 32.5	-70 55 11	13	10	5	3	23	17	27	22	-	0.11	0.22	2.5	10.4	05415-7055	10	C
1558	05 41 33.1	-68 47 38	36	12	11	7	-	-	-	-	p	0.89	0.44	-	-	05415-6847	06	S
1559	05 41 34.3	-69 00 13	32	28	25	22	-	-	-	-	-	0.15:	0.33	C	C	05416-6900	10/06	M
1560	05 41 36.0	-69 56 32	-	-	12	10	80	70	-	-	-	-	0.22	4.1:	C	05416-6956	10	M
1561	05 41 39.6	-69 48 17	30	20	31	20	140	100	120	100	p:	0.37	1.22	16.6:	41.6:	05416-6948	10	C
1562	05 41 41.7	-70 25 02	30	21	12	7	47	32	50	40	-	0.33	0.56	6.2	20.8	05416-7025	10	C
1563	05 41 41.9	-68 46 11	11	9	7	5	55	40	40	35	-	0.07:	0.22:	6.2:	10.4:	05416-6846	06	C
1564	05 41 45	-69 33	18	14	12	9	-	-	-	-	-	0.15	0.33	C	C		10	M
1565	05 41 49	-67 26	5	-	-	-	-	-	-	-	-	0.19	-	-	-		06	S
1566	05 41 50.2	-68 54 08	20	15	-	-	-	-	-	-	p	0.19:	C	C	C	05418-6854	06	S
1567	05 41 58.8	-70 03 20	90	10	13	6	-	-	-	-	-	2.96	0.78	-	C	05419-7003	10	S
1568	05 42 00	-67 02	5	-	2	-	13	6	14	7	-	0.19	0.22	2.9	14.6		06	C
1569	05 42 00	-70 39	18	13	-	-	-	-	-	-	-	0.19	-	-	-		10	S
1570	05 42 01.5	-69 43 33	22	12	18	15	-	-	-	-	-	0.37	0.33:	C	C	05420-6943	10	S

No.	T	n	Designation	V1	V2	R1	R2	f	d1	d2	d3	d4	d5	d6	d7	d8	RA	Dec
1571	M	14	05420-7118	C	8.3	0.67	0.33	p	-	-	15	35	10	4	17	8	05 42 03.7	-71 18 45
1572	M	14	05421-7116	C	4.1	0.61	0.26	p:	-	-	15	25	9	3.5	13	6	05 42 06.9	-71 16 52
1573	M	10/06	-	C	C	1.00	0.44	p:	40	43	-	-	27	18	32	20	05 42 15	-68 59
1574	C	10	05422-7032	6.2	1.7	0.22:	0.19	-	-	-	30	34	7	5	17	12	05 42 16.4	-70 32 18
1575	S	06	-	-	-	0.11:	0.19	-	-	-	-	-	5	4	11	6	05 42 20	-68 44
1576	C	06	05423-6719	2.1:	0.8	-	-	-	4	5	4	6	-	-	-	-	05 42 21.6	-67 19 22
1577	C	14	05423-7120	124.8	53.0	5.22	1.15	p	15	75	12	140	49	2	35	4	05 42 21.8	-71 20 33
1578	C	14	X0542-713	10.4	1.2	0.11:	0.15	-	20	25	12	15	3	2	10	6	05 42 24	-71 13
1579	C	06	05424-6813	2.1:	0.4:	0.28	0.19:	p:	4	5	6	7	4	1.5	7	2	05 42 27.4	-68 13 25
1580	C	06	-	4.2	0.8	0.22	0.07:	-	5	7	6	8	2	-	2	-	05 42 30	-67 25
1581	C	10	05425-6910	41.6:	20.7	1.66	0.56	-	160	180	250	250	40	25	45	30	05 42 30.4	-69 10 39
1582	S	10	05425-6914	C	C	-	0.37	-	-	-	-	-	-	-	30	20	05 42 32.1	-69 14 23
1583	C	10	05426-6929	4.2:	1.2	0.22	0.19	p:	32	34	31	34	6	4	14	9	05 42 38.4	-69 29 10
1584	C	02	-	6.2:	0.8	0.11:	0.19:	-	6	9	7	9	1	7	5	-	05 42 40	-66 44
1585	C	10	05426-6935	6.2:	1.2	0.22	0.07:	-	29	32	30	33	7	5	11	9	05 42 41.4	-69 36 00
1586	C	10	-	41.6:	20.7	0.78	0.74	-	150	170	200	250	42	35	60	40	05 42 45	-69 08
1587	S	06	-	-	-	-	0.15	-	-	-	-	-	5	4	4	11	05 42 46	-67 10
1588	C	10	05427-7006	16.6	1.7	0.11:	0.19	-	30	38	30	34	5	4	16	-	05 42 46.1	-70 06 31
1589	M	10	05429-7011	C	0.8	0.11:	-	-	-	-	21	23	5	4	-	-	05 42 54.3	-70 11 43
1590	C	06	05429-6814	8.3	1.7	0.22:	0.15:	-	8	12	8	12	2	-	4	-	05 42 59.5	-68 14 50
1591	C	02	-	4.2	1.2	0.11:	0.11:	-	10	12	6	9	1	-	3	-	05 43 00	-66 27
1592	C	02	-	6.2	1.2	-	-	-	9	12	7	10	15	11	-	20	05 43 00	-66 37
1593	C	10	05430-6913	20.8:	8.3:	0.44:	0.30	-	80	90	90	110	54	35	28	35	05 43 03.4	-69 13 01
1594	C	10	-	20.8:	20.7	2.11	0.74	-	140	150	200	250	3	1	55	9	05 43 10	-69 06
1595	C	10	-	2.1:	0.4:	0.22:	0.15	-	30	31	21	22	-	-	13	-	05 43 10	-70 24
1596	C	06	05431-6742	2.1:	0.4:	0.11:	0.07:	-	9	10	8	9	3	2	2	15	05 43 12.0	-67 42 26
1597	M	10/06	05432-6858	C	6.2:	0.44:	0.19:	-	-	-	110	125	15	11	20	-	05 43 12.6	-68 58 03
1598	C	06	05432-6735	2.1	1.2	0.22:	-	-	8	9	7	10	2	-	-	-	05 43 13.7	-67 35 42
1599	M	10	05432-6915	C	5.4	0.33:	-	-	-	20	80	93	13	10	-	4	05 43 14.6	-69 15 04
1600	M	14	05432-7118	4.2:	6.6	1.00	0.19	-	18	20	9	25	11	2	9	-	05 43 16.3	-71 18 44
1601	C	10/14	05432-7057	4.2:	0.8:	0.11:	0.07:	-	18	20	10	12	4	3	11	9	05 43 16.6	-70 57 27
1602	M	06	05433-6728	C	0.8:	0.22:	0.15	-	-	-	7	9	2	-	4	-	05 43 18.6	-67 28 56
1603	C	10	-	C	C	C	0.22:	-	-	-	-	-	-	-	26	20	05 43 20	-69 17
1604	S	10	X0543-678	-	-	0.11:	0.11	-	-	-	-	-	3	2	13	10	05 43 20	-70 32
1605	C	06	05433-6750	20.8	12.0	1.33	0.30	p	15	25	15	44	15	3	14	6	05 43 20.2	-67 50 55
1606	C	02	-	10.4	1.7:	0.22:	0.22	-	8	13	7	11	2	-	6	8	05 43 21	-66 48
1607	C	10	05433-6958	8.3:	1.7:	0.22:	0.07:	-	30	34	28	32	4	2	10	-	05 43 21.3	-69 58 30
1608	C	14	-	18.7	2.1	0.22:	0.19	-	18	27	10	15	4	2	11	6	05 43 25	-71 04
1609	C	10	05434-6946	124.8	58.0	3.88	1.29	p	60	120	60	200	45	10	55	20	05 43 26.0	-69 46 26
1610	C	14	-	12.5:	0.8	0.22:	0.30	p	15	21	9	11	4	2	13	5	05 43 27	-71 14

Infrared Sources in the LMC (continued 24).

(1)	(2)		(3)		(4)		(5)		(6)		(7)	(8)	(9)	(10)	(11)	(12)	(13)	(14)
Number LI-LMC	Position RA(1950) h m s	DEC(1950) o ' "	12 μm Peak	Bg	25 μm Peak	Bg	60 μm Peak	Bg	100 μm Peak	Bg	Size arcmin	F 12μm Jy	F 25μm Jy	F 60μm Jy	F 100μm Jy	IRAS-Id	DPM field	Spectrum
					1.0E-8 Watt / (m*m sr)													
1611	05 43 30	-67 57	13	6	5	3	25	18	23	18	p	0.26	0.22	2.9	10.4		06	C
1612	05 43 30	-70 11	11	8	4	3	22	20	28	26	-	0.11	0.11	0.8	4.2		10	C
1613	05 43 31.3	-66 19 43	11	2	10	1	32	6	32	9	p:	0.33	1.00	10.8	47.8:	05435-6619	02/01	C
1614	05 43 40	-66 22	9	2	6	2	25	12	-	-	p:	0.26:	0.44	5.4	C	X0543-663	02/01	W
1615	05 43 41.0	-70 08 02	-	-	-	-	21	19	22	20	-	-	-	0.8	4.2:	05436-7008	10	C
1616	05 43 43.5	-68 29 27	22	3	3	1.5	11	8	9	8	p	0.70	0.17	1.2:	2.1:	05437-6829	06	C
1617	05 43 45	-71 13	10	6	2	1.5	-	-	-	-	-	0.15	0.06:	-	-		14	S
1618	05 43 50	-68 06	10	-	3	2	13	11	14	13	p	0.37	0.11:	0.8:	2.1:	05439-6807:	06	C
1619	05 43 50	-68 49	15	10	7	5	68	55	65	55	-	0.19	0.22	5.4	20.8		06	C
1620	05 43 50	-69 41	12	9	7	5	40	35	37	35	-	0.11:	0.22	2.1	4.2:		10	C
1621	05 43 52.0	-67 28 30	16	6	8	3	38	14	25	12	p	0.37	0.56	9.9	27.0:	05438-6728	06	C
1622	05 43 52.0	-69 26 05	40	30	24	14	110	80	-	-	-	0.37:	1.11	12.4:	C	05438-6926	10	W
1623	05 43 54.4	-67 43 10	225	4	43	2	11	9	-	-	-	8.18	4.55	0.8:	C	05439-6743	06	S
1624	05 43 58.6	-65 55 11	5	-	2	-	1	-	1	-	p:	0.19	0.22	0.4:	2.1:	05439-6555	01/02	C
1625	05 44 00	-71 31	2	-	1.5	-	6	3	12	8	-	0.07:	0.17:	1.2	8.3		14	C
1626	05 44 00.9	-70 29 09	16	8	4	3	19	17	27	25	-	0.30	0.11:	0.8	4.2:	05440-7029	10	C
1627	05 44 10	-68 17	10	6	4	2	22	15	24	18	-	0.15	0.22	2.9	12.5		06	C
1628	05 44 10	-70 16	12	9	4	3	20	18	28	27	-	0.11:	0.11:	0.8:	2.1:		10	C
1629	05 44 10.7	-69 16 52	40	25	26	20	150	100	-	-	p	0.56	0.67	20.7	C	05441-6916	10	W
1630	05 44 11.7	-68 53 12	12	8	7	6	65	60	-	-	-	0.15:	0.11:	2.1:	C	05441-6853	06	W
1631	05 44 12.7	-67 48 29	5	-	2	-	17	12	16	14	-	0.19:	0.22:	2.1	4.2:	05442-6748	06	C
1632	05 44 13.2	-68 23 51	6	-	1	-	18	16	20	17	-	0.22:	0.11:	0.8	6.2:	05442-6823	06	C
1633	05 44 17.6	-69 23 19	50	35	25	20	150	125	140	110	-	0.56	0.56	10.3:	62.4:	05442-6923	10	C
1634	05 44 18	-66 18	14	4	4	2	15	10	20	15	1x1:	0.55	0.31:	2.4:	10.9		01/02	C
1635	05 44 18	-66 22	9	3	5	2	-	-	-	-	-	0.22	0.33:	C	C		01/02	W
1636	05 44 20	-70 18	11	9	4	3	23	20	28	27	-	0.07:	0.11	1.2	2.1:		10	C
1637	05 44 27	-71 12	9	5	3	1.5	10	8	18	15	-	0.15	0.17	0.8:	6.2:		14	C
1638	05 44 30	-67 26	11	7	5	3	32	18	25	20	p:	0.15	0.22	5.8	10.4:		06	C
1639	05 44 30	-67 43	11	6	3	1.5	14	10	16	12	-	0.19	0.17	1.7	8.3		06	C
1640	05 44 30	-69 01	13	10	8	5	80	70	65	55	-	0.11	0.33	4.1	20.8		10/06	C

No.	RA (1950)	Dec (1950)									flag					Ident.	Ep.	Cl.
1641	05 44 30	−70 23	13	10	23	21	3	4	−	−	−	0.11	0.11	0.8	C		10	M
1642	05 44 35	−68 49	18	13	−	−	5	8	−	−	−	0.19	0.33	C	C		06	M
1643	05 44 35	−69 21	44	33	150	130	20	24	130	120	−	0.41	0.44	8.3:	20.8:		10/09	C
1644	05 44 39.5	−65 45 19	33	−	1	130	−	−	120	−	p	1.22	0.33	0.4:	−	05446-6545	01	S
1645	05 44 40	−69 39	14	10	54	50	7	9	−	−	−	0.15:	0.22	1.7	C		10	M
1646	05 44 40.6	−69 45 25	24	14	−	−	7	10	−	−	p	0.37	0.33:	−	−	05446-6945	10	S
1647	05 44 42.2	−68 33 10	9	5	27	20	2	3	27	22	−	0.15:	0.11:	2.9	10.4	05447-6833	06	C
1648	05 44 45	−69 43	13	9	53	50	6	8	54	50	p	0.15	0.22	1.2	8.3		10	C
1649	05 44 45	−70 07	−	−	18	16	−	−	21	19	−	−	−	0.8	4.2		10	C
1650	05 44 46.3	−67 18 58	12	2	42	7	3	11	25	14	p	0.37	0.89	14.5	22.9:	05447-6718	06/05	C
1651	05 44 49.9	−70 25 07	11	9	22	20	3	4	27	25	−	0.07:	0.11:	0.8	4.2:	05448-7025	10	C
1652	05 44 50	−68 40	10	7	34	30	5	4	−	−	−	0.11:	0.11:	1.7:	C		06	M
1653	05 44 53.8	−66 41 09	3	−	8	3	−	1	10	4	−	0.11:	0.11:	2.1	12.5	05448-6641	01	C
1654	05 44 58.1	−68 49 15	15	9	−	−	4	6	−	−	p:	0.22	0.22	C	C	05449-6849	06	C
1655	05 45 00	−67 22	11	5	30	16	3	7	−	−	−	0.22	0.44	5.8	C		06/05	M
1656	05 45 01.3	−69 29 59	20	15	67	57	11	9	65	55	−	0.19:	0.22:	4.1:	20.8:	05450-6929	10	C
1657	05 45 03.0	−70 39 12	20	14	20	15	4	6	−	−	−	0.22:	0.22:	2.1	C	05450-7039	10	M
1658	05 45 10	−69 51	21	16	65	55	10	8	60	50	−	0.19	0.22	4.1	20.8:		10	C
1659	05 45 10	−70 32	−	−	22	20	3	2	23	21	−	−	0.11:	0.8	4.2		10	C
1660	05 45 10	−70 46	12	9	13	16	4	3	23	21	−	0.11	0.11	0.4:	4.2:		10	C
1661	05 45 11.4	−66 29 23	−	−	4	3	−	5	5	4	−	−	−	0.4	2.1	05451-6629	01	C
1662	05 45 14.2	−69 24 16	26	20	75	65	12	18	−	−	p	0.22	0.67	4.1:	C	05452-6924	10/09	M
1663	05 45 15	−68 12	7	5	22	17	2	3	21	18	−	0.07:	0.11:	2.1	6.2:		06	C
1664	05 45 20	−67 38	2	−	12	10	1	1	13	11	−	0.07:	0.11:	0.8	4.2		06/05	C
1665	05 45 20	−68 05	−	−	13	10	−	3	14	12	−	−	−	1.2	4.2:		06	C
1666	05 45 21.6	−65 22 05	4	1	4	−	5	5	−	−	p	0.15:	0.56	−	−	05453-6522	01	C
1667	05 45 30	−67 08	13	3	31	9	7	7	−	−	p:	0.44	0.56	9.1	C		05/01	M
1668	05 45 30	−68 12	5	−	15	13	2	2	16	15	−	0.07:	0.11:	0.8	2.1:		06	C
1669	05 45 35	−69 37	30	18	80	65	14	10	70	65	−	0.44:	0.44	6.2	10.4		10/09	C
1670	05 45 38.0	−72 34 07	−	−	2	−	−	−	1	−	−	−	−	0.8:	2.1:	05456-7234	14	C
1671	05 45 44.2	−69 22 27	23	19	4	−	11	11	−	−	−	0.15:	0.33	C	C	05457-6922	10/09	C
1672	05 45 45.8	−68 36 50	10	6	25	20	3	3	28	24	−	0.15	0.11:	2.1	8.3	05457-6836	06	C
1673	05 45 48.5	−67 10 49	16	1	56	9	18	18	37	6	p	0.56	1.89	19.5	64.5	05458-6710 / X0545-671	05	C
1674	05 45 53.1	−66 23 20	2	−	4	2	1	1	6	4	−	0.07:	0.11:	0.8	4.2	05458-6623	01	C
1675	05 45 53.2	−69 47 37	32	17	150	70	50	50	86	70	p	0.56	4.44	33.1	33.3:	05458-6947	10/09	C
1676	05 45 55	−70 39	16	11	17	14	4	3	26	23	−	0.19	0.11:	1.2:	6.2:	05459-6951	10/09	C
1677	05 45 55.4	−69 51 28	23	16	60	50	8	12	−	−	p:	0.26	0.44	4.1:	C	05459-6715	10/09	M
1678	05 45 57.0	−67 15 35	2	−	9	6	−	15	7	−	−	0.07:	1.66	1.2	C	05458-6651:	05	M
1679	05 46 00	−66 53	2	−	5	3	−	1	7	2	−	0.07:	0.11:	0.8	10.4		01	C
1680	05 46 00	−67 52	3	−	10	6	−	2	11	9	p:	0.11	0.22	1.7	4.2:		05/06	C

Infrared Sources in the LMC (continued 25).

(1)	(2)		(3)		(4)		(5)		(6)		(7)	(8)	(9)	(10)	(11)	(12)	(13)	(14)
Number LI-LMC	Position RA(1950) h m s	DEC(1950) o ' "	12 μm Peak	Bg	25 μm Peak	Bg	60 μm Peak	Bg	100 μm Peak	Bg	Size arcmin	F 12μm Jy	F 25μm Jy	F 60μm Jy	F 100μm Jy	IRAS-Id	DPM field	Spectrum
					1.0E-8 Watt / (m²·m sr)													
1681	05 46 00	-69 32	20	15	9	7	-	-	-	-	p:	0.19	0.22:	C	C		09/10	W
1682	05 46 00	-70 16	4	-	2	1	17	14	20	17	-	0.15:	0.11:	1.2	6.2		10/09	C
1683	05 46 00.2	-69 57 40	9	7	2	-	27	24	-	-	-	0.07:	0.22:	1.2	C	05460-6957	10/09	W
1684	05 46 06.3	-70 08 35	3	-	1	-	18	14	21	18	-	0.11:	0.11:	1.7	6.2	05461-7008	10/09	C
1685	05 46 10	-69 07	15	10	8	5	60	50	60	40	-	0.19	0.33	4.1	41.6:		09/10	C
1686	05 46 15	-67 44	4	-	-	-	-	-	-	-	p	0.15:	-	1.2:	-		05	S
1687	05 46 20	-68 10	4	-	2	-	11	8	18	15	-	0.15:	0.22:	1.7	6.2		05	C
1688	05 46 20	-68 26	-	-	2	-	17	13	17	14	-	-	0.22	1.7	6.2		05/06	C
1689	05 46 20.3	-66 12 37	-	-	1	-	3	1	-	-	-	-	0.11:	0.4	-	05463-6612	01	W
1690	05 46 27	-71 07	8	3	2	1	6	4	13	10	-	0.19	0.11:	0.8	6.2		14/13	C
1691	05 46 28.2	-68 51 44	10	4	6	3	43	25	43	28	p:	0.22	0.33	7.5	31.2:	05464-6851	06/05	C
1692	05 46 30	-69 40	30	17	9	7	85	65	-	-	-	0.48	0.22:	8.3	C		09/10	W
1693	05 46 34.8	-69 17 14	12	8	7	4	64	45	-	-	-	0.15	0.33	7.9	C	05465-6917	09/10	W
1694	05 46 40	-69 05	16	8	8	4	66	40	-	-	-	0.30	0.44	10.8	C		05/09	W
1695	05 46 45	-70 31	-	-	-	-	15	13	23	22	-	-	-	0.8	2.1:		10/09	C
1696	05 46 47.0	-69 35 45	29	19	18	10	93	66	80	65	p	0.37	0.89	11.2	31.2	05467-6935	09/10	C
1697	05 46 50	-67 22	4	-	1	-	10	5	12	6	-	0.15:	0.11:	2.1	12.5		05	C
1698	05 46 52.7	-67 10 15	-	-	2	-	8	5	-	-	-	-	0.22:	1.2:	C	05468-6710	05	W
1699	05 47 00	-68 11	4	-	2	-	13	7	15	10	-	0.15	0.22:	2.5	10.4		05	C
1700	05 47 00	-69 26	20	15	7	6	-	-	-	-	-	0.19	0.11:	C	C		09	S
1701	05 47 00	-69 45	20	15	10	7	75	65	-	-	-	0.19:	0.33	4.1:	C		09/10	W
1702	05 47 14	-71 16	8	2	1	-	-	-	-	-	-	0.22	0.11:	-	-		14/13	S
1703	05 47 20	-70 15	9	6	2	1.5	-	-	-	-	-	0.11	0.06:	-	C		09/10	S
1704	05 47 21.4	-70 08 01	-	-	3	2	23	18	30	17	-	-	0.11:	2.1	27.0	05473-7008	09/10	C
1705	05 47 24	-71 05	6	2	1	-	-	-	-	-	-	0.15	0.11:	C	C		14/09	S
1706	05 47 25	-69 09	14	7	6	4	43	30	42	30	-	0.26	0.22	5.4	25.0		09/05	C
1707	05 47 25	-69 28	23	15	9	6	75	55	74	60	-	0.30	0.33	8.3	29.1		09	C
1708	05 47 28.6	-68 42 20	1	-	2	-	14	10	16	12	-	0.04:	0.22:	1.7	8.3	05474-6842	05	C
1709	05 47 30	-67 04	-	-	-	-	4	2	6	2	-	-	-	0.8	8.3		01/05	C
1710	05 47 31.0	-67 46 29	75	-	8	-	-	-	-	-	p	2.77	0.89	-	-	05475-6746	05	S

No.	T	Ref	ID	m1	m2	v1	v2	P	c1	c2	c3	c4	c5	c6	c7	c8	RA	Dec
1711	C	05	05475-6752	6.2	1.2	0.11:	0.11:	-	4	7	2	5	-	1	-	3	05 47 31.3	-67 52 29
1712	M	09/10	05475-7004	C	2.1	0.22:	0.22	P	4	-	21	26	2	4	7	13	05 47 31.5	-70 04 14
1713	C	14/13	05475-7128	4.2:	0.8:	0.22	0.30	P	-	6	1	3	2	2	7	8	05 47 31.5	-71 28 54
1714	C	14/13	05475-7135	4.2:	0.8:	-	-	-	4	4	-	2	-	-	-	-	05 47 32.5	-71 35 43
1715	C	05	05475-6743	8.3	1.7	-	-	-	2	7	2	6	-	-	-	-	05 47 35.4	-67 43 01
1716	C	05	05476-6829	4.2:	0.8:	0.11:	-	-	9	11	7	9	-	1	-	-	05 47 38.7	-68 29 17
1717	M	09		C	4.1:	0.33	0.11	-	22	24	30	40	5	8	15	18	05 47 40	-69 49
1718	C	09/10	05477-7040	4.2	1.7	0.33	0.30	-	17	19	9	13	2	5	11	19	05 47 40	-70 35
1719	C	09/10		4.2	1.2	0.22	0.15	-	-	-	7	10	2	4	7	11	05 47 42.5	-70 40 50
1720	M	09		C	1.7	0.33	0.30	-	-	-	33	37	6	9	14	22	05 47 50	-69 54
1721	C	09/10	05478-7045	10.4	0.4	0.44	0.26	P	12	17	7	8	2	6	7	14	05 47 51.8	-70 45 14
1722	M	09		C	4.1	0.56	0.26	-	-	-	40	50	4	9	15	22	05 48 00	-69 45
1723	C	05	05480-6839	6.2	0.8:	0.22	0.30	P:	10	13	9	9	-	2	-	8	05 48 03.4	-68 39 50
1724	C	01	05481-6631	4.2	0.8	-	0.07:	-	2	4	1	3	-	1	-	2	05 48 08.4	-66 31 43
1725	S	09/14		-	-	0.11:	0.15	-	-	-	-	-	-	1	-	4	05 48 14	-71 00
1726	C	09	05482-6856	12.5	1.7	0.22	0.26	-	20	26	9	13	2	4	7	14	05 48 15	-70 30
1727	C	05		10.4	1.2:	0.11:	0.07:	-	20	25	20	23	1	2	-	2	05 48 15.2	-68 56 19
1728	S	01		-	6.2	0.11:	0.11	-	-	30	-	-	1	1	-	6	05 48 18	-66 39
1729	C	09		20.8:	0.8:	0.11:	0.11	-	20	30	15	30	3	3	7	10	05 48 20	-69 13
1730	C	09		10.4:	-	0.11:	0.15	-	18	23	14	16	1	2	6	10	05 48 20	-70 16
1731	S	09	05484-6945	-	-	0.33:	1.11	P	-	-	-	-	4	7	8	38	05 48 26.6	-69 45 53
1732	C	09/14	05484-7101	6.2	0.8:	0.11:	-	P	4	7	2	4	-	1	-	-	05 48 29.6	-71 01 28
1733	C	09	05486-6953	27.0	15.3	1.55	0.63	-	45	58	25	62	4	18	8	25	05 48 36.7	-69 53 53
1734	C	01		4.2	0.8	-	-	-	-	2	-	2	-	-	-	-	05 48 40	-66 53
1735	C	05		4.2:	1.2	-	-	-	9	11	6	9	-	-	-	-	05 48 40	-68 16
1736	S	14/13	05488-7243	-	-	-	0.78:	P	-	-	-	16	-	-	4	25	05 48 49.0	-72 43 04
1737	C	05	05488-6850	10.4	2.9:	0.33:	0.07:	-	15	20	9	16	-	3	-	2	05 48 49.2	-68 50 58
1738	S	05		-	-	0.11:	0.19	-	-	-	-	-	-	1	-	6	05 48 50	-68 10
1739	C	09	05489-7002	20.8	3.3	0.11:	0.19	-	18	28	12	20	2	3	3	8	05 48 55	-68 58
1740	M	09		C	C	2.44	0.30	P:	-	-	-	-	8	30	12	20	05 48 57.6	-70 02 29
1741	M	09	05489-7009	C	6.2	0.78	0.52	P	-	-	20	35	5	12	12	26	05 48 58.7	-70 09 44
1742	M	09	05491-7006	C	0.8	0.22	0.19	-	-	-	6	8	1	3	5	10	05 49 00	-70 37
1743	M	09	X0549-700	C	C	2.22:	0.78	P	-	-	-	-	10	30	12	33	05 49 06.2	-70 06 24
1744	C	09	05494-7004	208.0	58.4	4.99	0.96	P	20	120	24	165	6	51	10	36	05 49 24.6	-70 04 14
1745	M	09	05495-7034	-	1.2	0.89	0.15	P	-	-	3	6	-	8	-	4	05 49 34.3	-70 34 09
1746	M	05	05495-6859	C	1.2:	0.11:	-	-	-	-	13	16	2	2	-	-	05 49 35.1	-68 59 11
1747	C	09		10.4	2.1	0.22	0.15:	-	10	15	8	13	1	2	-	4	05 49 50	-69 19
1748	S	01	05499-6654	-	3.3	-	0.26	P	-	-	10	18	-	-	-	7	05 49 55.5	-66 54 53
1749	M	09	05499-7010	C	0.8:	-	-	-	-	-	-	2	-	1	-	-	05 49 59.9	-70 10 49
1750	C	13/14	05500-7234	2.1:	0.8:	0.11:	0.11:	-	-	1	-	2	-	1	-	3	05 50 00.8	-72 34 04

Infrared Sources in the LMC (continued 26).

(1)	(2)		(3)		(4)		(5)		(6)		(7)	(8)	(9)	(10)	(11)	(12)	(13)	(14)
Number LI–LMC	Position RA(1950) h m s	DEC(1950) o ' "	12 μm Peak	Bg	25 μm Peak	Bg	60 μm Peak	Bg	100 μm Peak	Bg	Size arcmin	F 12μm Jy	F 25μm Jy	F 60μm Jy	F 100μm Jy	IRAS-Id	DPM field	Spectrum
					1.0E-8 Watt / (m²m sr)													
1751	05 50 20	-68 17	2	-	2	-	8	6	13	9	-	0.07:	0.22	0.8:	8.3	-	05	C
1752	05 50 24.1	-69 41 48	10	3	1	-	-	-	-	-	-	0.26:	0.11:	-	-	05504-6941	09	S
1753	05 50 30	-68 27	6	-	2	-	15	9	17	12	-	0.22:	0.22	2.5	10.4	-	05	C
1754	05 50 30	-68 47	-	-	-	-	8	6	13	10	-	-	-	0.8	6.2	-	05	C
1755	05 50 30	-69 40	4	-	2	-	9	7	14	9	-	0.15:	0.22	0.8	10.4	-	09	C
1756	05 50 36.1	-70 53 58	17	-	4	-	-	-	-	-	p	0.63	0.44	-	-	05506-7053	09	S
1757	05 50 45.8	-67 51 08	4	-	2	-	-	-	-	-	-	0.15	0.22:	-	-	05507-6751	05	W
1758	05 50 53.8	-71 46 43	3	-	-	-	-	-	-	-	-	0.11:	-	-	-	05508-7146	13	S
1759	05 50 57.9	-69 56 53	13	-	7	-	7	4	10	8	p	0.48	0.78	1.2:	4.2:	05509-6956	09	C
1760	05 51 01.5	-71 15 14	-	-	1	-	5	1	3	1.5	-	-	0.11:	1.7	3.1	05510-7115	13	C
1761	05 51 02.5	-69 05 02	2	-	1	-	8	6	14	10	-	0.07:	0.11:	0.8	8.3	05510-6905	09/05	C
1762	05 51 12.1	-69 28 53	3	-	1	-	9	6	10	8	-	0.11:	0.11:	1.2	4.2	05512-6928	09	C
1763	05 51 30.1	-71 03 45	4	-	1	-	-	-	-	-	-	0.15	0.11:	-	C	05515-7103	13/09	S
1764	05 51 38.3	-71 24 25	5	-	1	-	-	-	2	1	-	0.19	0.11:	-	2.1:	05516-7124	13	C
1765	05 51 54	-71 54	4	-	-	-	-	-	2	1	-	0.15	-	-	-	-	13	S
1766	05 52 15.3	-65 45 23	3	-	1	-	-	-	-	-	-	0.11:	0.11:	-	-	05522-6545	01	S
1767	05 52 15.4	-69 56 42	-	-	-	-	5	2	6	3	-	-	-	1.2	6.2:	05522-6956	09	C
1768	05 52 15.9	-71 20 10	5	-	2	-	-	-	-	-	-	0.19	0.22	-	-	05522-7120	13	W
1769	05 52 29.2	-72 36 44	-	-	-	-	-	-	2	1	-	-	-	-	4.2:	05524-7236	13	C
1770	05 52 38.8	-65 20 34	4	-	-	-	-	-	-	-	-	0.15:	-	-	-	05526-6520	01	S
1771	05 53 10	-67 17	6	-	1	-	8	6	0.5	-	-	0.22	0.11:	0.8:	1.0:	-	05	C
1772	05 53 15	-68 24	6	-	2	-	10	6	-	-	-	0.22:	0.22	1.7	C	-	05	W
1773	05 53 40	-70 34	5	-	1	-	-	-	-	-	-	0.19	0.11:	-	-	-	09	S
1774	05 53 42	-66 39	-	-	2	-	-	-	-	-	-	-	0.22	-	-	-	01	W
1775	05 53 42	-71 37	12	-	-	-	-	-	-	-	p	0.44:	-	-	-	-	13	S
1776	05 53 42.1	-66 53 14	-	-	4	-	4	-	1.5	-	p	-	0.44	1.7	3.1	05537-6653	01	C
1777	05 53 44.5	-70 15 52	3	-	5	-	1	-	-	-	p	0.11:	0.56	0.4:	-	05537-7015	09	W
1778	05 53 56	-68 14	5	-	1	-	11	5	16	9	-	0.19:	0.11:	2.5	14.6	05539-6811:	05	C
1779	05 54 00.7	-68 21 42	6	-	2	-	10	6	14	10	-	0.22:	0.22	1.7	8.3	05540-6821	05	C
1780	05 54 01.8	-65 33 37	8	-	3	1	-	-	-	-	p	0.30	0.22	-	-	05540-6533	01	S

No.	RA (1950)	Dec (1950)	(1)	(2)	(3)	(4)	(5)	(6)	(7)	(8)	flag	m1	m2	m3	m4	Name	Code	Type
1781	05 54 12.2	–69 08 32	2	–	–	–	2	–	4	2	–	0.07:	–	0.8	4.2	05542–6908	09	C
1782	05 54 17.4	–69 14 55	3	–	–	–	1	–	–	–	–	0.11:	–	0.4:	–	05542–6914	09	M
1783	05 54 40.7	–69 49 59	–	–	3	–	3	–	3	1	P	–	0.33:	1.2	4.2	05546–6949	09	C
1784	05 54 43.2	–65 15 27	4	–	4	–	–	–	–	–	P	0.15:	0.22	–	–	05547–6515	01	M
1785	05 55 06.2	–65 28 39	–	2	–	–	–	2	–	–	P	–	–	–	–	05551–6528	01	M
1786	05 55 12.4	–68 47 05	–	–	–	–	1	–	2	–	–	–	–	0.4:	4.2	05552–6847	05	C
1787	05 55 25	–68 07	–	–	2	–	7	–	–	–	p:	–	0.22:	2.9	C	–	05	M
1788	05 55 38.0	–67 34 31	5	–	–	–	–	–	–	–	p:	0.19	0.11:	–	–	05556–6734	05	S
1789	05 55 50.4	–68 03 15	12	–	1	–	–	–	–	–	P	0.44	0.67	–	–	05558–6803	05	S
1790	05 55 50.7	–70 00 24	20	–	6	–	–	–	–	–	P	0.74	–	–	–	05558–7000	09	S
1791	05 56 04.7	–68 11 46	12	–	7	2	26	2	22	2	p:	0.33	0.67	9.9	41.6	05560–6811 (X0556–682)	05	C
1792	05 56 10.1	–68 21 25	5	–	2	–	5	–	6	–	p:	0.19:	0.22:	2.1	12.5	05561–6821	05	C
1793	05 56 10.7	–67 32 51	7	–	–	–	–	–	–	–	P	0.26	–	4.1	–	05561–6732	05	S
1794	05 56 12.1	–69 33 58	6	–	5	–	10	–	5	1	P	0.22	0.56	–	8.3	05562–6933	09	C
1795	05 56 49.6	–67 53 54	9	–	4	–	–	–	1	1	P	0.33	0.44	–	–	05568–6753	05	M
1796	05 56 49.6	–70 25 03	–	–	–	–	2	–	1	–	–	0.11:	–	0.8:	2.1:	05568–7025	09	C
1797	05 56 51.3	–66 18 42	3	–	–	–	–	–	–	–	P	0.15:	–	–	–	05568–6618	01	S
1798	05 56 51.6	–65 28 39	4	–	–	–	–	–	–	–	P	0.33	0.11:	–	–	05568–6528	01	S
1799	05 57 07.3	–68 27 44	9	–	–	–	–	–	–	–	P	0.19	–	–	–	05571–6827	05	S
1800	05 57 12.5	–70 07 01	5	2	1	–	–	–	–	–	–	–	–	–	–	05572–7007	09	S
1801	05 57 19.5	–69 51 26	–	–	–	–	1	–	1	–	–	0.26:	–	0.6:	2.1:	05573–6951	09	C
1802	05 58 30.9	–69 01 29	7	–	5	–	–	–	–	–	P	0.19	–	–	–	05585–6901	09/05	S
1803	05 58 52.1	–69 44 31	5	–	–	–	–	–	–	–	P	0.81:	0.56	–	–	05588–6944	09	M
1804	05 58 59.7	–69 51 29	22	–	–	–	–	–	–	–	P	0.19:	0.17:	–	–	05589–6951	09	S
1805	05 59 03	–71 20	6	1	–	1.5	–	–	–	–	–	–	–	–	–	–	13	S
1806	06 00 12.1	–66 20 05	–	–	–	–	1	–	0.7	–	–	0.11:	0.22:	0.4:	1.5	06002–6620A	01	C
1807	06 01 08.9	–66 36 34	3	–	2	–	–	–	–	–	p:	–	–	–	–	06011–6636A	01	M
1808	06 01 26.0	–66 28 59	–	–	–	–	1.5	–	0.7	–	p:	0.74	0.22	0.6:	1.5:	06014–6628	01	C
1809	06 02 14.3	–70 06 41	20	–	2	–	–	–	–	–	P	–	0.11:	–	–	06022–7006	09	S
1810	06 02 17.1	–67 43 03	–	–	1	–	–	–	–	–	–	–	–	–	–	06022–6743	05	M
1811	06 02 25.4	–70 35 29	75	–	15	–	–	–	–	–	P	2.77:	1.66:	–	–	06024–7035	09	S
1812	06 02 25.5	–66 45 54	4	–	4	–	–	–	–	–	–	0.15:	0.44:	–	–	06024–6645A	01	S
1813	06 02 35.3	–67 12 43	12	–	3	–	–	–	–	–	P	0.44:	0.33:	–	–	06025–6712	05	S
1814	06 02 38.2	–72 08 44	40	1	5	–	–	–	–	–	P	1.37:	0.56	–	–	06026–7208	13	S
1815	06 02 40.4	–70 40 22	42	–	–	–	–	–	–	–	–	1.55	–	–	–	06026–7040	09	S
1816	06 02 51	–71 03	4	–	–	–	–	–	–	–	–	0.15:	–	–	–	–	13	S
1817	06 02 51.1	–67 22 15	18	–	3	–	–	–	–	–	–	0.67:	0.33:	–	–	06028–6722	05	S
1818	06 03 07.4	–72 27 10	18	4	6	–	12	1	3	1	P	0.52:	0.56:	–	4.2	06031–7227	13	M
1819	06 03 34.6	–71 02 58	–	–	6	–	–	–	3	–	P	–	0.67	5.0	–	06035–7102	13	M
1820	06 04 13.0	–69 42 22	24	–	5	–	–	–	–	–	P	0.89:	0.56:	–	–	06042–6942	09	S

Infrared Sources in the LMC (continued 27).

(1)	(2)		(3)		(4)		(5)		(6)		(7)	(8)	(9)	(10)	(11)	(12)	(13)	(14)
Number LI-LMC	Position RA(1950) h m s	DEC(1950) o ' "	12 μm Peak	Bg	25 μm Peak	Bg	60 μm Peak	Bg	100 μm Peak	Bg	Size arcmin	F 12μm Jy	F 25μm Jy	F 60μm Jy	F 100μm Jy	IRAS-Id	DPM field	Spec-trum
					1.0E-8 Watt / (m*m sr)													
1821	06 04 32.6	-67 22 54	12	–	3	–	–	–	–	–	–	0.44:	0.33:	–	–	06045-6722	05	S
1822	06 04 47.0	-67 36 59	4	–	–	–	–	–	–	–	–	0.15:	–	–	–	06047-6736	05	S
1823	06 06 55.7	-72 38 26	13	6	–	–	–	–	–	–	–	0.26:	–	–	–	06069-7238	13	S

Infrared Sources in the LMC (continued 28), Additions from Co-added IRAS Survey data.

(1)	(2)		(3)	(4)	(5)	(6)	(7)	(8)	(9)	(10)	(11)	(12)	(13)	(14)
Number LI-LMC	Position RA(1950) h m s	DEC(1950) o ' "	12 µm Peak Bg	25 µm Peak Bg	60 µm Peak Bg	100 µm Peak Bg	Size arcmin	F 12µm Jy	F 25µm Jy	F 60µm Jy	F 100µm Jy	IRAS-Id	CoAdd field	Spectrum
					1.0E-8 Watt / (m*m sr)									
1824	04 27 23.0	-71 00 38	46	6	-	-	p	1.70	0.67	-	-	04273-7100	23	S
1825	04 28 41.9	-69 37 15	6	1.5	-	-	-	0.22	0.17	-	-	04286-6937	22	S
1826	04 29 56.2	-68 51 07	-	-	-	0.5	-	-	-	-	1.0:	04299-6851	22	C
1827	04 30 01.2	-69 04 42	-	-	1	0.5	-	-	-	0.4	1.0:	04300-6904	22	C
1828	04 30 10	-67 59	5	-	-	-	-	0.19	-	-	-	-	21	S
1829	04 31 35.2	-72 29 31	8	-	-	-	p	0.30	-	-	-	04315-7229	23	S
1830	04 31 42.0	-71 09 48	-	-	1	-	-	-	-	0.4	-	04317-7109	23	W
1831	04 32 16.6	-65 06 26	9	1	-	-	-	0.33	0.11:	-	-	04322-6506	20	S
1832	04 32 29.5	-71 12 43	-	-	1.5	-	-	-	-	0.6	-	04324-7112	23	W
1833	04 32 30	-65 21	4	-	-	-	-	0.15	-	-	-	-	20	S
1834	04 32 42.2	-68 53 58	-	-	1	0.5	-	-	-	0.4	1.0	04327-6853	22	C
1835	04 33 03.7	-67 25 09	12	1	-	-	-	0.44	0.11:	-	-	04330-6725	21	S
1836	04 33 34.0	-71 37 06	-	-	1	-	-	-	-	0.4	1.0	04335-7137	23	W
1837	04 34 25.0	-71 56 09	-	-	1	0.5	-	-	-	0.4	-	04344-7156	23	C
1838	04 35 14.5	-66 43 59	-	-	1	-	-	-	-	0.4	-	04352-6643	21	W
1839	04 35 31.3	-69 09 58	-	-	1	-	-	-	-	0.4	-	04355-6909	22	W
1840	04 35 35.1	-70 08 03	22	2	-	-	p	0.81	0.22	-	-	04355-7008	22	S
1841	04 37 00	-66 28	-	-	1	0.7	-	-	-	0.4	1.5	-	20	C
1842	04 37 08.5	-70 24 38	10	1.5	-	-	-	0.37	0.17	-	-	04371-7024	22	S
1843	04 37 14.4	-68 45 51	-	-	1	0.5	-	-	-	0.4:	1.0:	04372-6845	22	C
1844	04 37 27.3	-68 31 10	5	1.5	-	-	-	0.19	0.17	-	-	04374-6831	21/22	S
1845	04 37 31.2	-69 18 00	-	-	3	0.7	-	-	-	1.2	1.5	04375-6918	22	C
1846	04 37 40	-66 16	-	-	1	0.7	-	-	-	0.4	1.5	-	20	C
1847	04 38 15	-65 12	4	-	-	-	-	0.15	-	-	-	-	20	S
1848	04 39 03.3	-69 33 01	21	2	-	-	p	0.78	0.22	-	-	04390-6933	22	S
1849	04 39 05.4	-69 36 09	-	1.5	3	-	-	-	0.17:	1.2	3.1	04390-6936	22	C
1850	04 39 09.6	-71 54 02	-	-	-	0.5	-	-	-	-	1.0	04391-7154	23	C

Infrared Sources in the LMC (continued 29).

(1) Number LI-LMC	(2) RA(1950) h m s	(2) DEC(1950) ° ' "	(3) 12 μm Peak	Bg	(4) 25 μm Peak	Bg	(5) 60 μm Peak	Bg	(6) 100 μm Peak	Bg	(7) Size arcmin	(8) F 12μm Jy	(9) F 25μm Jy	(10) F 60μm Jy	(11) F 100μm Jy	(12) IRAS-Id	(13) CoAdd field	(14) Spectrum
			(1.0E-8 Watt / (m×m sr))															
1851	04 39 30	-65 46	5	-	-	-	1.5	-	0.5	-	-	0.19	-	-	-		20	S
1852	04 40 54.2	-64 54 55	-	-	-	-	1	-	2	1	-	-	-	0.6	1.0	04409-6454	20	C
1853	04 41 07.1	-69 38 47	-	-	-	-	-	-	-	-	-	-	-	0.4:	2.1:	04411-6938	22	C
1854	04 41 31.1	-66 59 46	10	-	1	-	-	-	-	-	-	0.37	0.11:	-	-	04415-6659	21	S
1855	04 42 18.3	-65 06 03	-	-	-	-	1.5	-	0.5	-	-	-	-	0.6	1.0:	04423-6506	20	C
1856	04 42 40	-67 15	-	-	-	-	3	-	1.5	-	-	-	-	1.2	3.1		21	C
1857	04 42 48.3	-65 53 49	-	-	-	-	1.5	-	1	-	-	-	-	0.6	2.1	04428-6553	20	C
1858	04 43 05.9	-68 01 02	-	-	1	-	2	-	2	1	-	-	0.11:	0.8	2.1	04430-6801	21	C
1859	04 43 45	-65 42	-	-	-	-	1.5	0.5	1.5	0.5	-	-	-	0.4	2.1		20	C
1860	04 44 42.4	-68 22 03	-	-	-	-	-	-	4	-	-	-	-	-	2.1	04447-6822	21	C
1861	04 45 20	-67 49	-	-	1	-	3	-	5	3	-	-	0.11:	0.4	4.2		21	C
1862	04 45 40	-69 08	-	-	-	-	11	8	11	8	-	-	-	1.2	6.2		22	C
1863	04 45 49.0	-66 22 50	-	-	4	-	4	-	-	-	-	0.19	0.44	1.7	-	04458-6622	20	W
1864	04 45 50	-66 10	5	-	-	-	-	-	-	-	-	-	-	-	-		20	S
1865	04 46 03.4	-66 48 06	-	-	2	-	2	-	3	2	-	-	0.22	0.8	2.1	04460-6648	21	C
1866	04 48 30	-64 28	4	-	3	-	1	-	1	0.5	-	0.15:	-	0.4	1.0		20	S
1867	04 49 21.3	-64 41 15	-	-	-	-	1	-	2	1	-	-	0.33	0.4:	2.1	04493-6441	20	C
1868	04 49 50	-71 48	-	-	-	-	1	-	2	1	-	-	-	0.4:	2.1		23	C
1869	04 51 04.2	-70 35 23	-	-	-	-	-	-	-	-	-	-	-	-	-	04510-7035	23	C
1870	04 55 50	-64 40	4	-	-	-	-	-	-	-	-	0.15	-	-	-		19	S
1871	04 57 33.0	-64 40 21	6	-	-	-	-	-	-	-	-	0.22	-	-	-	04575-6440	19	S
1872	05 00 14.2	-64 27 47	24	-	1.5	-	-	-	-	-	P	0.89	0.17	-	-	05002-6427	19	S
1873	05 02 57.0	-64 36 11	16	-	1	-	-	-	-	-	P	0.59	0.11	-	-	05029-6436	19	S
1874	05 03 41.9	-65 04 45	130	-	13	-	0.5	-	-	-	P	4.81	1.44	0.2:	-	05036-6504	19	S
1875	05 04 10.9	-64 33 24	-	-	-	-	1.5	-	0.4	-	-	-	-	0.6	0.8:	05041-6433	19	C
1876	05 04 30	-64 37	5	-	-	-	-	-	-	-	-	0.19	-	-	-		19	S
1877	05 06 47.9	-65 14 03	-	-	-	-	1	0.5	3	2	-	-	-	0.2:	2.1:	05067-6514	19	C
1878	05 07 21.9	-64 46 30	-	-	-	-	-	-	0.7	-	-	-	-	0.6	1.5	05073-6446	19	C
1879	05 11 50	-65 14	5	-	-	-	1.5	-	-	-	-	0.19	-	-	-		18	S
1880	05 12 50.1	-64 55 03	4	-	2	-	-	-	-	-	-	0.15	0.22	-	-	05128-6455	18	W

No.	RA	Dec	N				i	P	g	f	e	d	IRAS	mag	Sp
1881	05 12 58.2	−65 03 28	–	–	–	–	–	–	–	–	0.4	–	05129-6503	18	M
1882	05 16 57.1	−65 00 36	9	1	1	2	–	–	–	0.11:	2.1	4.2	05169-6500	18	C
1883	05 20 50.6	−64 59 30	8	1	5	–	–	–	0.33	0.22	–	–	05208-6459	18	S
1884	05 30 53.4	−65 09 25	–	2	–	–	–	P	0.30	–	–	–	05308-6509	17	S
1885	05 34 08.4	−65 09 24	–	–	2	1.1	0.4	P	–	–	0.8	1.5	05341-6509	17	C
1886	05 35 36.8	−65 08 39	–	–	1.5	0.5	–	–	–	–	0.6	1.0	05356-6508	17	C
1887	05 37 11.4	−65 05 49	10	–	1	0.3	–	–	–	–	0.4	0.6	05371-6505	17	C
1888	05 39 03.8	−64 49 13	6	–	–	–	–	P	0.37	–	–	–	05390-6449	17	S
1889	05 40 26.4	−64 58 47	7	–	–	–	–	–	0.22	–	–	–	05404-6458	17	S
1890	05 47 12.7	−64 35 24	–	–	–	–	–	–	0.26	–	–	–	05472-6435	17	S
1891	05 48 24.7	−65 10 54	12	–	–	–	–	P	0.44	–	–	–	05484-6510	17	S

ATLAS OF INFRARED

EMISSION OF THE

MAGELLANIC CLOUDS

4. Atlas of Infrared Emission of the Magellanic Clouds

In this section, we present maps of the infrared emission of the LMC and SMC in the four IRAS wavelength bands at 12, 25, 60 and 100 microns. For clarity of presentation, the DPM maps constructed from IRAS Additional Observations (*cf*. Part 1) have been subdivided, with some overlap, in individual fields of dimensions roughly 2° x 2°. An overview of the SMC and the LMC in their totality is given in the overall maps. These maps show the DPM observation boundaries and rough field boundaries. As the DPM maps did not completely cover the LMC and the SMC, we have also used the IRAS survey database to construct 2 Co–Added fields on the SMC, and 7 Co–Added fields on the LMC to complete coverage of infrared emission at the Cloud edges. The accompanying Table 4.1 lists individual field centers, and corresponding map page numbers, while Table 4.2 lists map properties of interest.

The Atlas proper is contained in the detail maps for both Clouds. For the SMC we include DPM map sets 2 (NS scans) and 3 (EW scans), while for the LMC we include map sets 1 (EW scans) and 2 (NS scans). Note that the LMC EW scanned maps show a bad detector scan at the SE end of the Bar. Maps are identified in the margin by wavelength and by field number. The numbers correspond to those in Table 4.1 and those of the transparent overlays, which can be used to read off equatorial coordinates (1950.0). On the overlays, source positions and numbers corresponding to the Catalogue entries of Part 3 are marked. Also marked are the positions of SAO (Galactic foreground) stars for comparison with optical images.

The maps are given in a combined contour and halftone representation. Halftones are used to enhance intermediate intensity levels. Darker tones correspond to higher intensities. The contour levels used in the Atlas are given in Table 4.3.

Table SMC–4.1: *SMC fields 1 – 6 presented in the Atlas.*

Field	Field centre		Page numbers				Comments
number	α_{1950}	δ_{1950}	12 μm	25 μm	60 μm	100 μm	
SMC Additional Observations DPM fields 1 – 4							
1	01^h11^m	$-72°00'$	106	107	108	109	NE–field
2	00^h45^m	$-72°00'$	110	111	112	113	NW–field
3	01^h12^m	$-74°00'$	114	115	116	117	SE–field
4	00^h43^m	$-74°00'$	118	119	120	121	SW–field
SMC Co–added IRAS Survey Data fields 5 – 6							
5	01^h28^m	$-73°35'$	122	122	123	123	SE–edge
6	00^h30^m	$-74°04'$	124	124	125	125	SW–edge

Note to Table SMC–4.1:

The orientation of the fields is with respect to the centre of the SMC–Bar (00^h58^m, $-73°00'$). Fields 1 to 4 have sizes of 2°11 x 2°11 around the the specified field centre. Fields 5 and 6 have sizes of 2° x 2°. There is ~ 1° overlap between fields 5 and 6 with fields 3 and 4. In the Atlas' DPM fields, the NS–scanned map is displayed above the EW–scanned map.

Table LMC–4.1: *LMC fields 1 – 23 presented in the Atlas.*

Field number	Field centre α_{1950}	δ_{1950}	12 μm	25 μm	60 μm	100 μm	Comments
LMC Additional Observations DPM fields 1 – 16							
1	05^h53^m	$-66°00'$	128	129	130	131	NE–corner
2	05^h34^m	$-66°00'$	132	133	134	135	SGS 4
3	05^h14^m	$-66°00'$	136	137	138	139	SGS 5
4	04^h55^m	$-66°00'$	140	141	142	143	SGS 1; NW
5	05^h57^m	$-68°00'$	144	145	146	147	
6	05^h35^m	$-68°00'$	148	149	150	151	SGS 4
7	05^h13^m	$-68°00'$	152	153	154	155	NW–Bar
8	04^h51^m	$-68°00'$	156	157	158	159	SGS 8
9	05^h58^m	$-70°00'$	160	161	162	163	
10	05^h35^m	$-70°00'$	164	165	166	167	30 Dor, SE–Bar
11	05^h13^m	$-70°00'$	168	169	170	171	SGS 8, Bar
12	04^h50^m	$-70°00'$	172	173	174	175	SGS 7, 8
13	06^h01^m	$-71°57'$	176	177	178	179	SE–corner
14	05^h37^m	$-72°00'$	180	181	182	183	SGS 9
15	05^h11^m	$-72°00'$	184	185	186	187	SGS 9
16	04^h47^m	$-71°57'$	188	189	190	191	SW–corner
LMC Co-added IRAS Survey Data fields 17 – 23							
17	05^h40^m	$-65°30'$	192	192	193	193	N–edge; SGS 4
18	05^h20^m	$-65°30'$	194	194	195	195	N–edge; SGS 4,5
19	05^h00^m	$-65°30'$	196	196	197	197	N–edge; SGS 1
20	04^h40^m	$-65°30'$	198	198	199	199	NW–corner
21	04^h40^m	$-67°30'$	200	200	201	201	W–edge
22	04^h40^m	$-69°30'$	202	202	203	203	W–edge
23	04^h40^m	$-71°30'$	204	204	205	205	W–edge

Note to Table LMC–4.1:

All fields have sizes of $2°11 \times 2°11$ around the field centre. Note that there is considerable overlap between fields 17 – 23 with fields 1 – 16. In the Atlas' DPM fields, the NS–scanned map is displayed above the EW–scanned map.

Table SMC–4.2: *Description of the SMC map characteristics (fields 1 – 6).*

Characteristic (Unit)	Wavelength band			
	12 μm	25 μm	60 μm	100 μm
General characteristics of data				
Effective frequency (10^{12} Hz)	25	12	5	3
Bandwidth (μm)	7.0	11.2	32.5	31.5
Bandwidth correction (10^{12} Hz) [a]	13.48	5.16	2.58	1.00
Zero–magnitude flux density				
$\quad$ f_ν (0.0 mag) (Jy)	28.3	6.73	1.19	0.43
Point Source Conversion factor				
$\quad$ (Jy / 10^{-8} Watt m^{-2} sr^{-1})	0.037	0.11	0.41	2.08
Positional accuracy ($''$)	15	15	15	15
Nominal detector size($'$x$'$)	0.75 x 4.5	0.75 x 4.7	1.5 x 4.8	3.0 x 5.0
Resolution ($'$x$'$) [b]	1.1 x 6.9	1.1 x 6.9	2.1 x 6.9	4.3 x 6.9
Absolute calibration (%)	10	10	10	10
SMC Additional Observations DPM fields 1 – 4				
Median noise (MJy/sr) [c]	0.096	0.097	0.12	0.20
$\quad$ (10^{-8} Watt m^{-2} sr^{-1})	1.3	0.5	0.3	0.2
Zero–level uncertainty (MJy/sr)	0.015	0.019	0.039	0.10
$\quad$ (10^{-8} Watt m^{-2} sr^{-1})	0.2	0.1	0.1	0.1
Stripe residuals (MJy/sr) [d]	0.052	0.058	0.12	0.30
$\quad$ (10^{-8} Watt m^{-2} sr^{-1})	0.7	0.3	0.3	0.3
Sensitivity (MJy/sr) [e]	0.3	0.3	0.4	0.6
$\quad$ (10^{-8} Watt m^{-2} sr^{-1})	4	2	1	0.5
SMC Co–added IRAS Survey Data fields 5 – 6				
Median noise (MJy/sr) [c]	0.04	0.04	0.03	0.05
$\quad$ (10^{-8} Watt m^{-2} sr^{-1})	0.6	0.2	0.09	0.05
Zero–level uncertainty (MJy/sr)	0.04	0.14	0.23	0.7
$\quad$ (10^{-8} Watt m^{-2} sr^{-1})	0.5	0.7	0.6	0.7
Stripe residuals (MJy/sr) [d]	0.04	0.06	0.07	0.2
$\quad$ (10^{-8} Watt m^{-2} sr^{-1})	0.5	0.3	0.2	0.2
Sensitivity (MJy/sr) [e]	0.3	0.3	0.4	0.6
$\quad$ (10^{-8} Watt m^{-2} sr^{-1})	4	2	1	0.5

Table LMC–4.2: *Description of the LMC map characteristics (fields 1 – 23).*

Characteristic (Unit)	Wavelength band			
	12 μm	25 μm	60 μm	100 μm

General characteristics of data (similar to table SMC–4.2)

LMC Additional Observations DPM fields 1 – 16

	12 μm	25 μm	60 μm	100 μm
Median noise (MJy/sr) [c]	0.096	0.097	0.16	0.6
(10^{-8} Watt m^{-2} sr^{-1})	1.3	0.5	0.4	0.6
Zero-level uncertainty (MJy/sr)	0.015	0.097	0.39	1.0
(10^{-8} Watt m^{-2} sr^{-1})	0.2	0.5	1.0	1.0
Stripe residuals (MJy/sr) [d]	0.15	0.3	0.8	2.0
(10^{-8} Watt m^{-2} sr^{-1})	2.0	1.5	2.0	2.0
Sensitivity (MJy/sr) [e]	0.3	0.3	0.4	0.6
(10^{-8} Watt m^{-2} sr^{-1})	4	2	1	0.5

LMC Co–added IRAS Survey Data fields 17 – 23

	12 μm	25 μm	60 μm	100 μm
Median noise (MJy/sr) [c]	0.040	0.037	0.035	0.07
(10^{-8} Watt m^{-2} sr^{-1})	0.54	0.19	0.091	0.07
Zero–level uncertainty (MJy/sr)	0.096	0.19	0.30	1.0
(10^{-8} Watt m^{-2} sr^{-1})	1.3	1.0	0.78	1.0
Stripe residuals (MJy/sr) [d]	0.045	0.12	0.078	0.2
(10^{-8} Watt m^{-2} sr^{-1})	0.6	0.6	0.2	0.2
Sensitivity (MJy/sr) [e]	0.3	0.3	0.4	0.6
(10^{-8} Watt m^{-2} sr^{-1})	4	2	1	0.5

Notes to Tables SMC–4.2 and LMC–4.2:
  a) Assuming an intrinsic spectrum $f_\nu \propto \nu^{-1}$.
  b) The Gaussian resolution is given in arc–minutes (in–scan x cross–scan). One arcminute at the distance of the SMC (63 kpc) corresponds to 18 pc. One arcminute at the distance of the LMC (53 kpc) corresponds to 15 pc.
  c) The value of the median noise is influenced somewhat by the extended emission of the Clouds itself. The real detector noise is lower, especially at 60 and 100 μm. This effect is relatively less for the Co–added Survey data. Average values for these edge fields are given. The individual LMC–fields show a 30 % spread. For the Co–added maps an improvement of a factor of 2 – 3 relative to the individual detector scans has been realized.
  d) Higher stripes than the average level given in these tables can occur.
  e) The sensitivity indicates the intensity limits of the Catalogue LI–SMC. Note that the IRAS PSC sensitivity limits are 0.25, 0.25, 0.40, 1.00 Jy.

Table SMC–4.3: *Contour levels of the SMC Infrared Atlas* (in 10^{-8} Watt m^{-2} sr^{-1}).

Band	Greyscales	Contour levels

Overall maps on pages 104–105 (4.3° x 4.3°)

Band	Greyscales	Contour levels
12 μm	9 – 33	5, 10, 15, 20, 50
25 μm	9 – 33	2, 4, 6, 8, 10, 30, 50, 100
60 μm	9 – 58	1, 2, 4, 6, 8, 10, 15, 20, 40, 80, 140, 200
100 μm	9 – 58	0.5, 1, 2, 4, 6, 8, 10, 15, 20, 40, 80

Detail maps on pages 106–125 (2.1° x 2.1°)

Band	Greyscales	Contour levels
12 μm	9 – 33	5, 10, 15, 20, 30, 40, 50, 60, 70
25 μm	9 – 33	2, 4, 6, 8, 10, 12, 20, 30, 40, 50, 100, 150
60 μm	9 – 58	1, 2, 4, 6, 8, 10, 15, 20, 25, 30, 35, 40, 45, 50, 60, 70, 80, 90, 100, 125, 150, 200, 250, 300
100 μm	9 – 58	0.5, 1, 2, 4, 6, 8, 10, 15, 20, 25, 30, 35, 40, 45, 50, 60, 70, 80, 90, 100

Table LMC–4.3: *Contour levels of the LMC Infrared Atlas* (in 10^{-8} Watt m^{-2} sr^{-1}).

Band	Greyscales	Contour levels

Overall maps on pages 126–127 (8.5° x 8.5°)

Band	Greyscales	Contour levels
12 μm	9 – 33	5, 10, 20, 40, 100, 200, 500, 1000, 2000
25 μm	9 – 33	2, 6, 10, 20, 40, 100, 200, 500, 1000, 2000
60 μm	9 – 58	2, 6, 10, 20, 40, 100, 200, 500, 1000, 2000, 3000, 4000, 5000, 6000
100 μm	9 – 58	2, 6, 10, 20, 40, 100, 200, 500, 1000, 2000

Detail maps on pages 128–205 (2.1° x 2.1°)

Band	Greyscales	Contour levels
12 μm	9 – 33	4, 8, 12, 20, 30, 40, 50, 60, 75, 100, 125, 200, 300, 400, 500, 750, 1000, 2000
25 μm	4 – 28	1.5, 3, 6, 9, 12, 15, 20, 25, 30, 40, 50, 75, 100, 125, 200, 300, 400, 500, 750, 1000, 2000, 3000, 4000, 5000
60 μm	9 – 58	2, 4, 6, 8, 12, 16, 20, 30, 40, 50, 60, 70, 80, 90, 100, 125, 150, 200, 250, 300, 400, 500, 750, 1000, 1250, 1500, 1750, 2000, 2500, 3000, 3500, 4000, 4500, 5000, 6000, 7000, 8000
100 μm	9 – 58	0.5, 1, 2, 4, 6, 8, 10, 15, 20, 25, 30, 35, 40, 45, 50, 60, 70, 80, 90, 100, 125, 150, 200, 250, 300, 350, 400, 500, 600, 700, 800, 1000, 1250, 1500, 1750, 2000

SMC
12 µm

SMC
25 µm

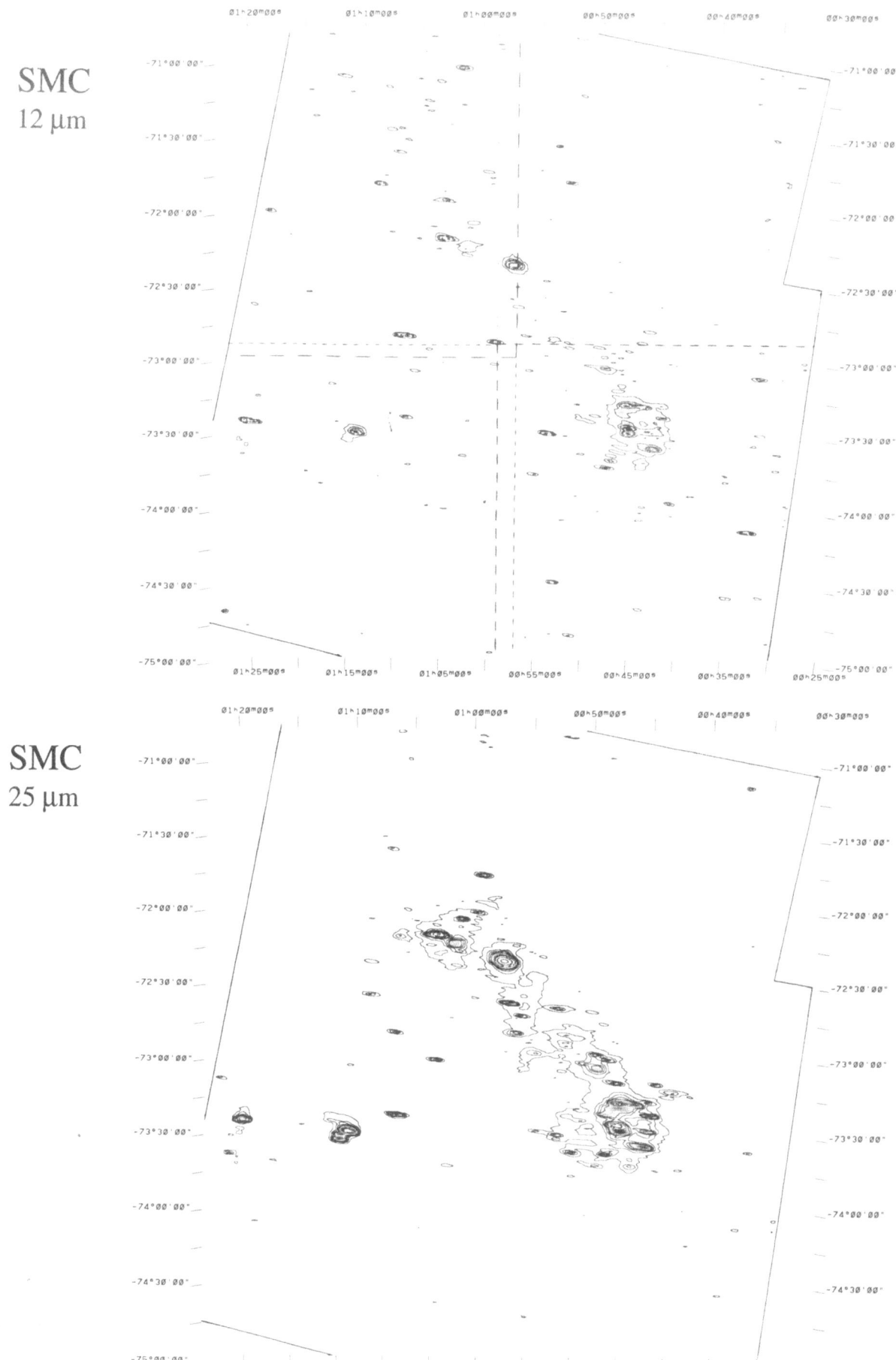

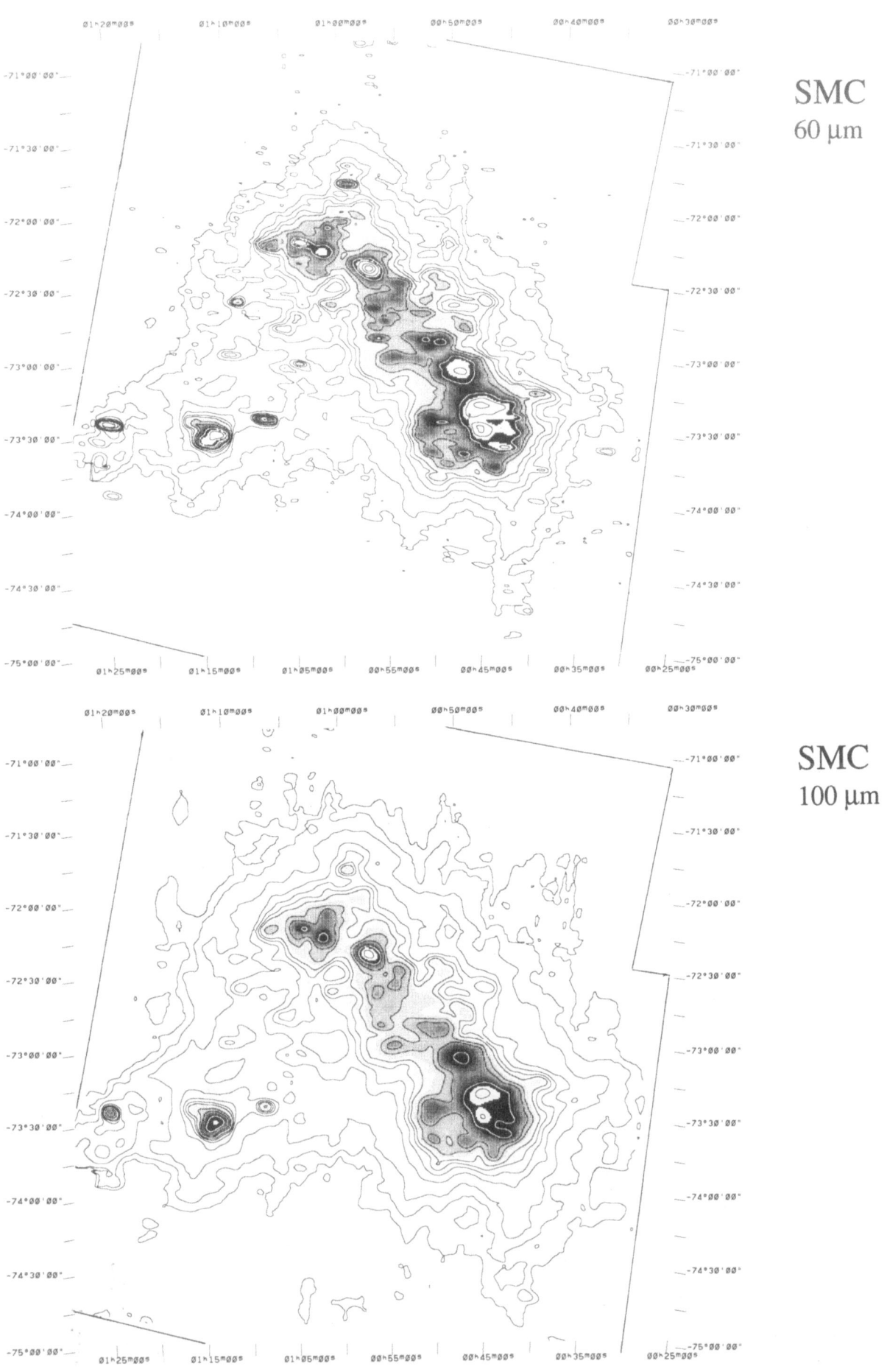

SMC
60 µm

SMC
100 µm

SMC
Field 1
12 μm

SMC
Field 1
25 μm

SMC
Field 1
60 μm

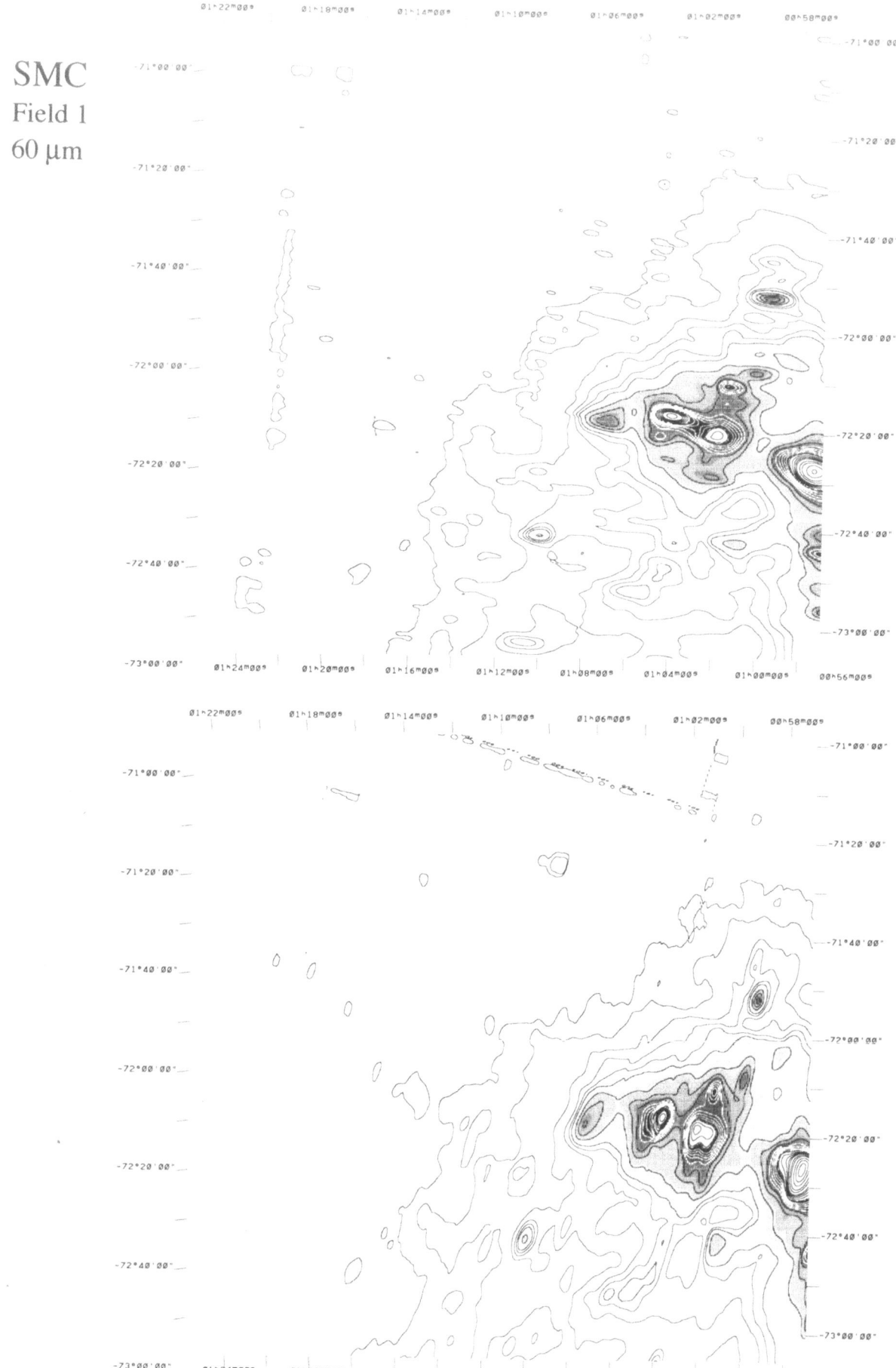

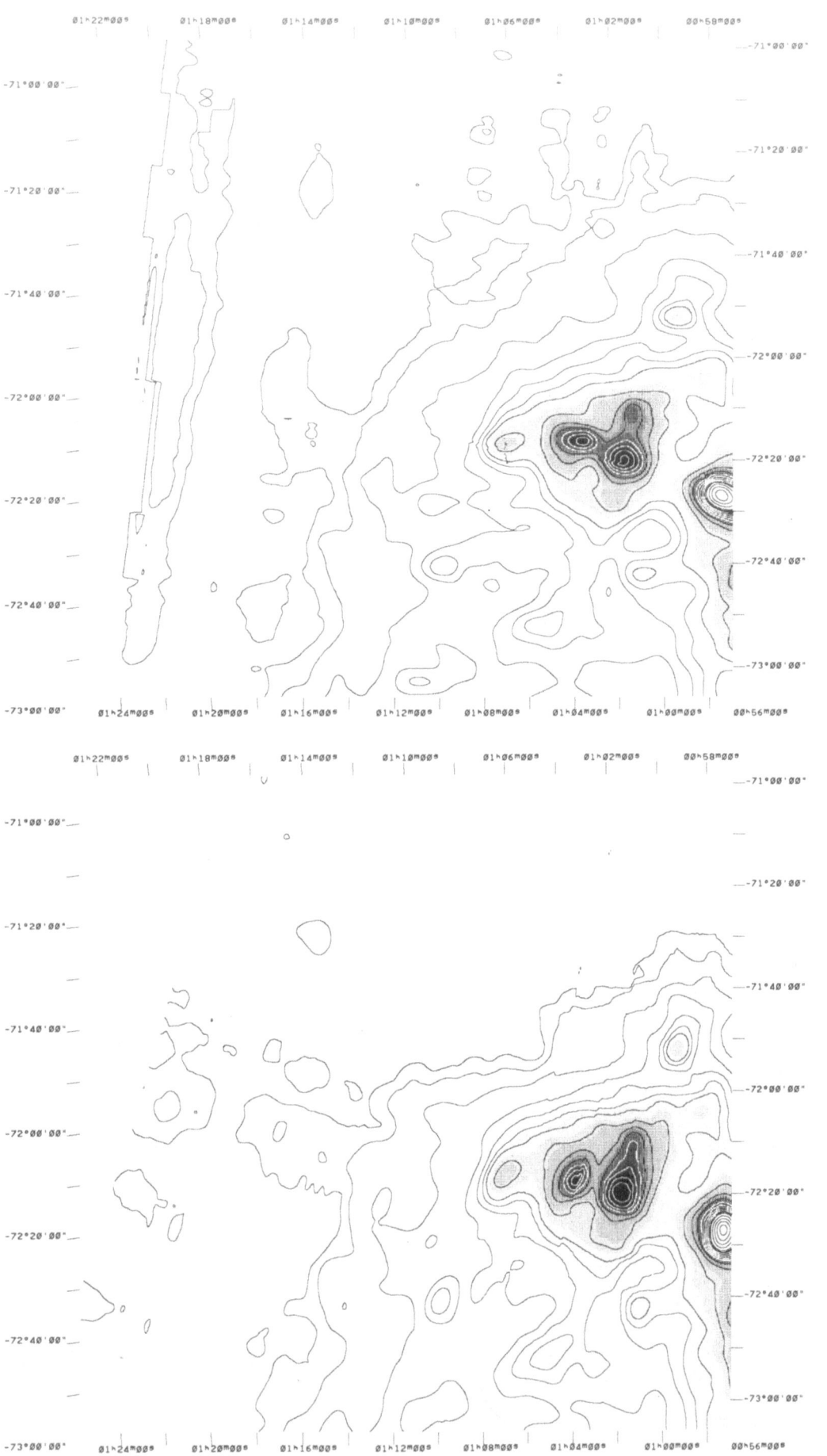

SMC
Field 1
100 μm

SMC
Field 2
12 μm

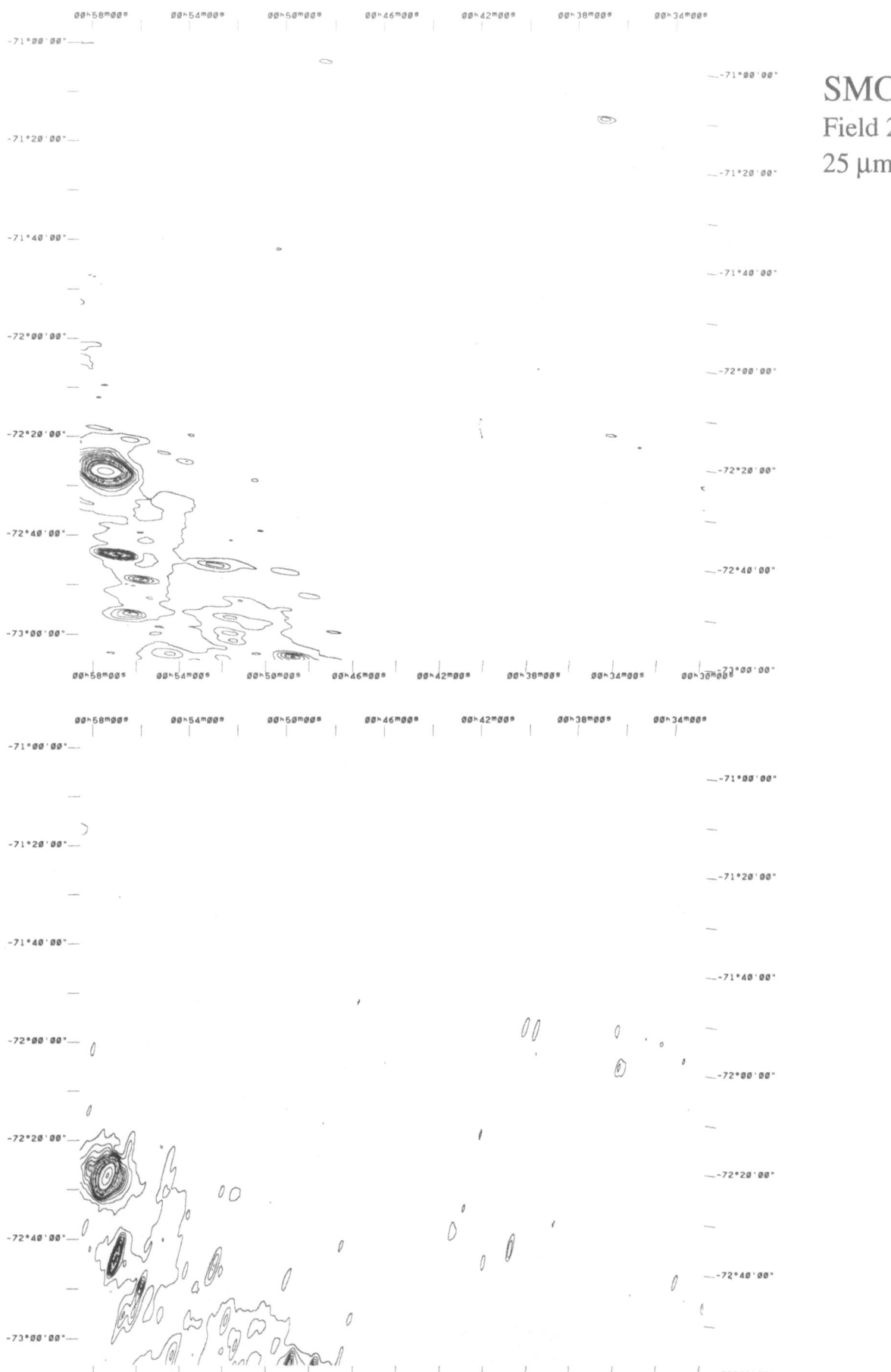

SMC
Field 2
25 μm

SMC
Field 2
60 μm

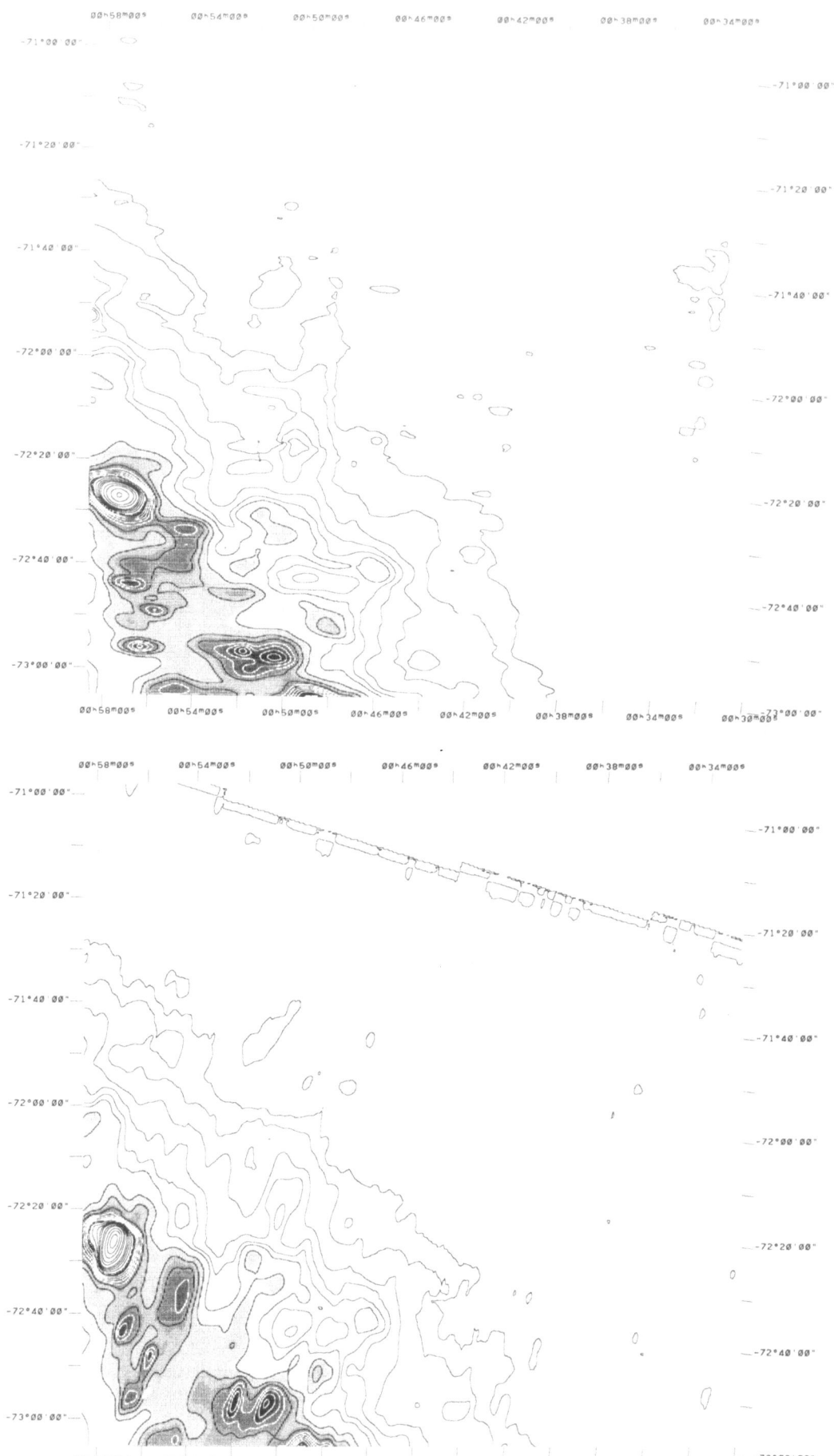

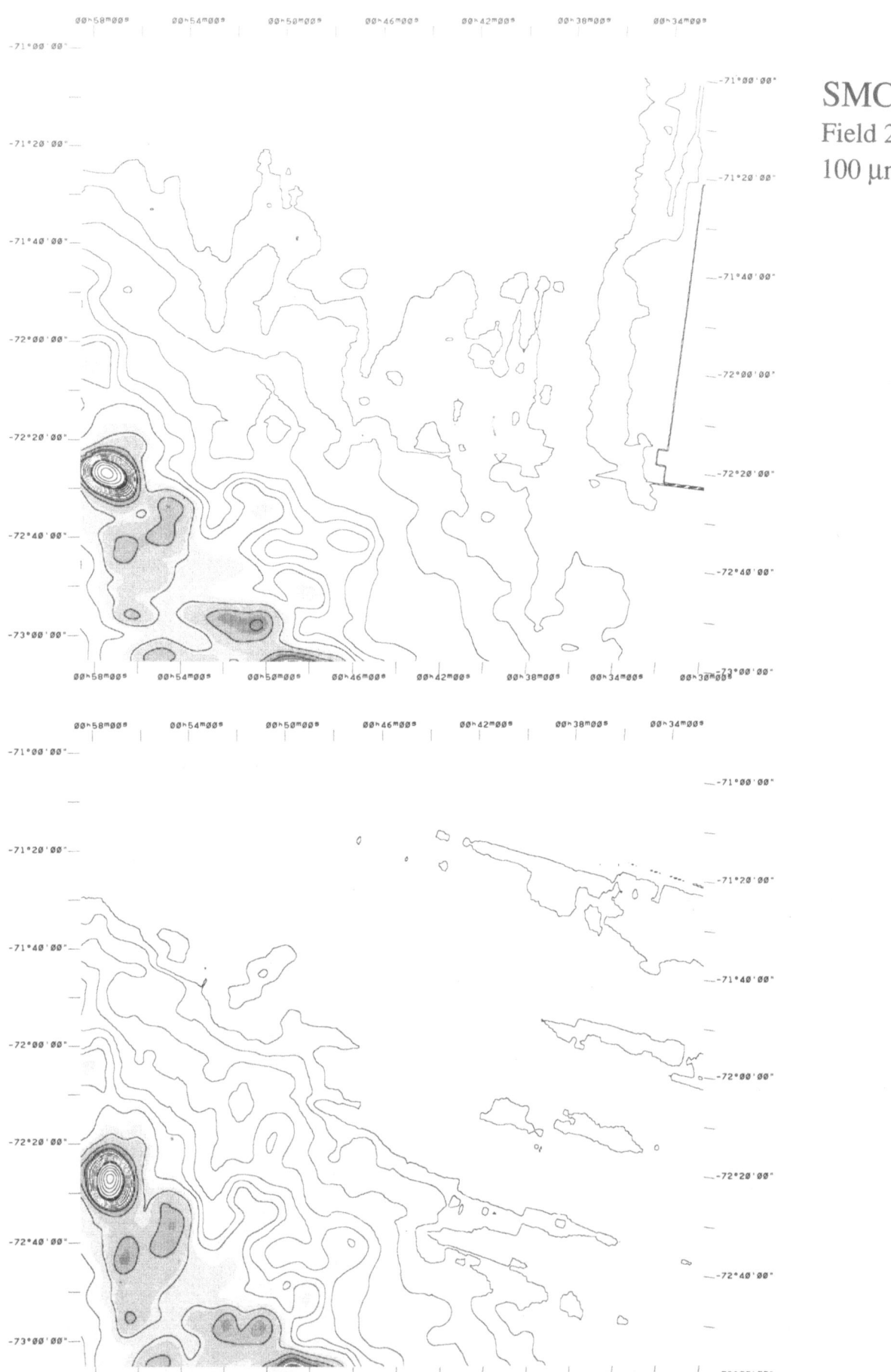

SMC
Field 2
100 μm

SMC
Field 3
12 μm

SMC
Field 3
25 μm

SMC
Field 3
60 µm

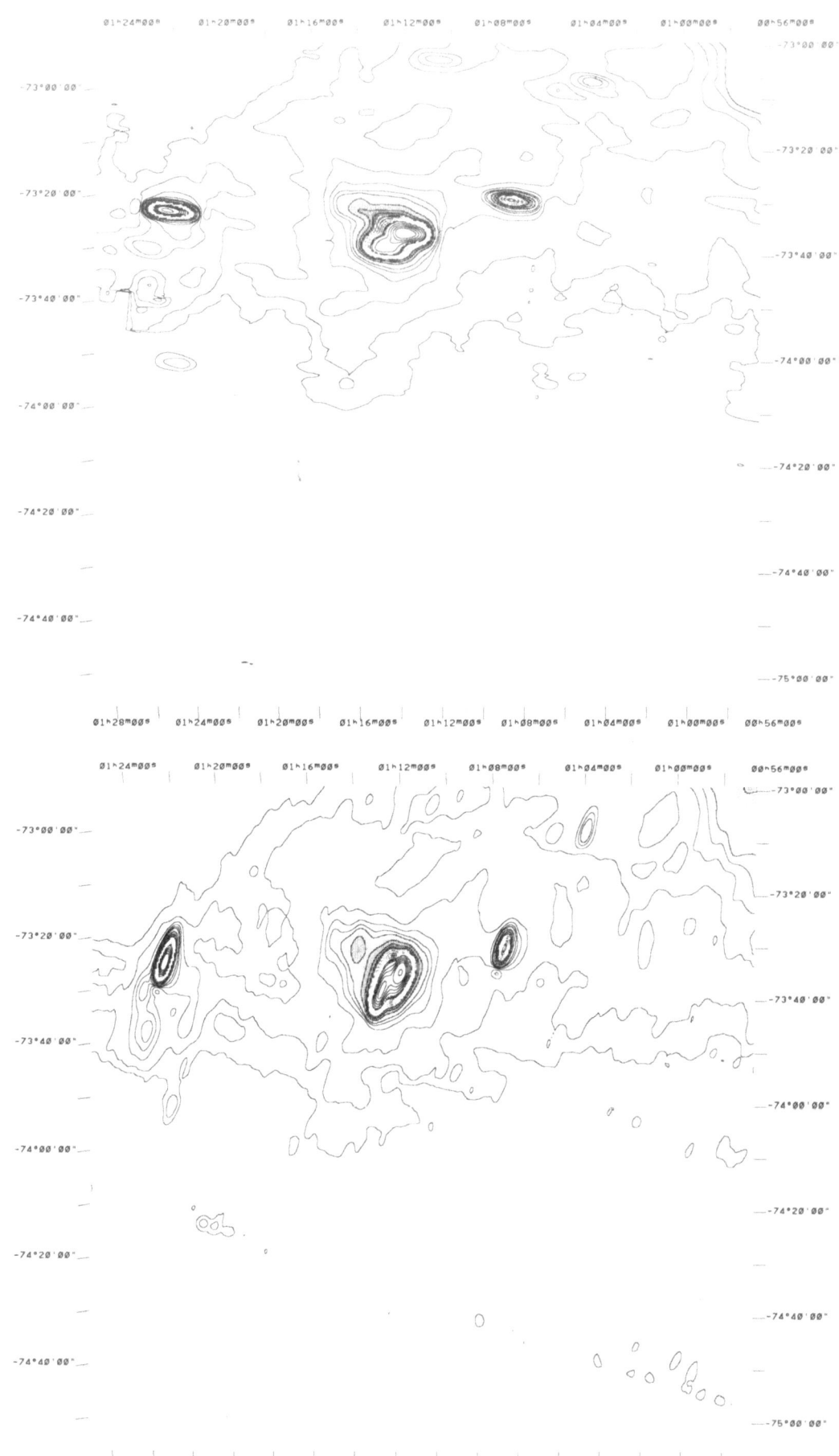

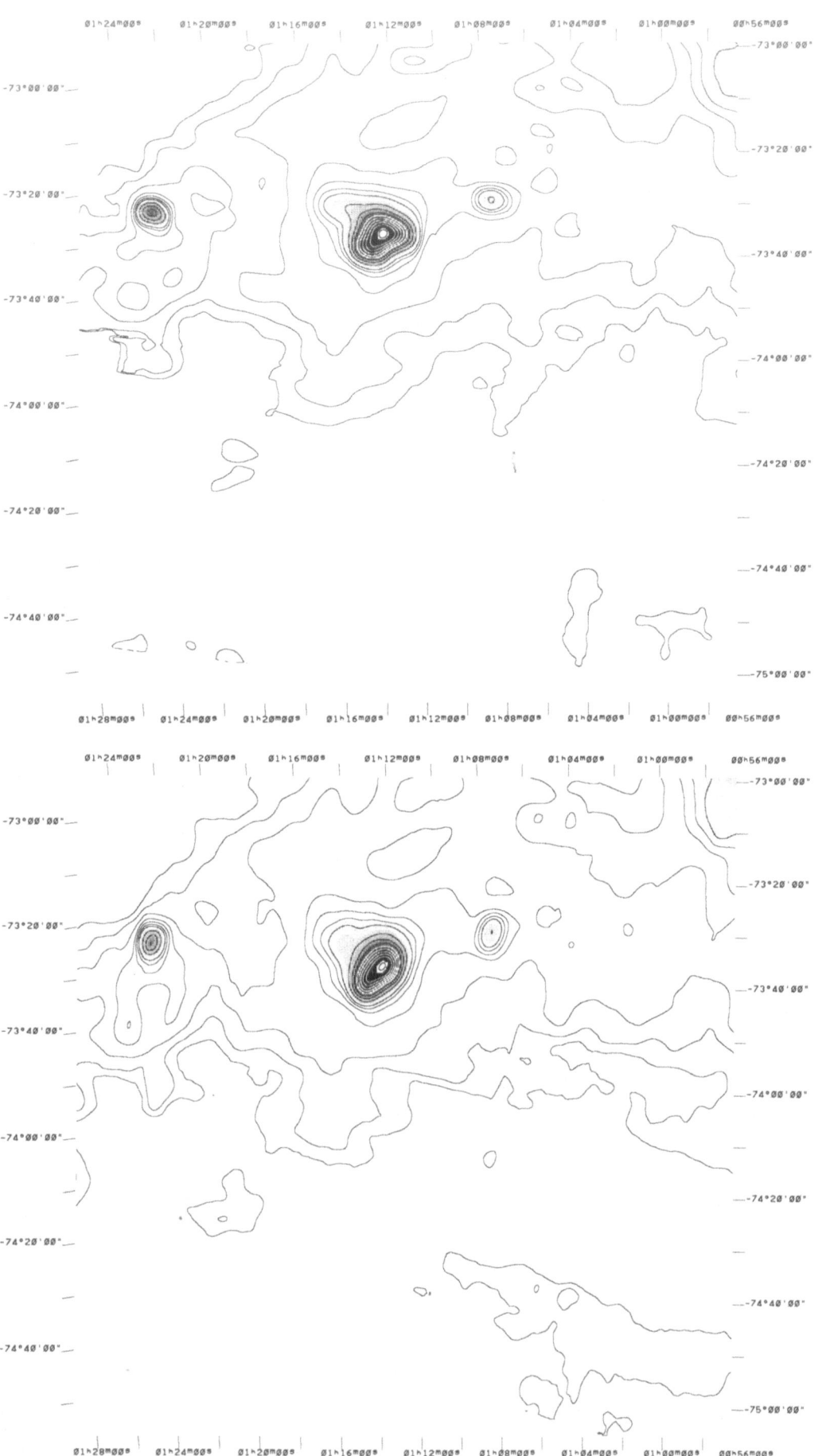

SMC
Field 3
100 μm

SMC
Field 4
12 μm

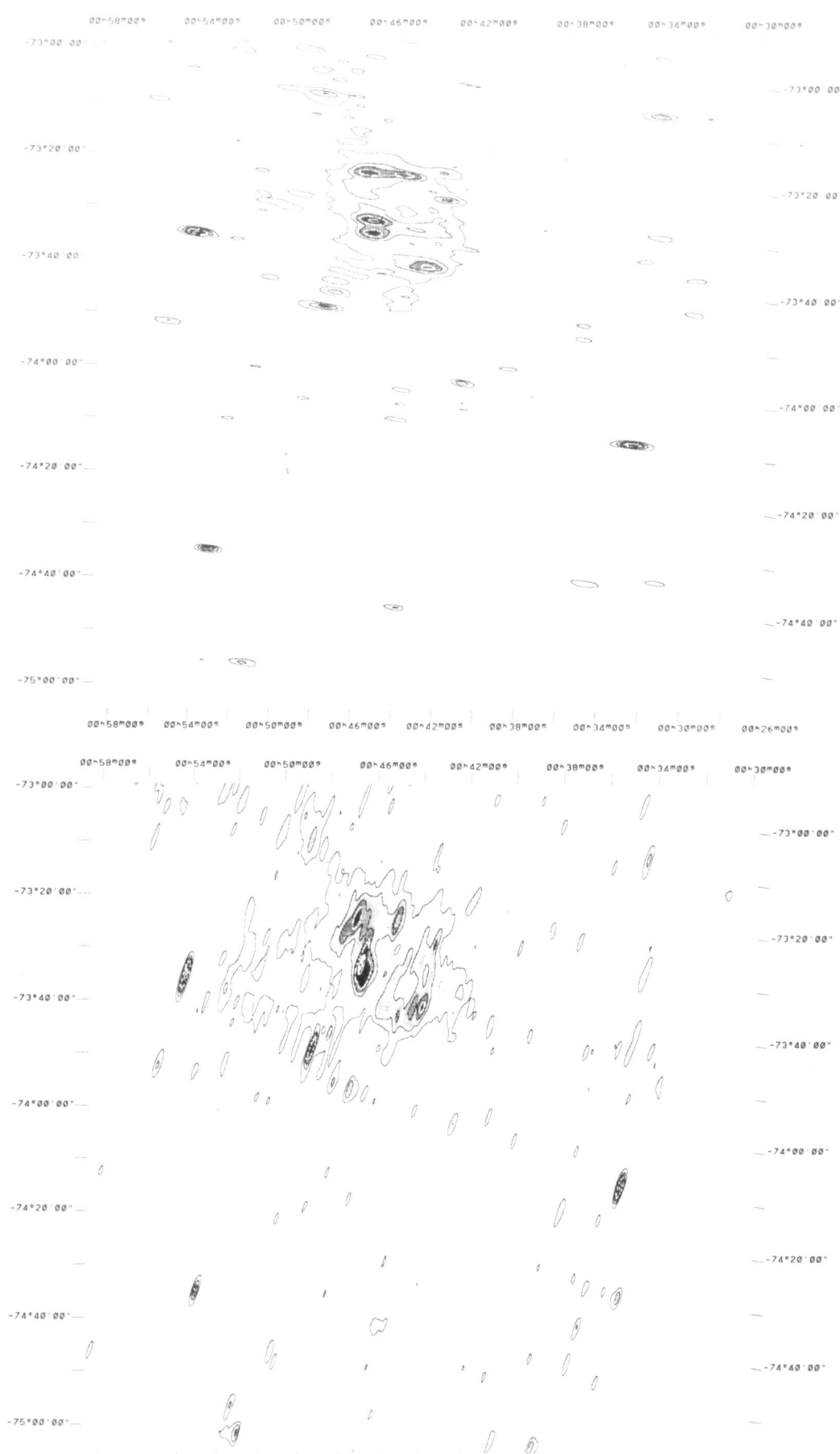

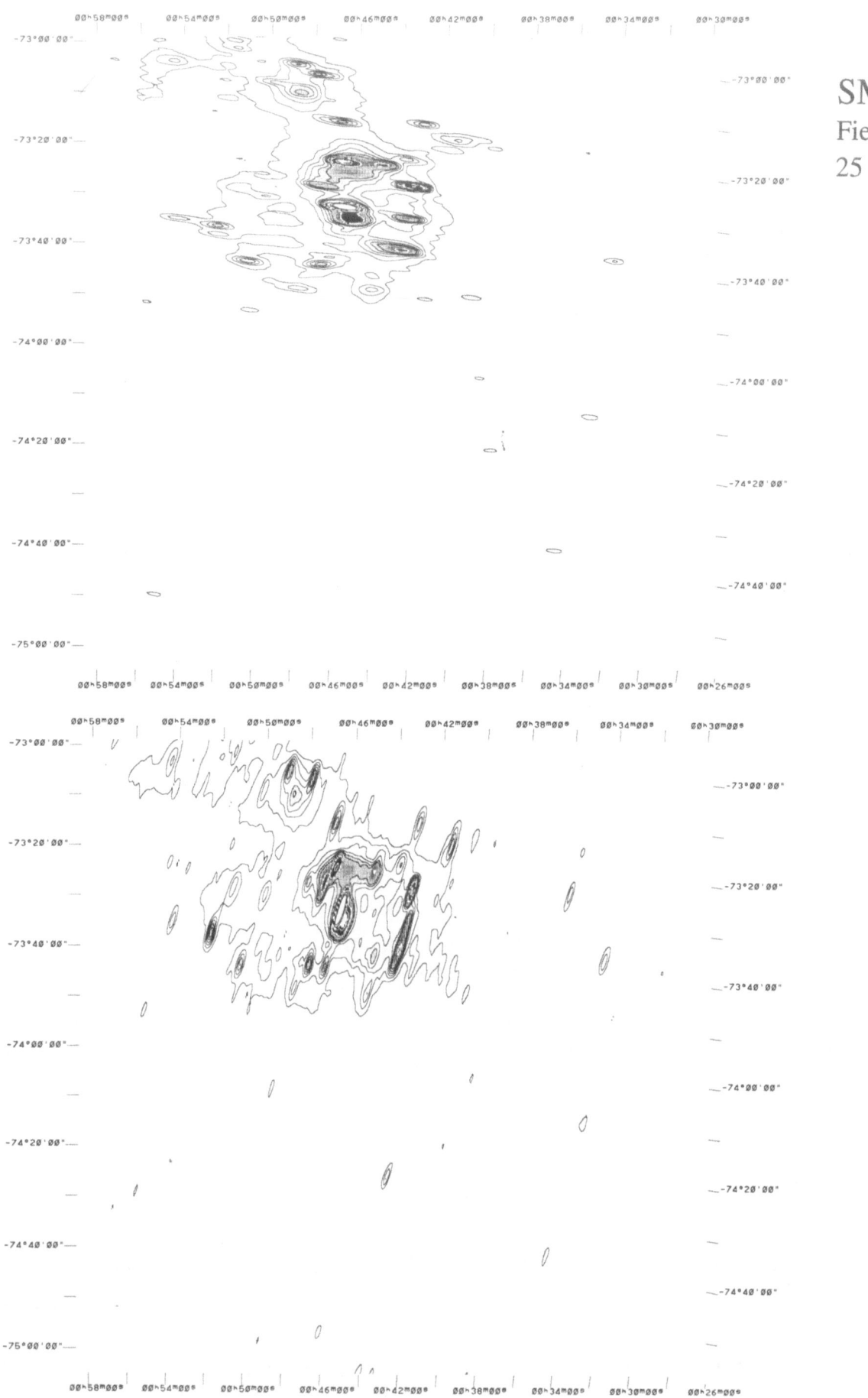

SMC

Field 4

25 μm

SMC
Field 4
60 μm

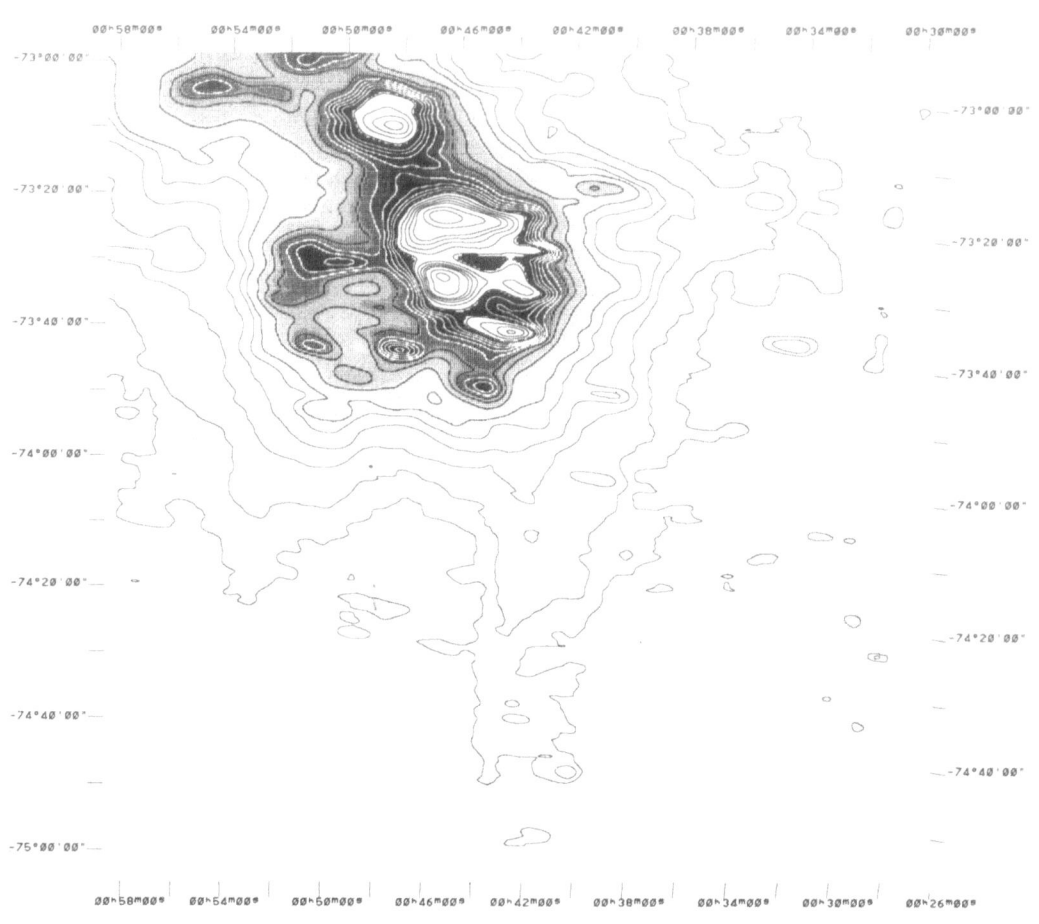

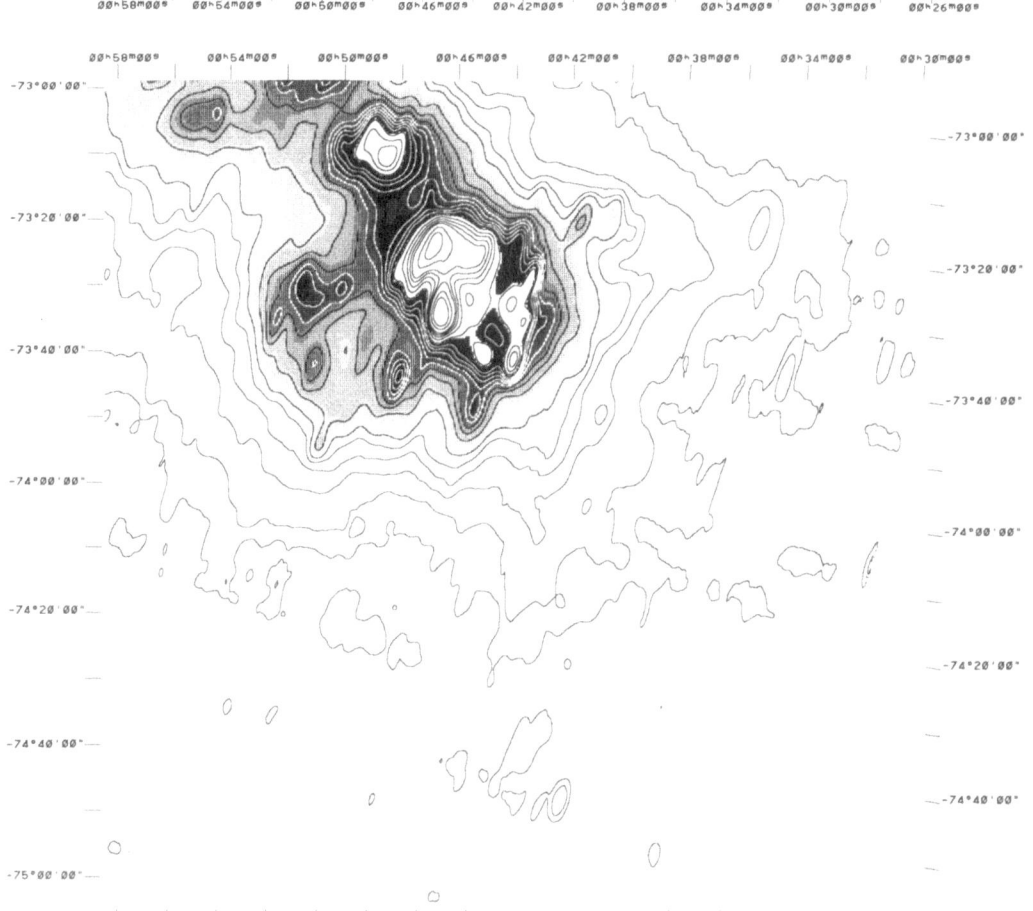

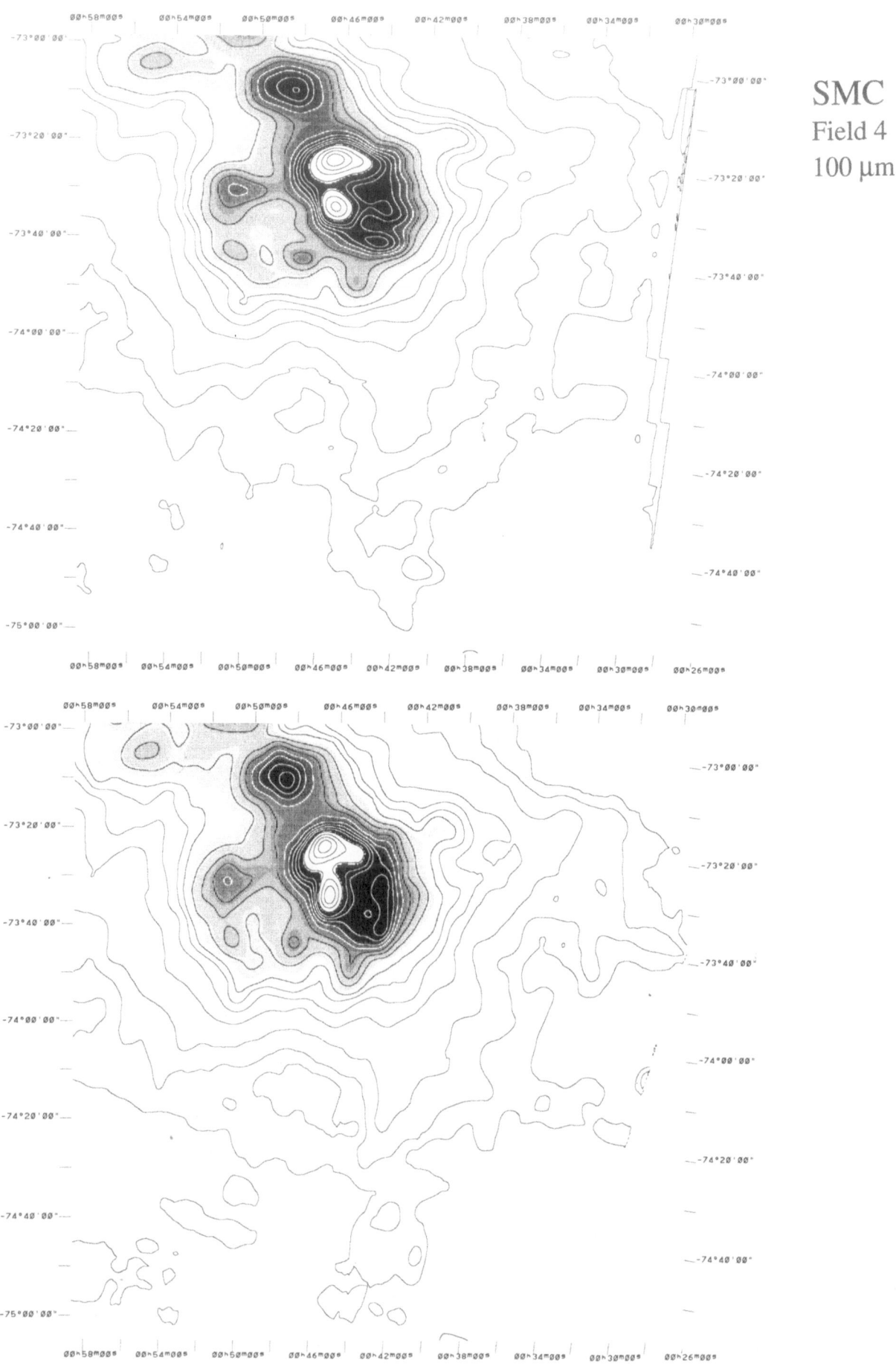

SMC
Field 4
100 μm

SMC
Field 5
12 μm

SMC
Field 5
25 μm

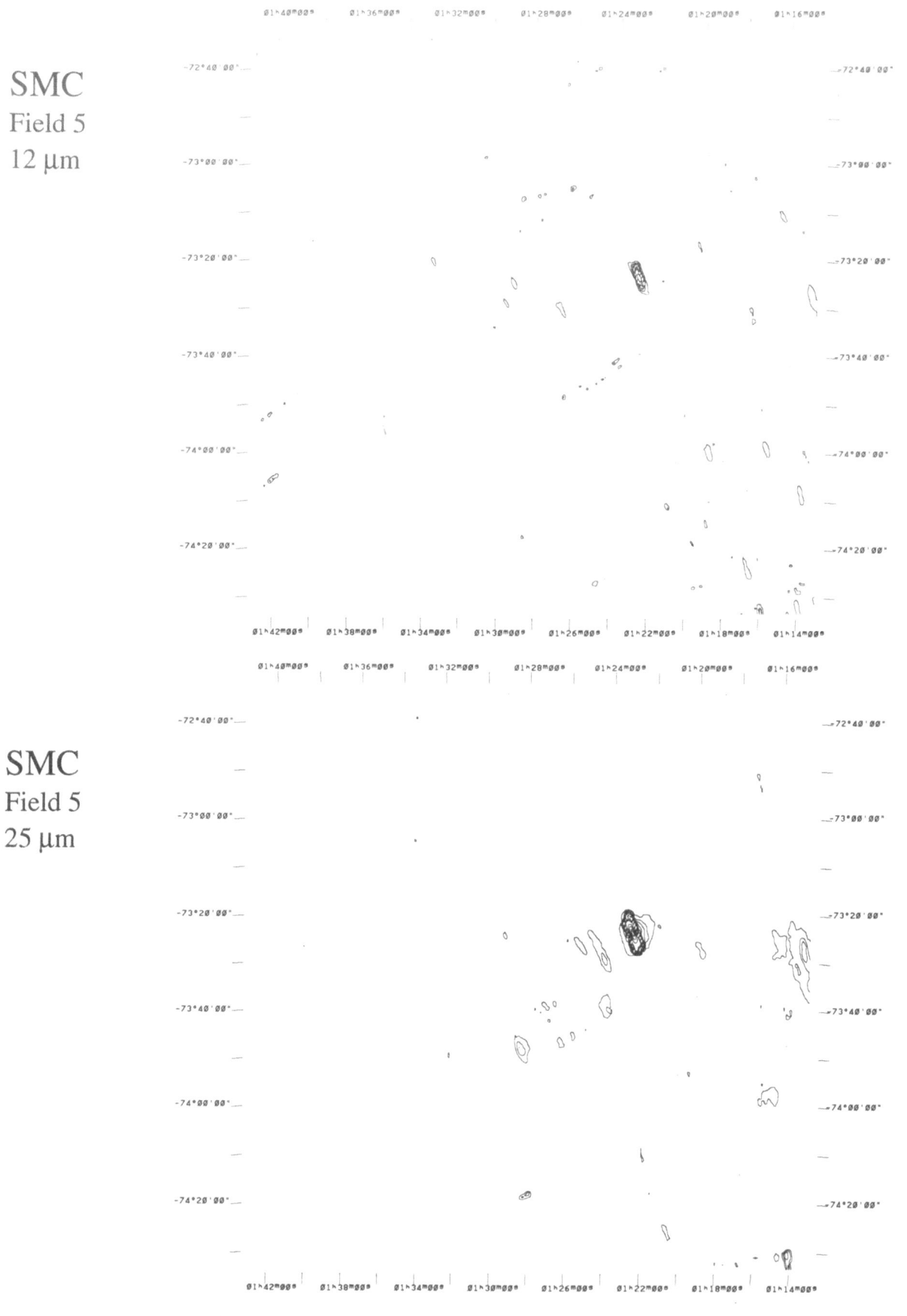

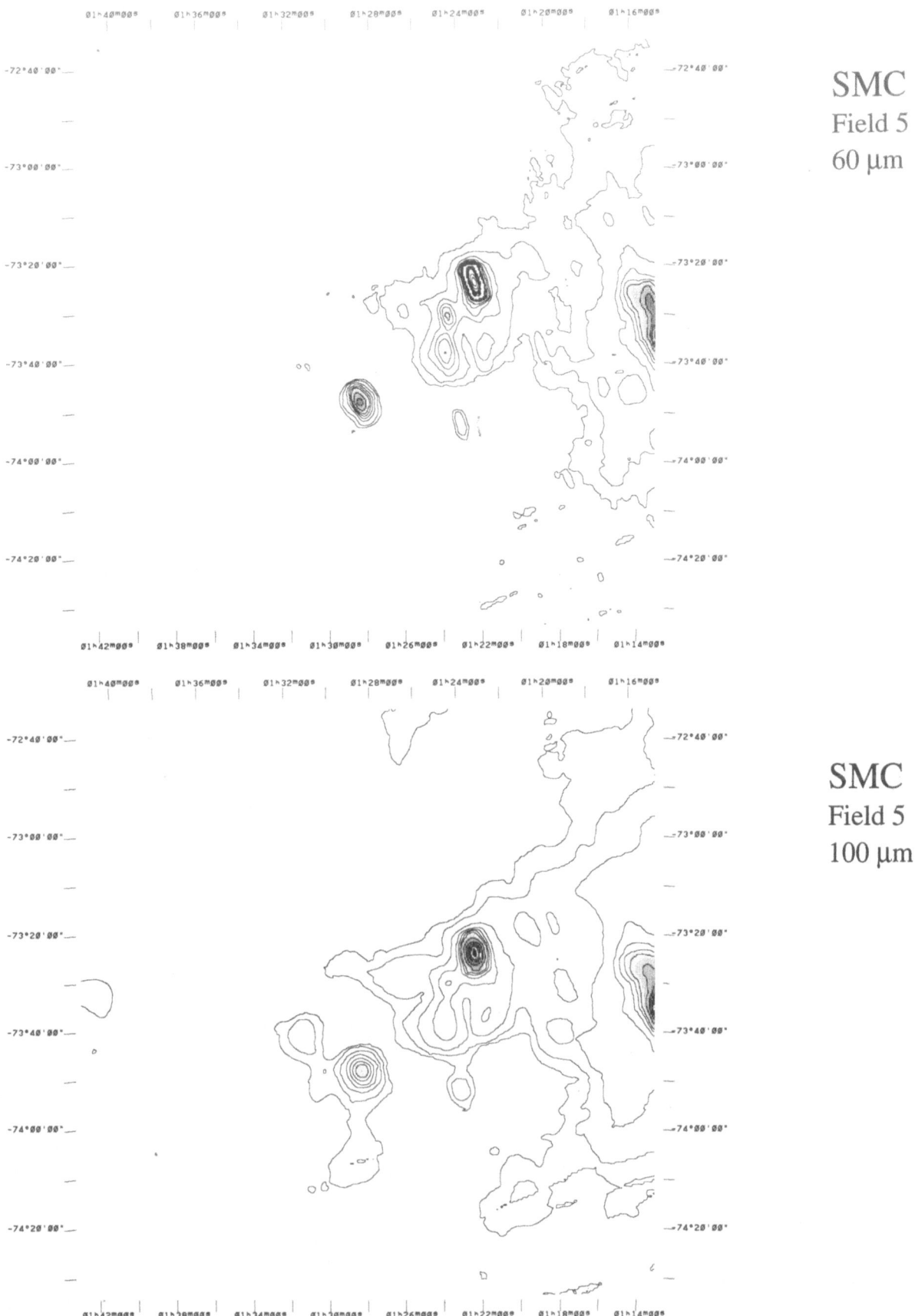

SMC
Field 5
60 μm

SMC
Field 5
100 μm

SMC
Field 6
12 μm

SMC
Field 6
25 μm

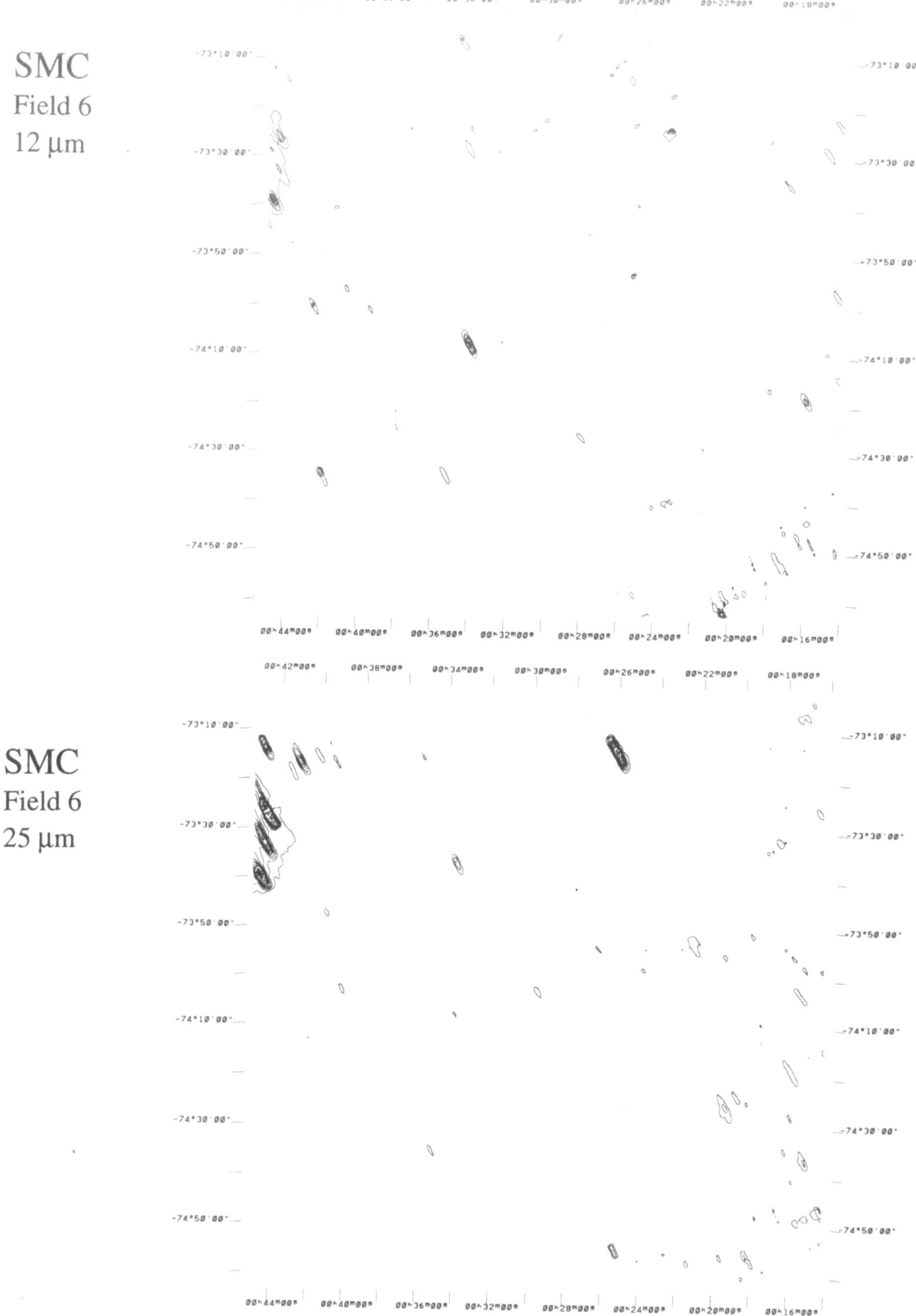

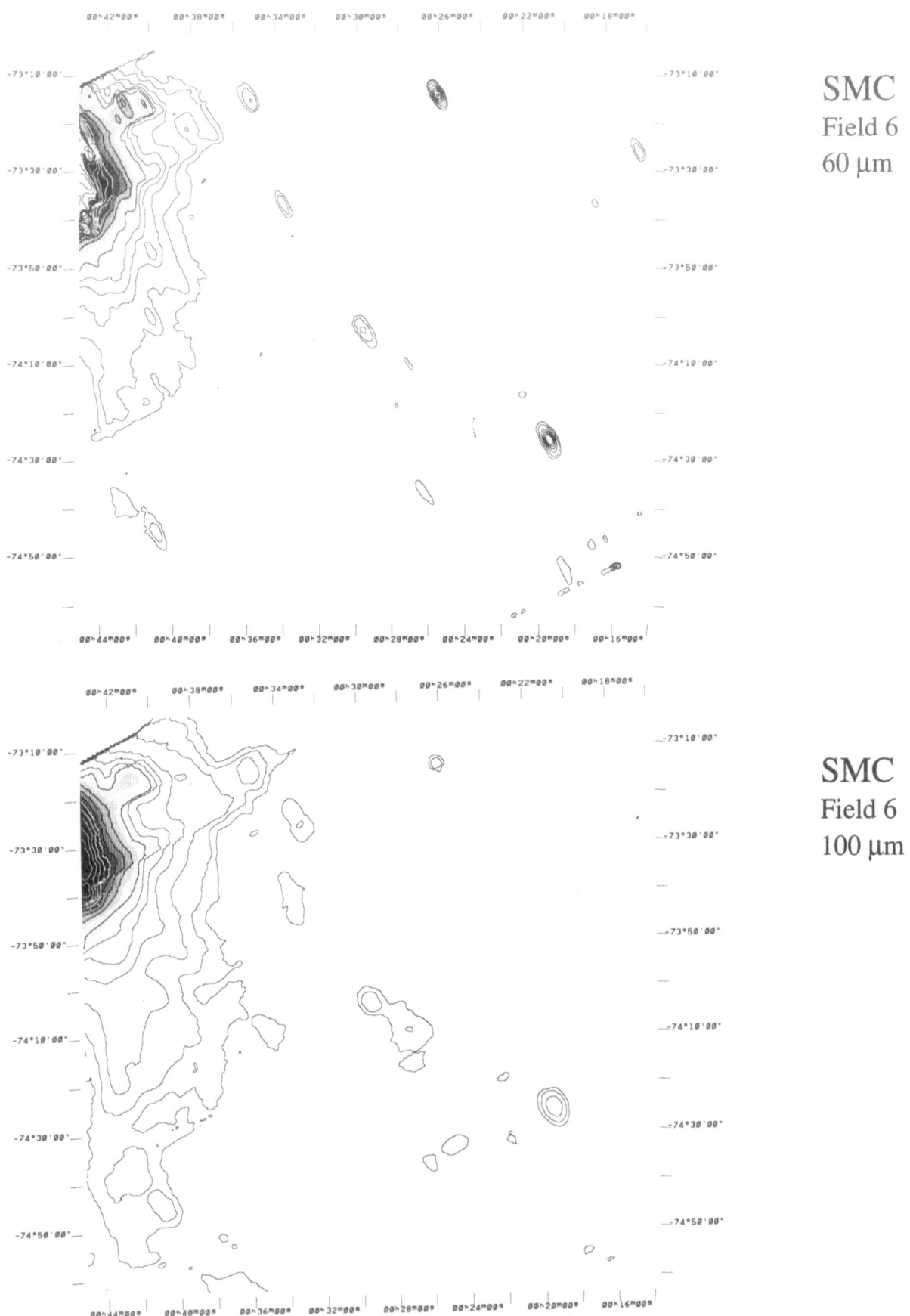

SMC
Field 6
60 μm

SMC
Field 6
100 μm

LMC
12 μm

LMC
25 μm

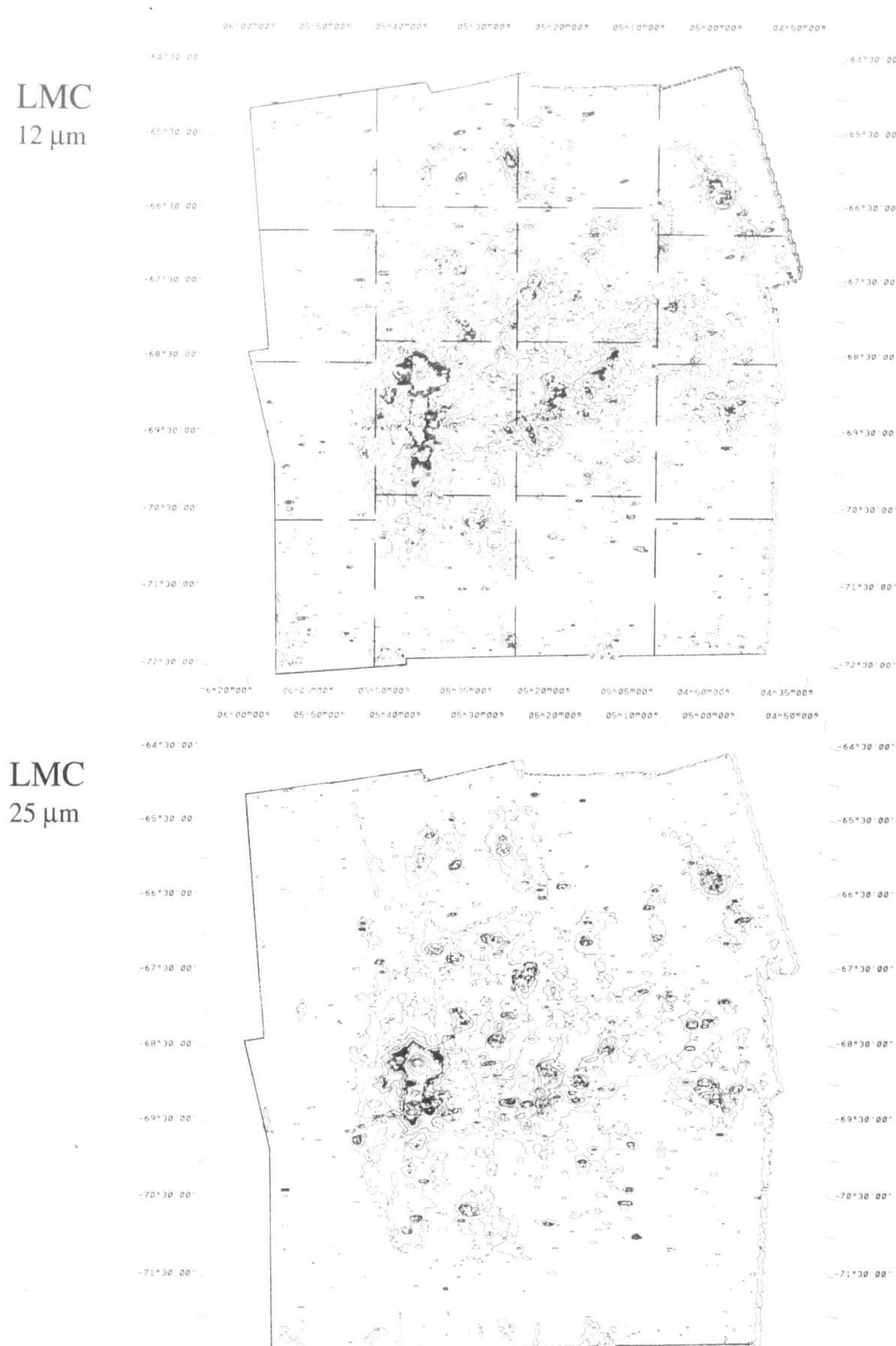

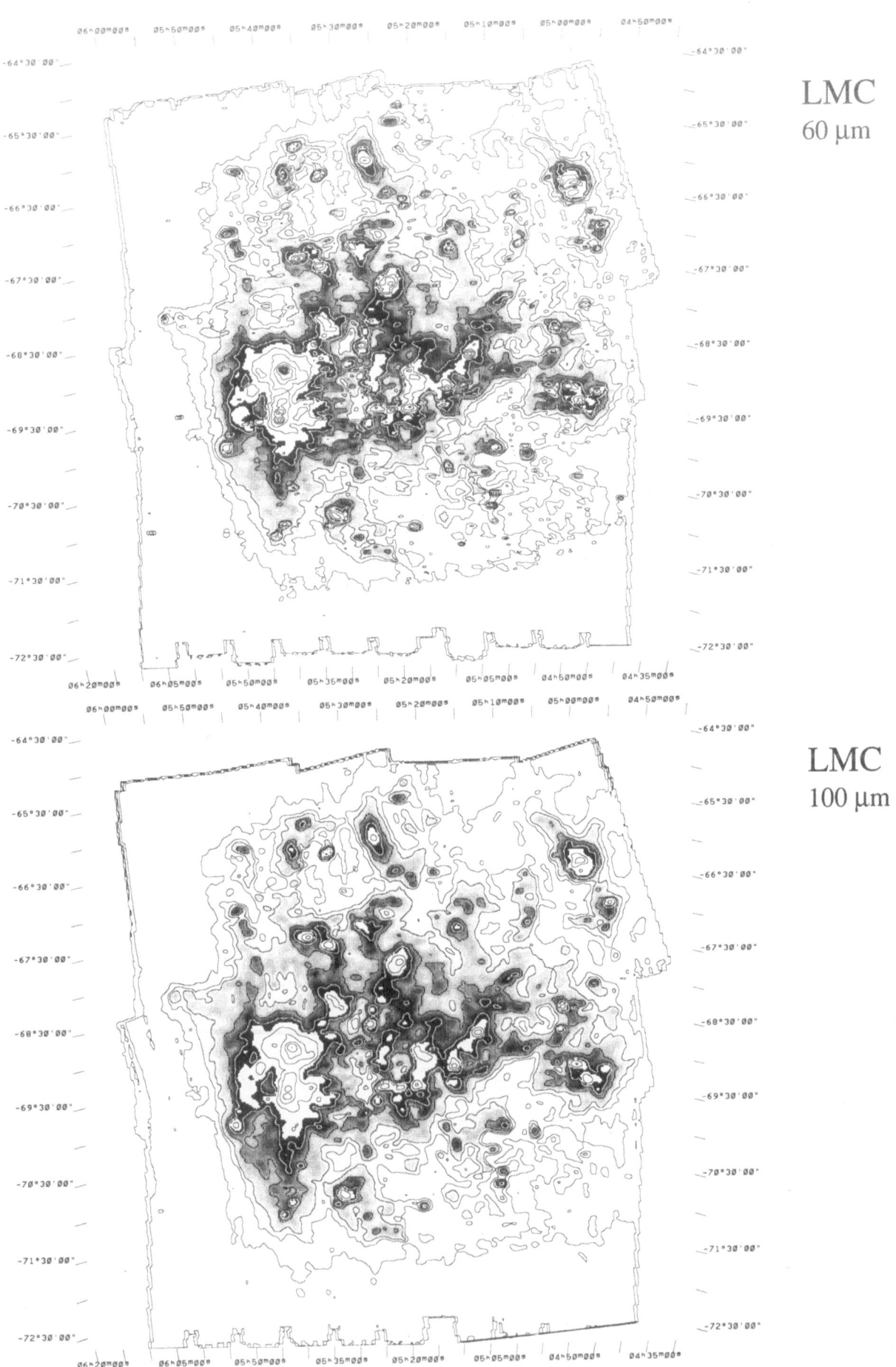

LMC
60 µm

LMC
100 µm

LMC
Field 1
12 μm

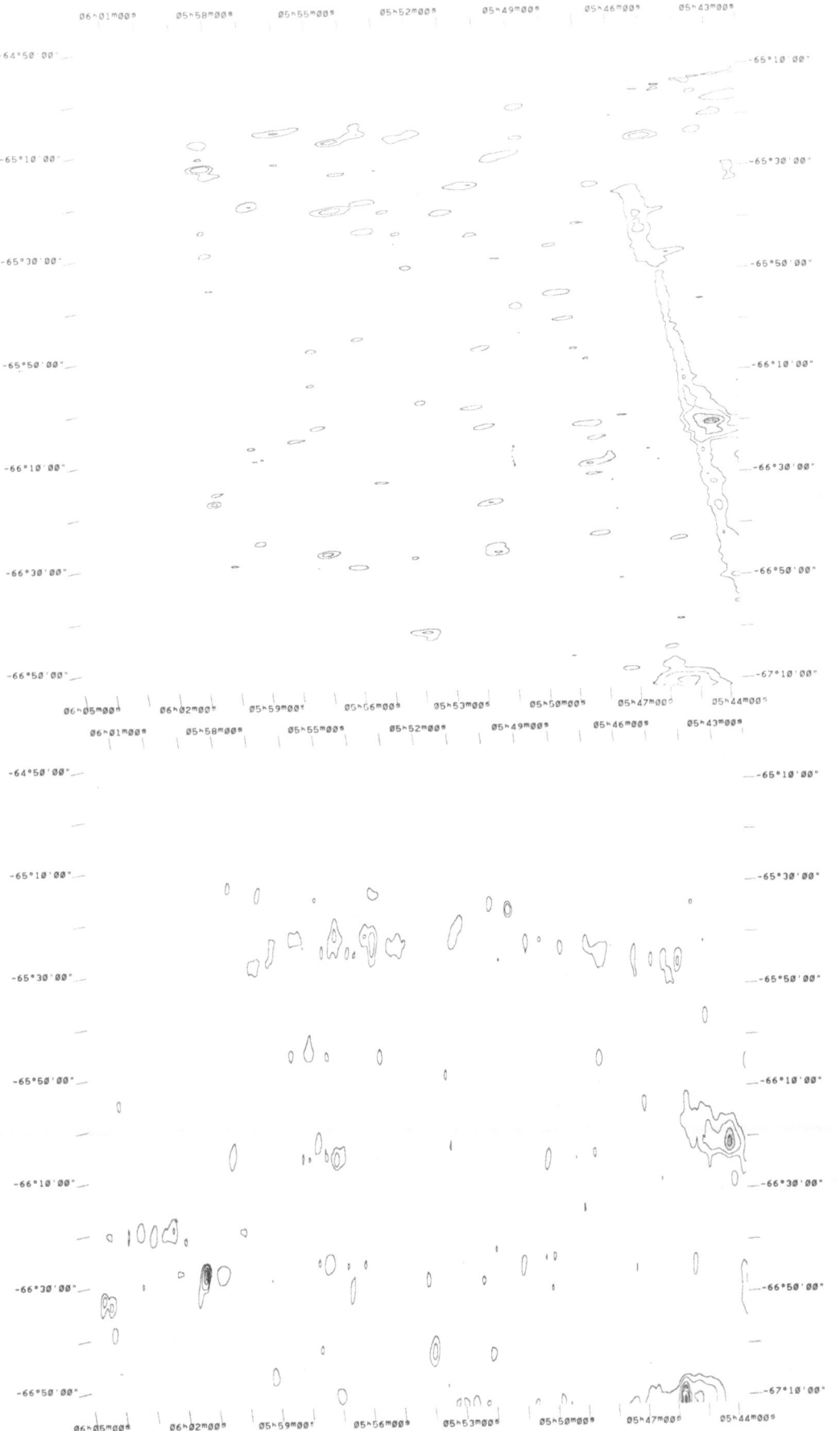

LMC
Field 1
25 μm

LMC
Field 1
60 μm

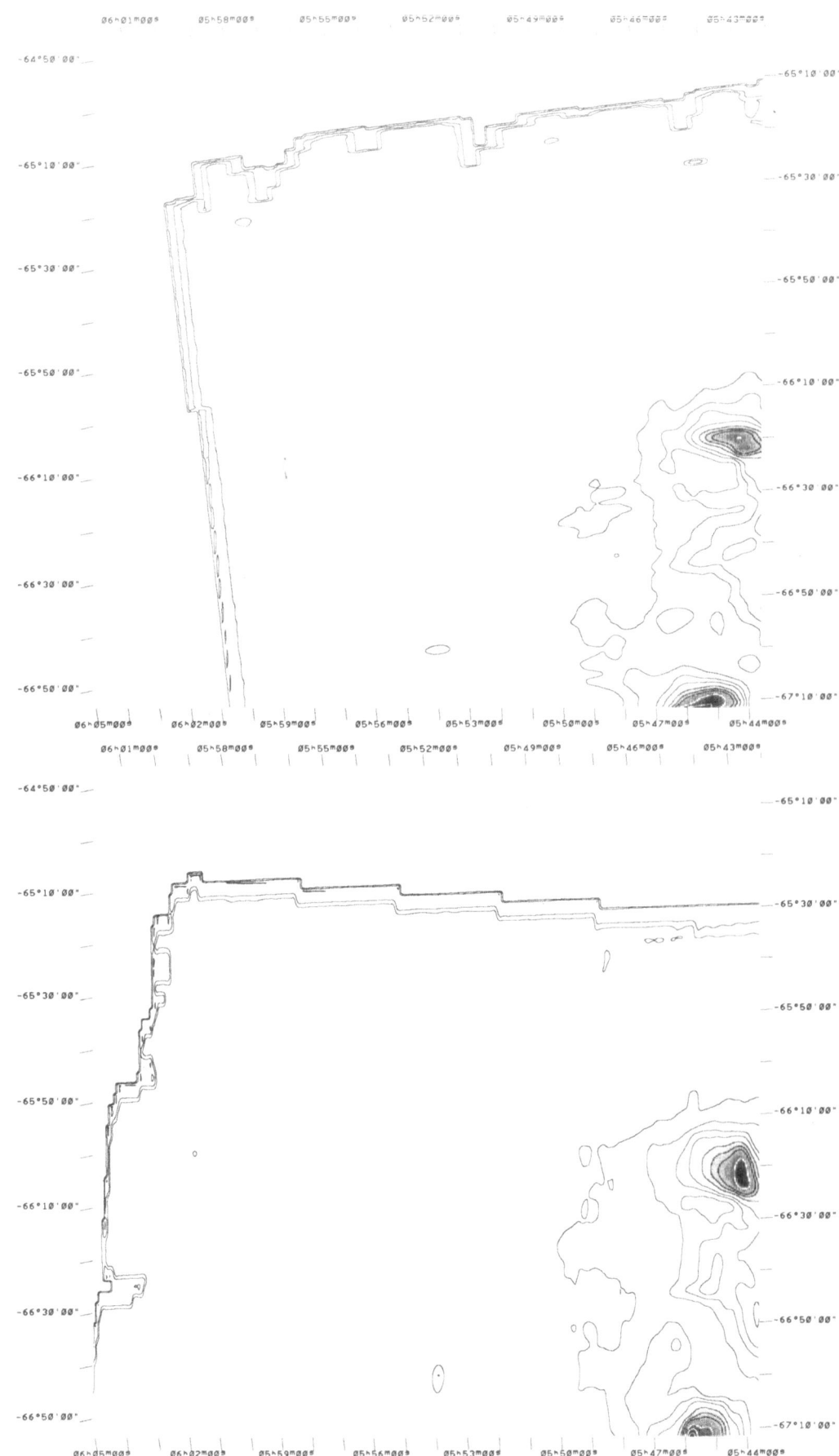

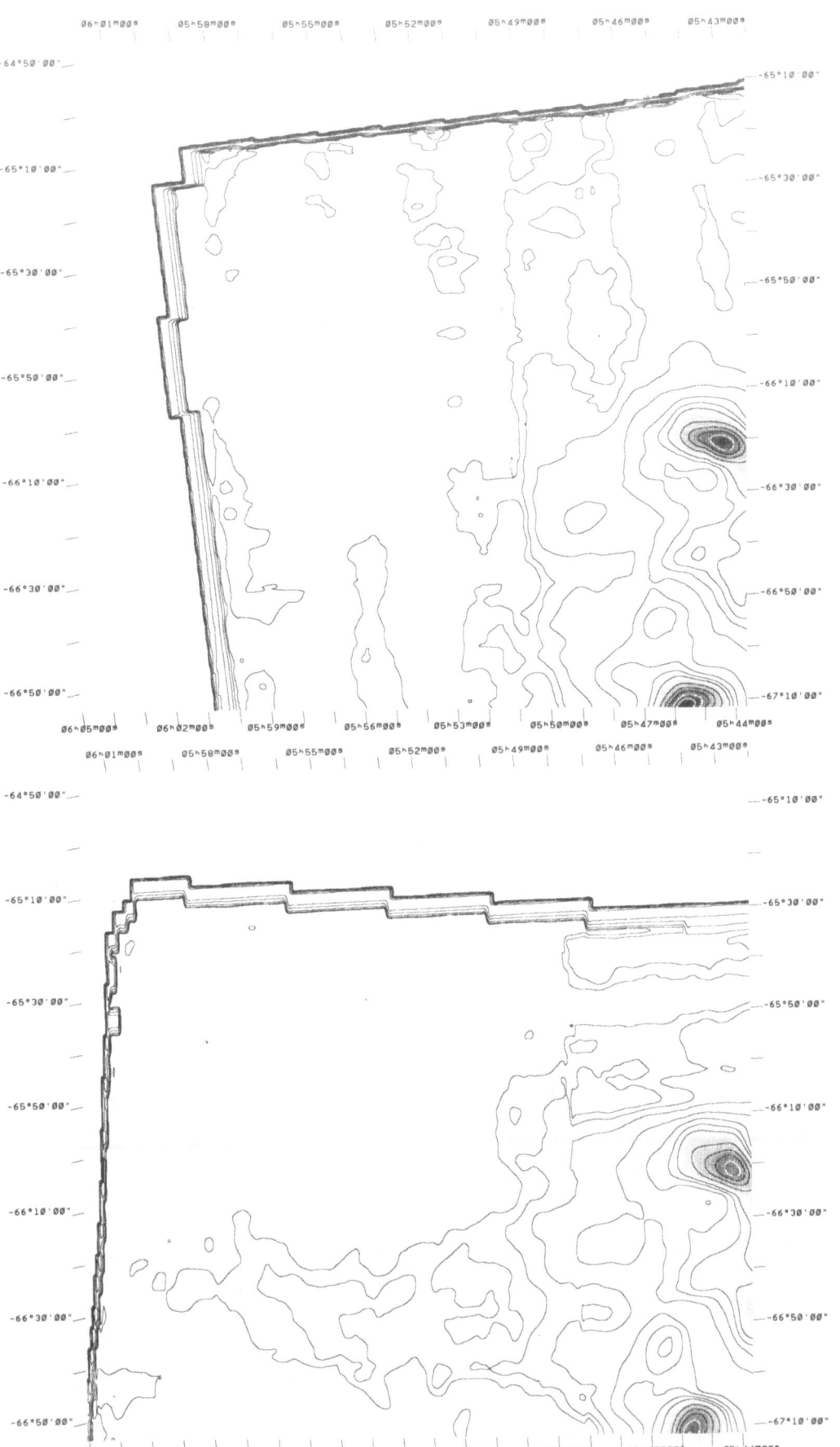

LMC
Field 1
100 μm

LMC
Field 2
12 μm

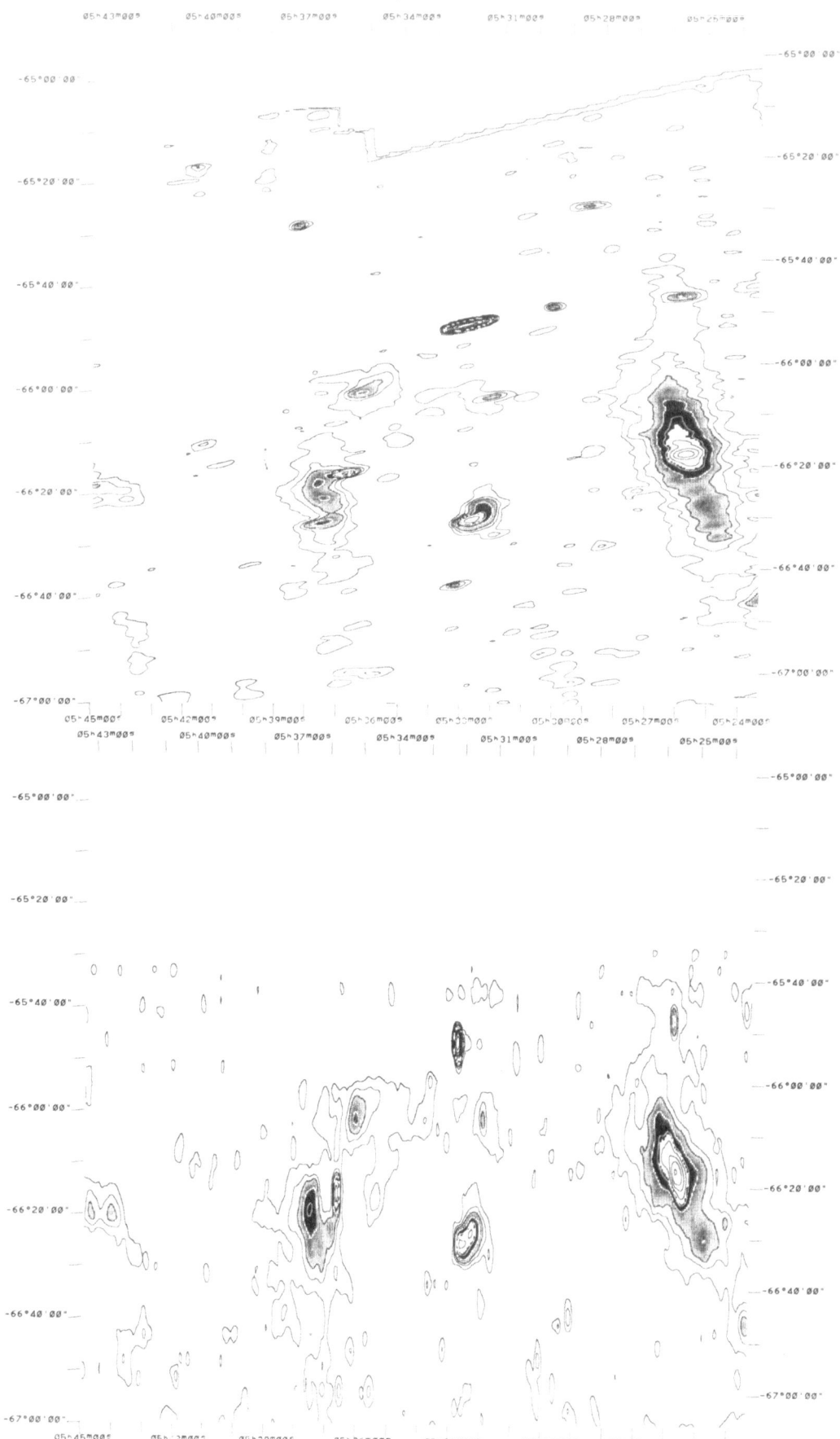

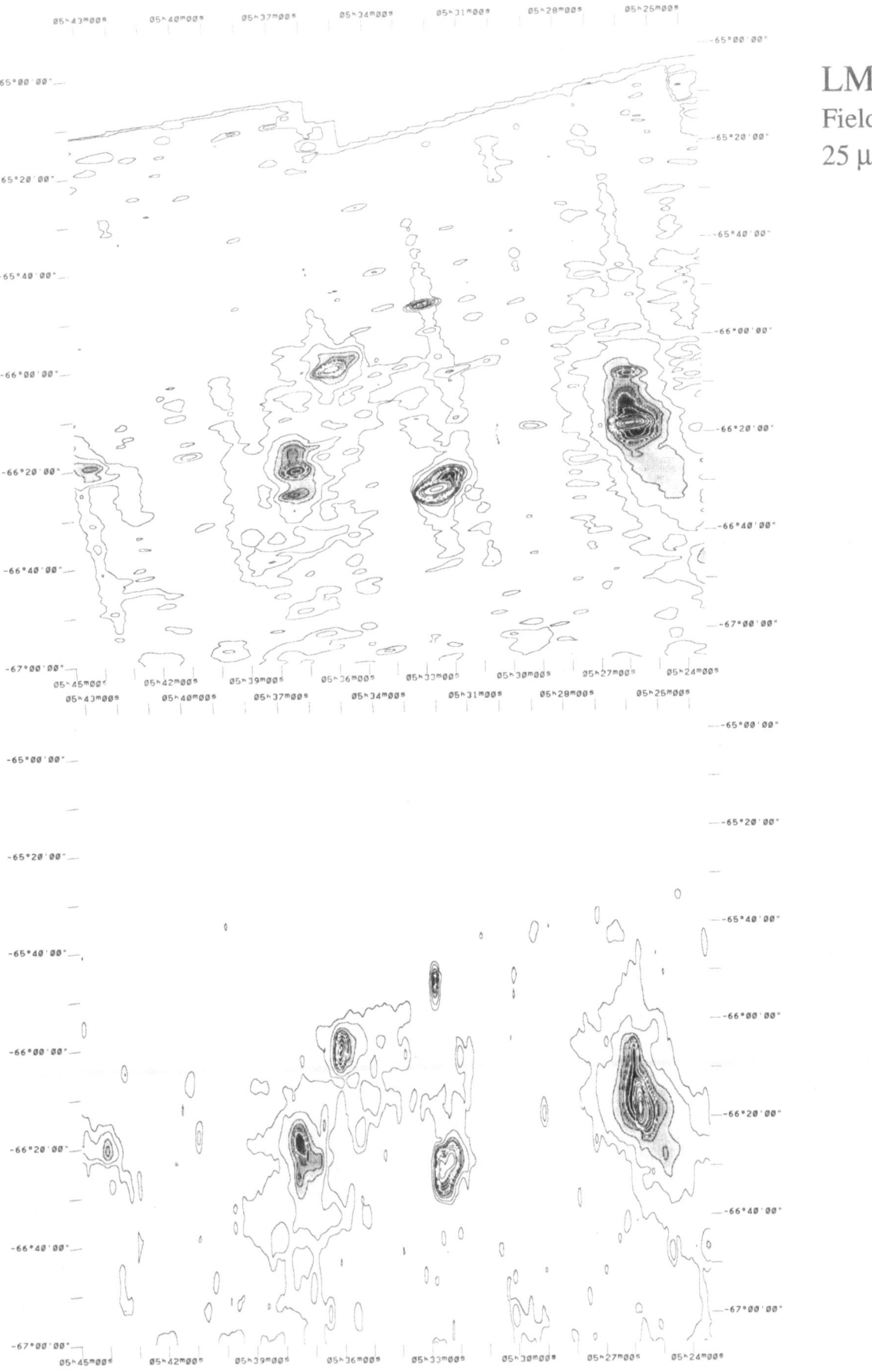

LMC
Field 2
25 μm

LMC
Field 2
60 μm

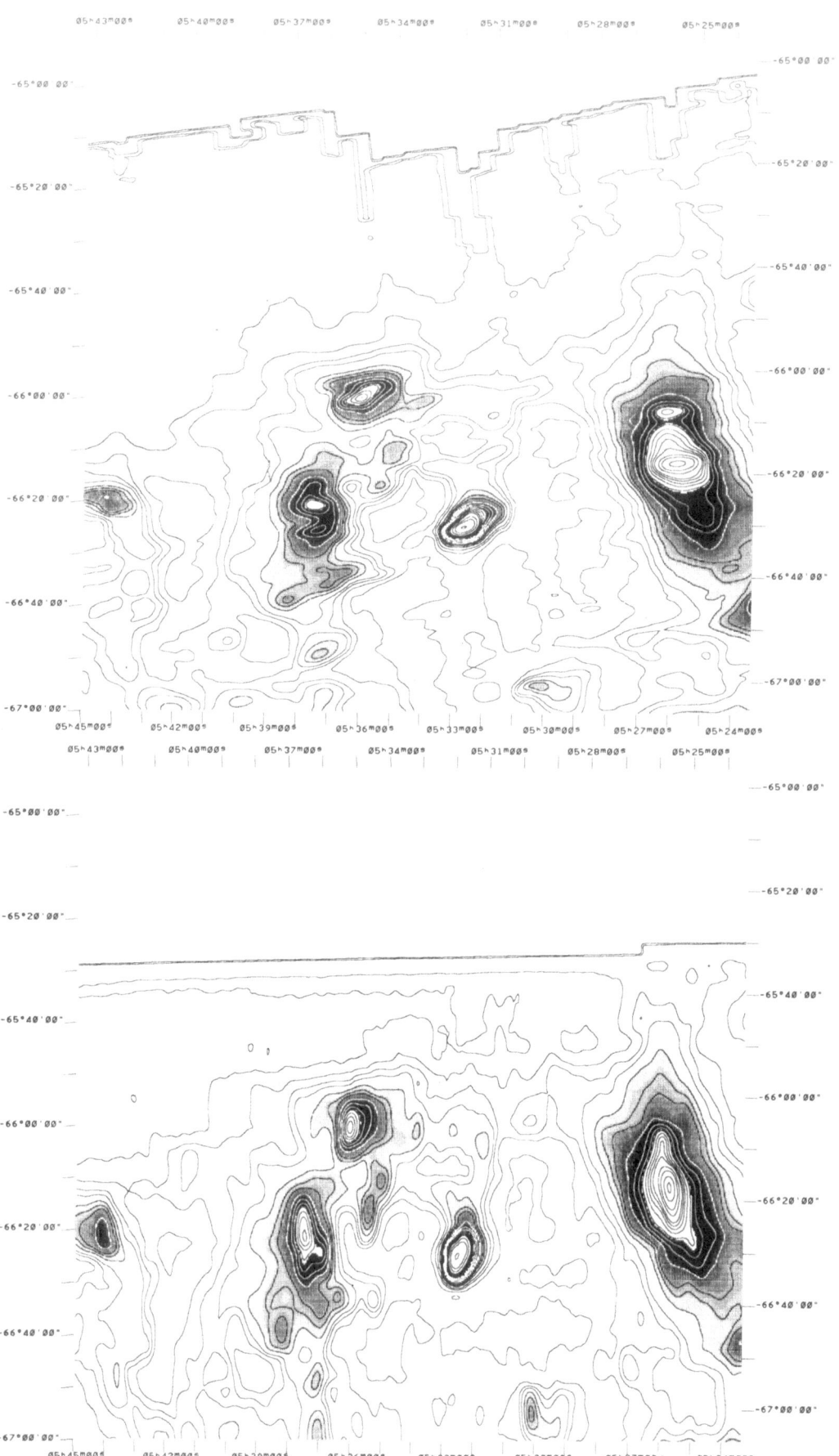

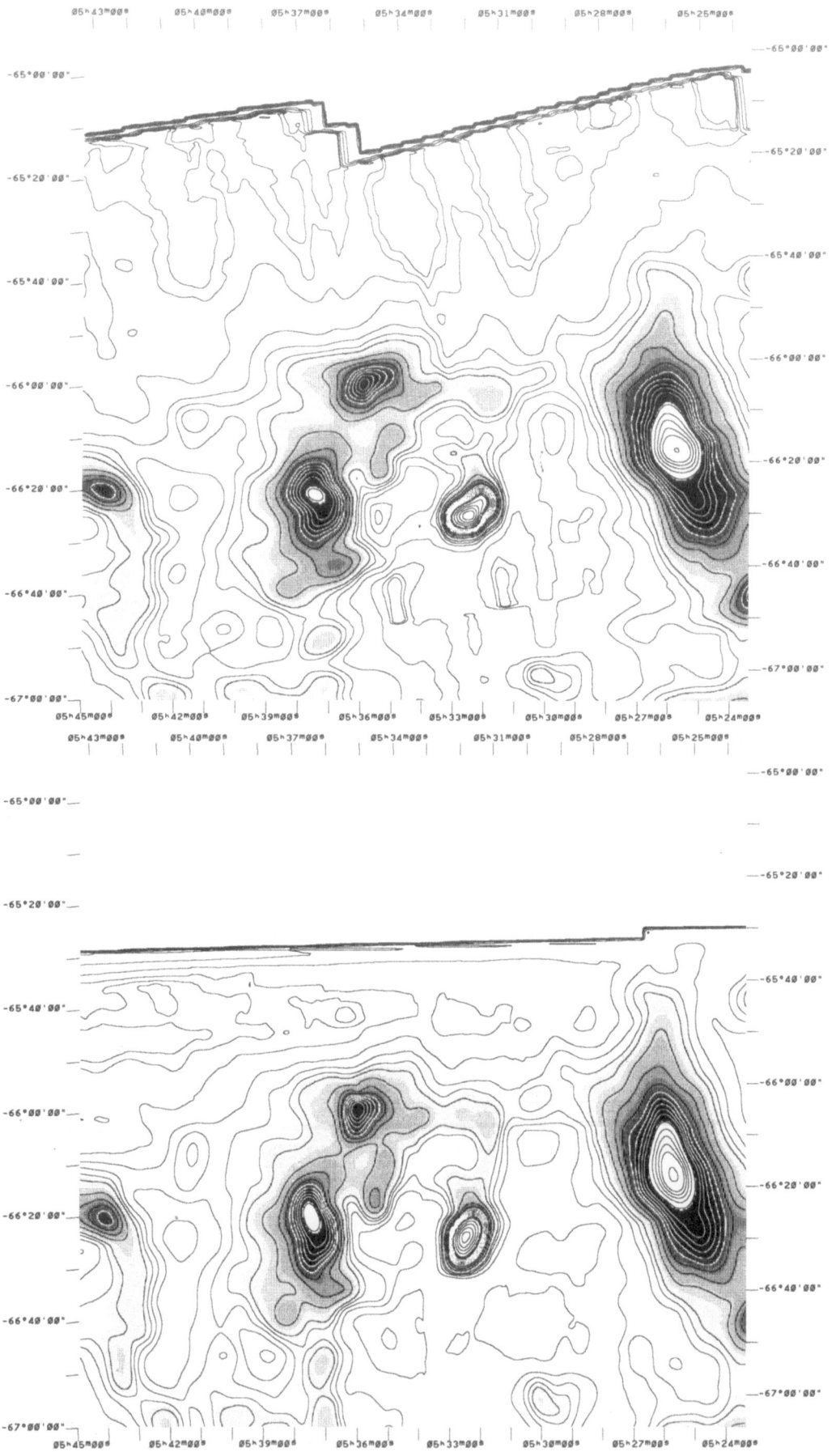

LMC
Field 2
100 μm

LMC
Field 3
12 μm

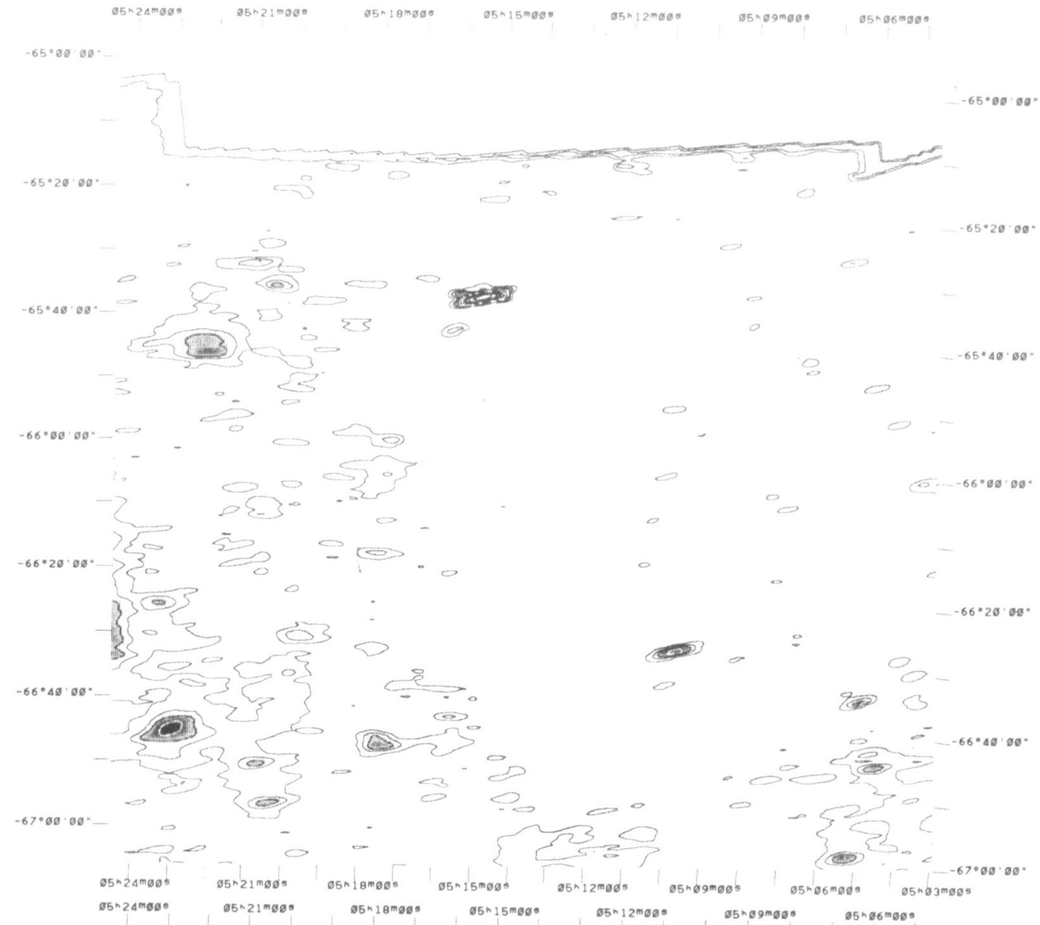

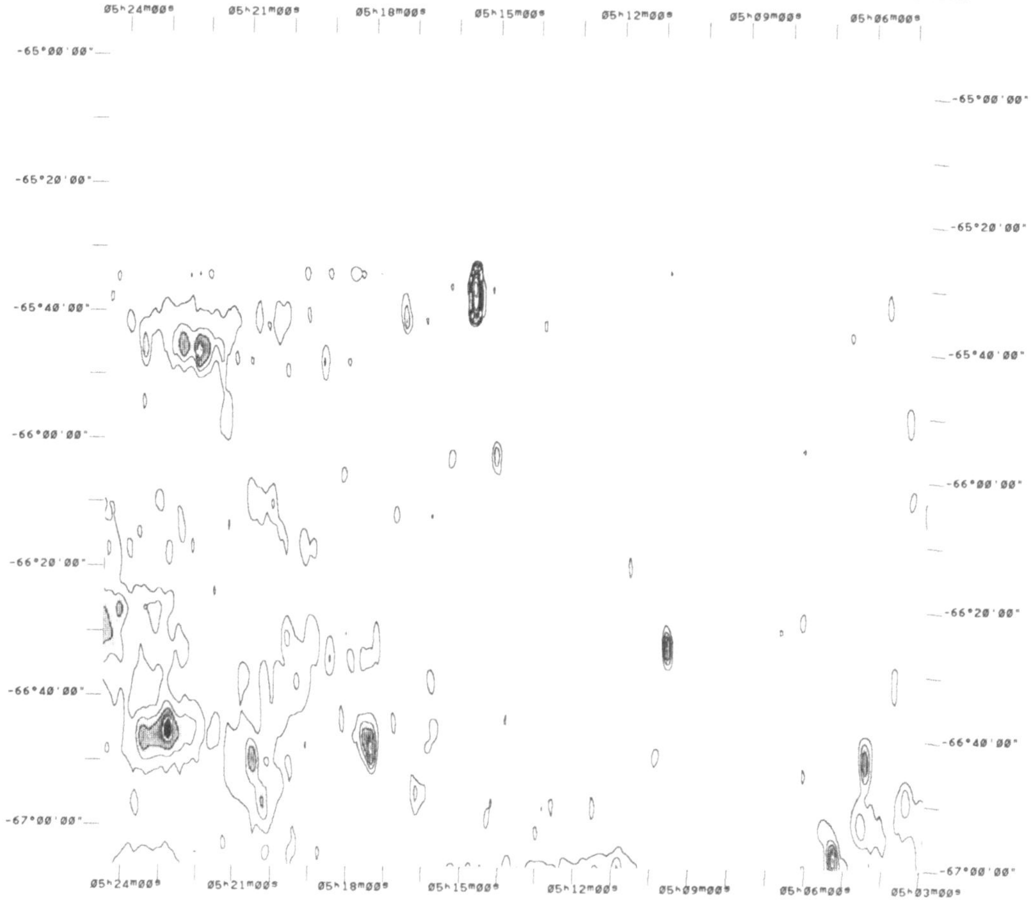

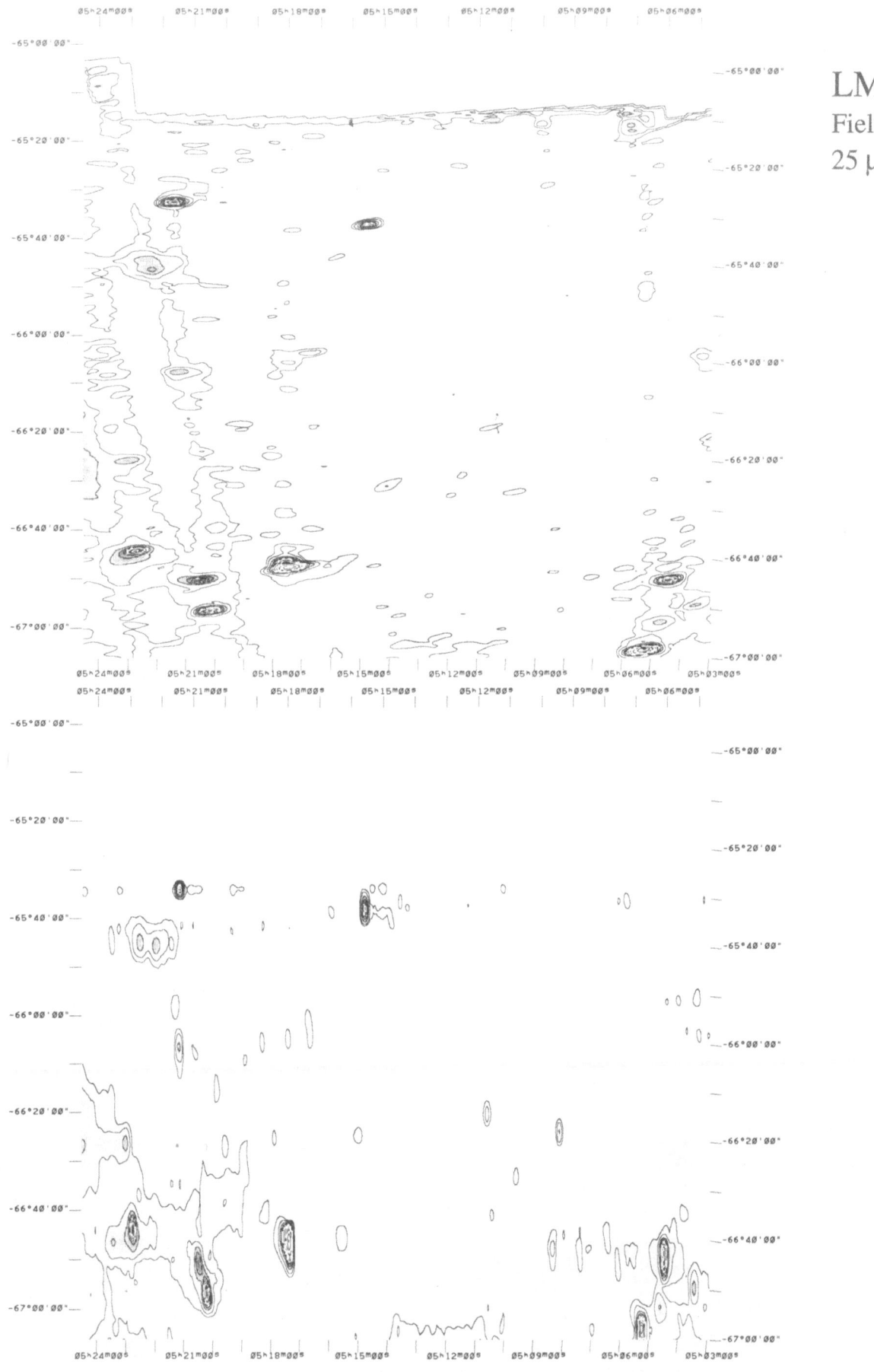

LMC
Field 3
25 μm

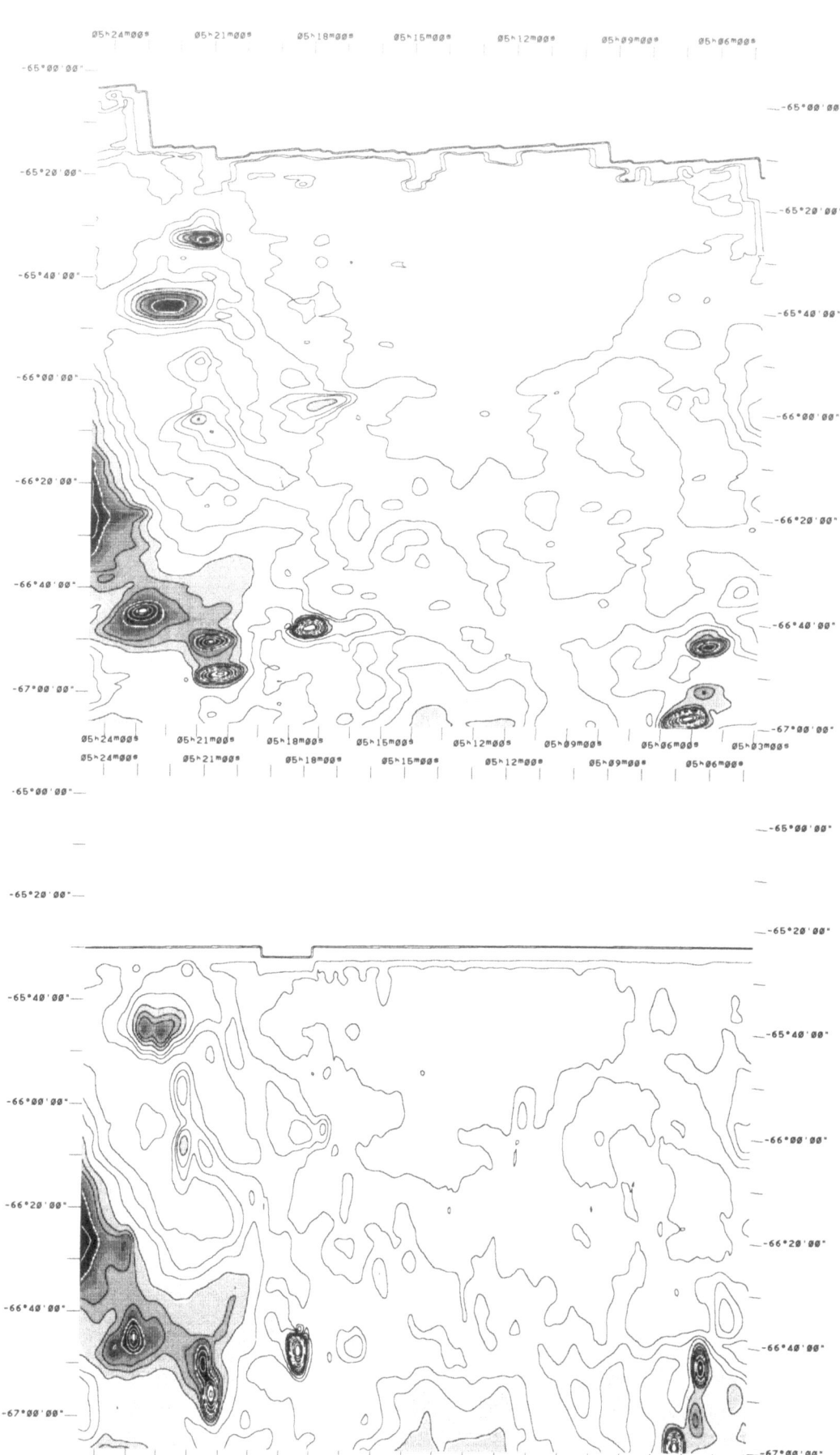

LMC
Field 3
60 μm

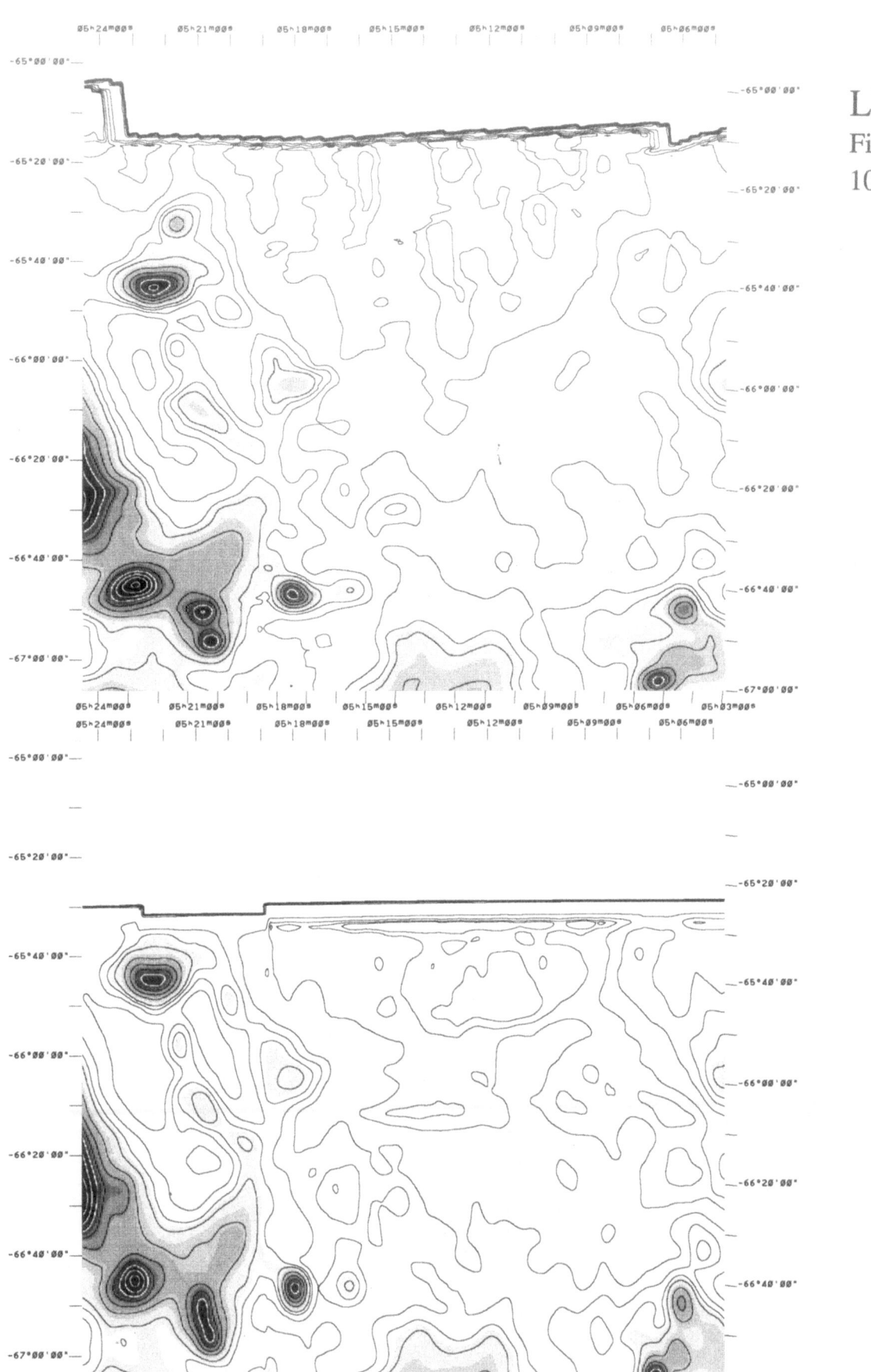

LMC
Field 3
100 μm

LMC
Field 4
12 µm

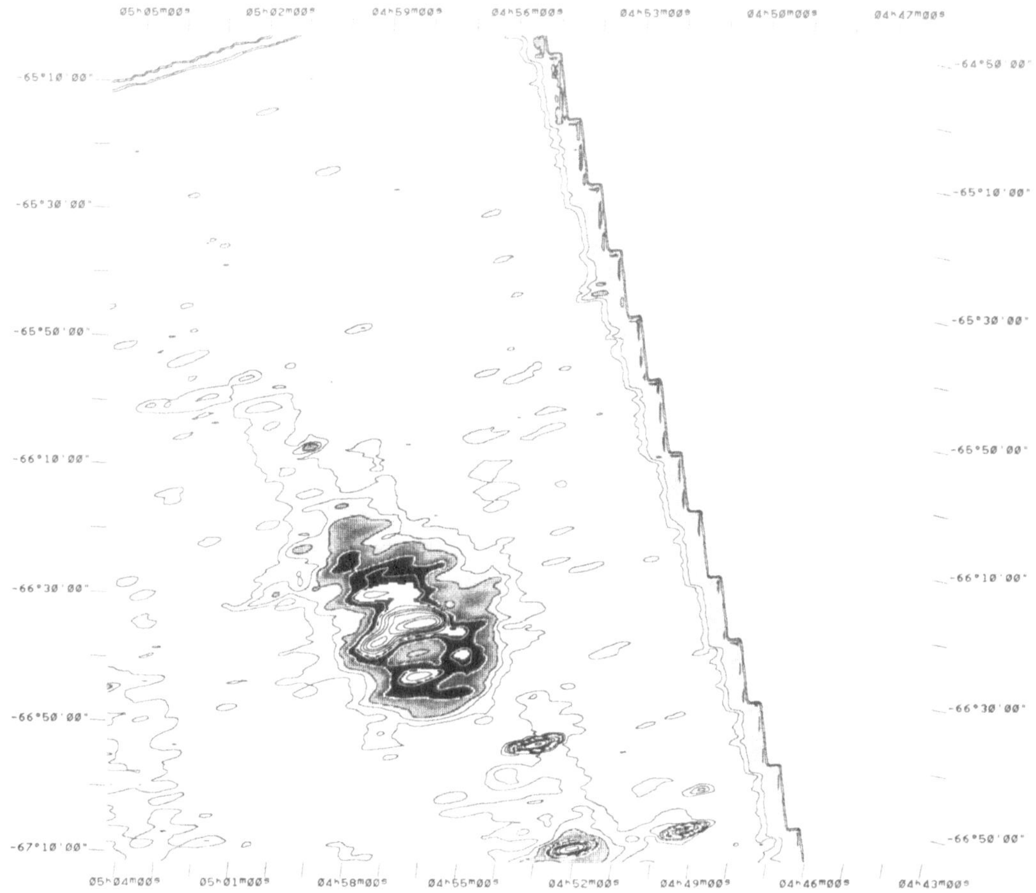

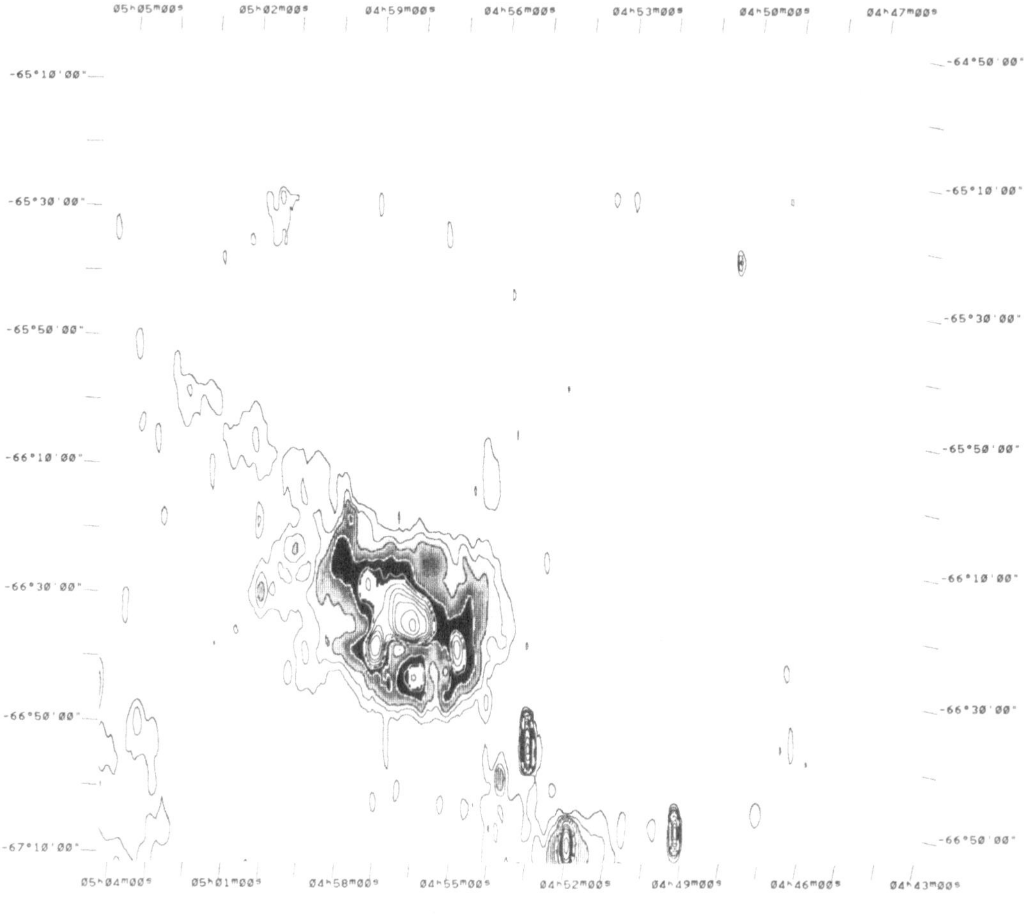

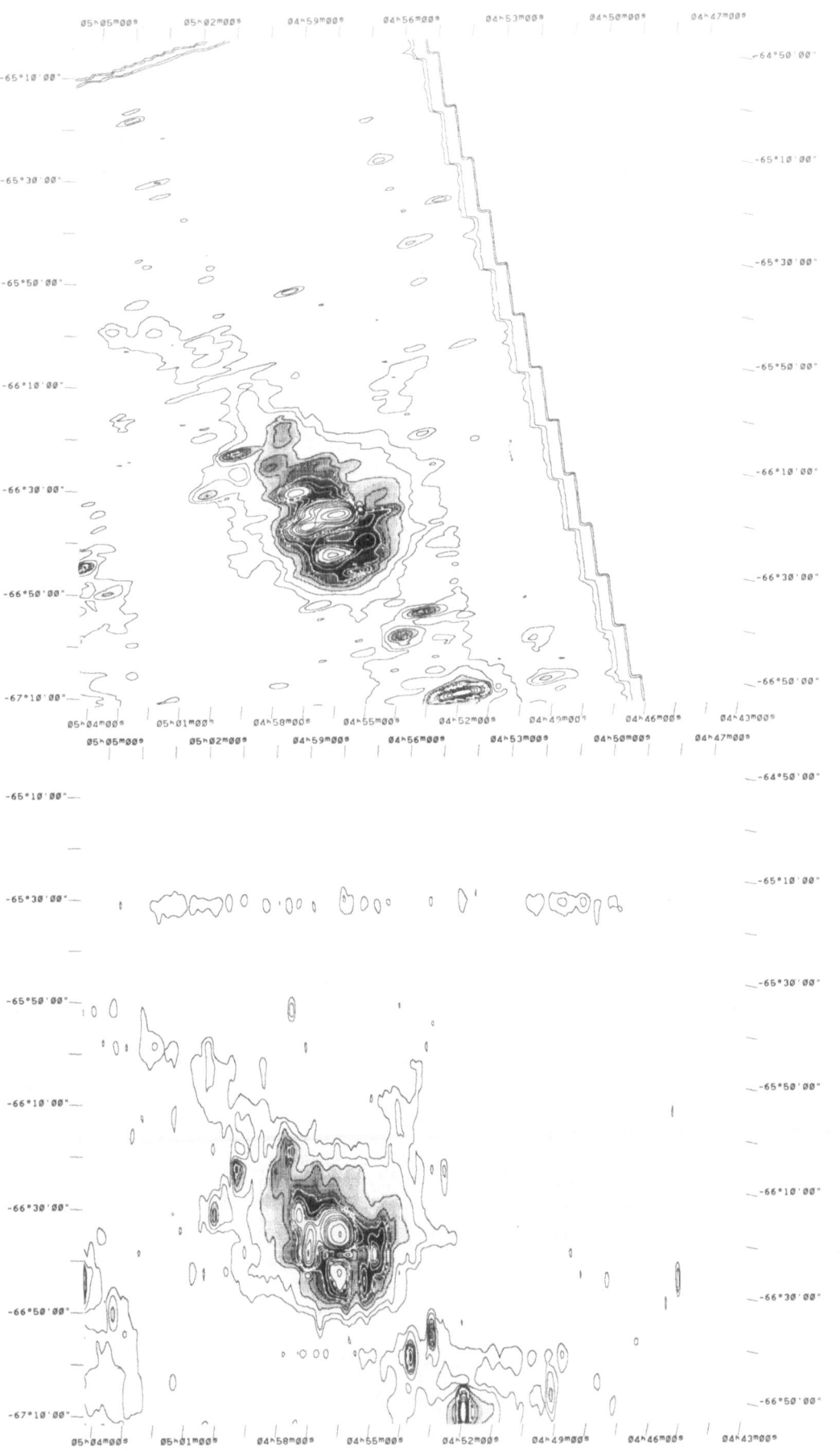

LMC
Field 4
25 μm

LMC
Field 4
60 μm

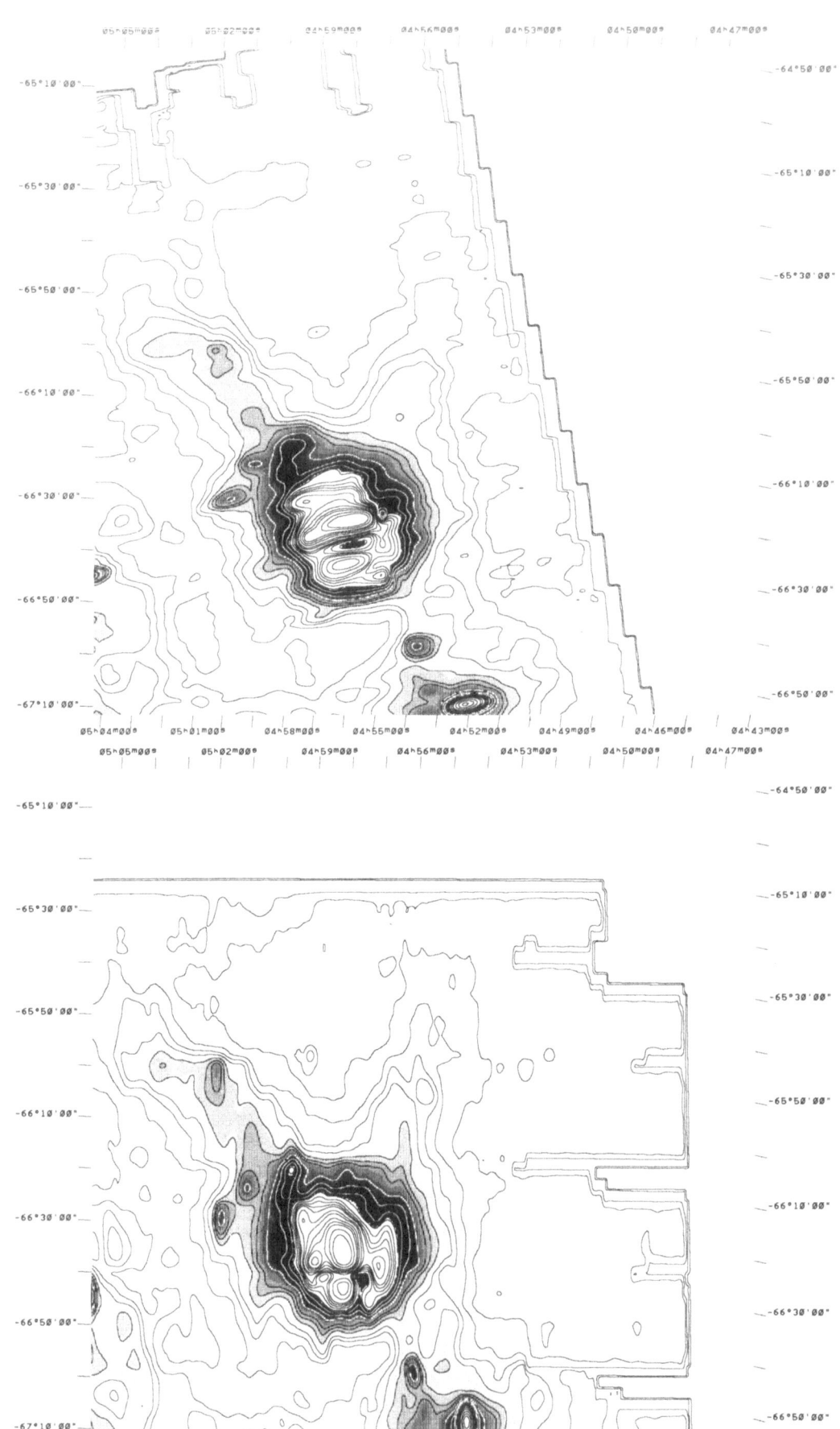

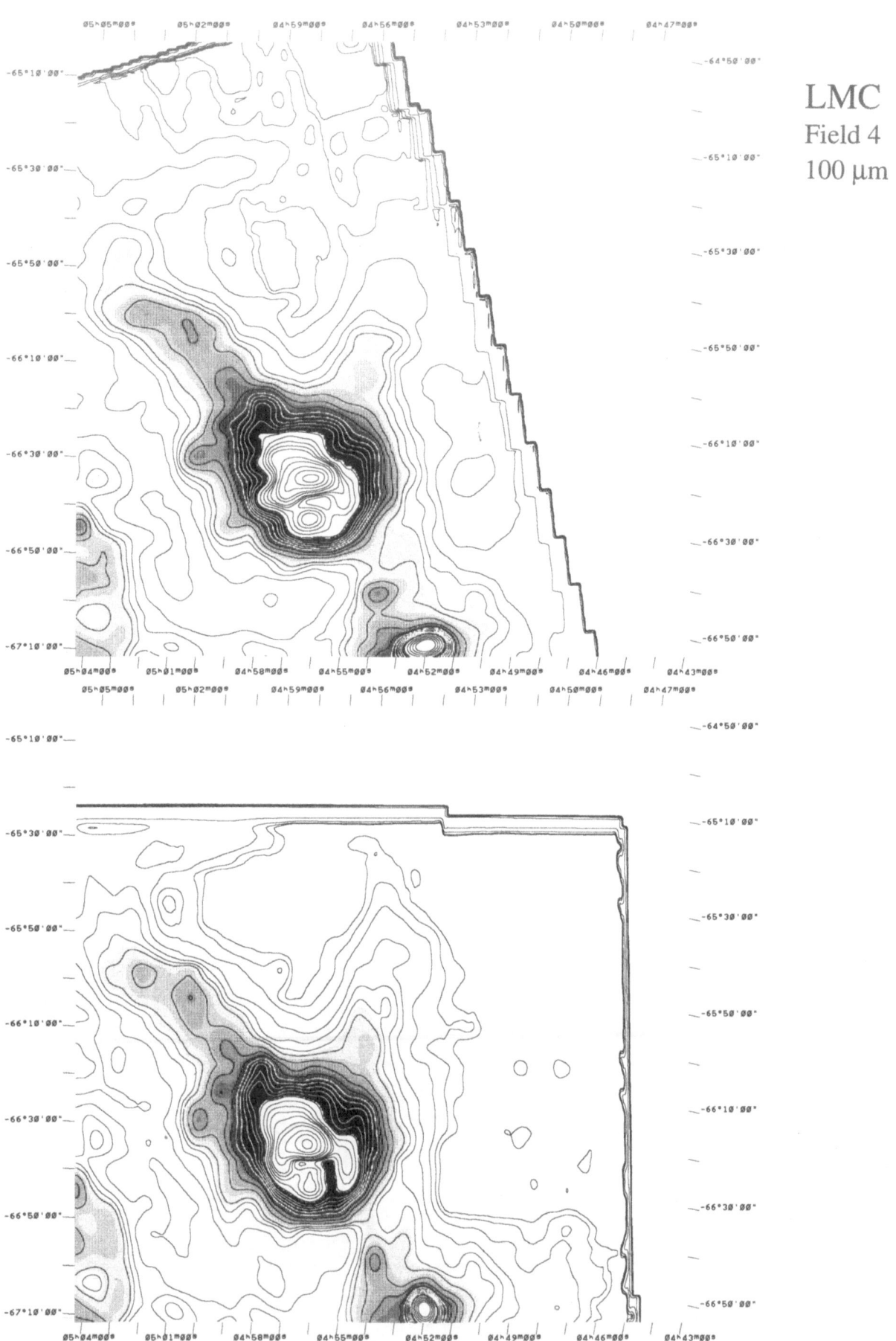

LMC
Field 4
100 µm

LMC
Field 5
12 μm

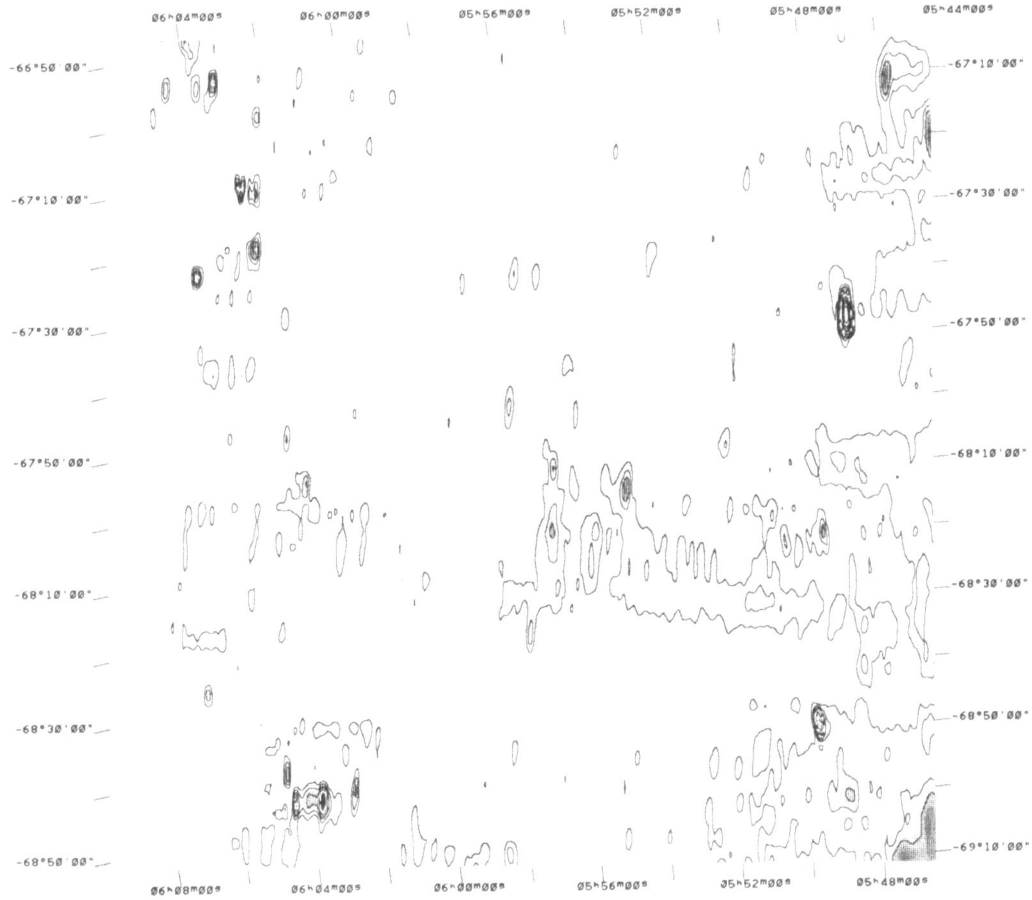

LMC
Field 5
25 µm

LMC
Field 5
60 µm

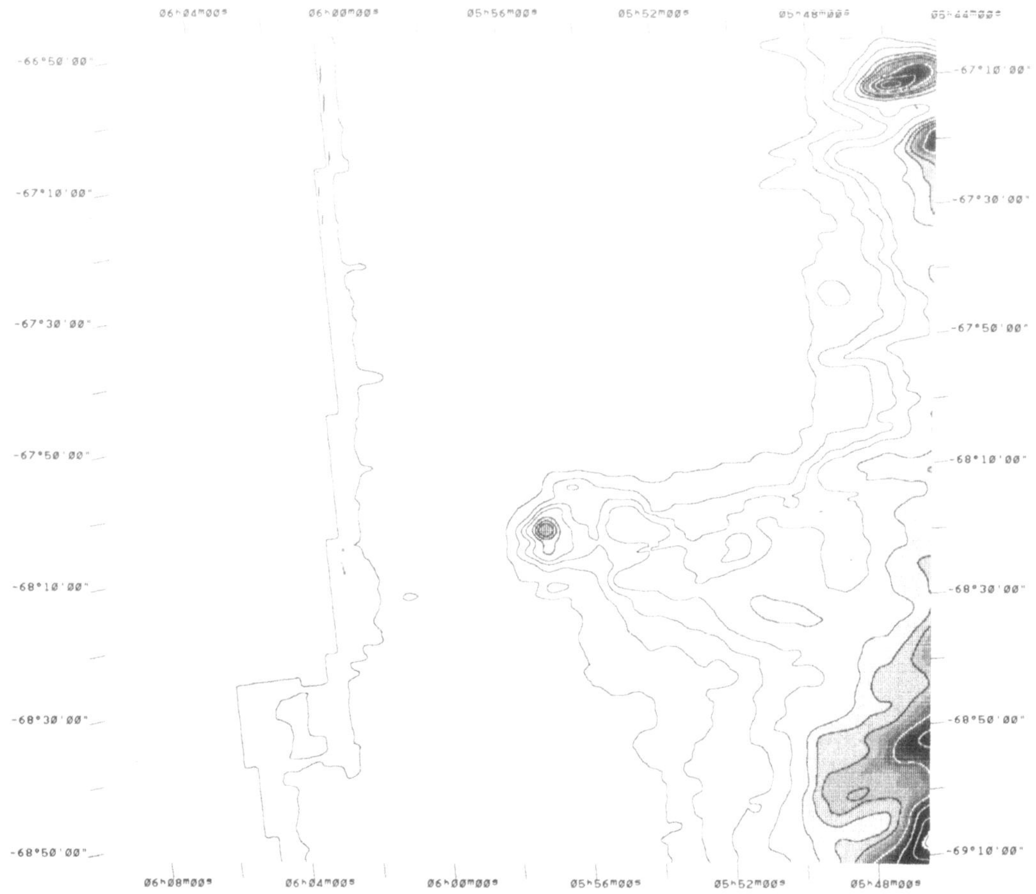

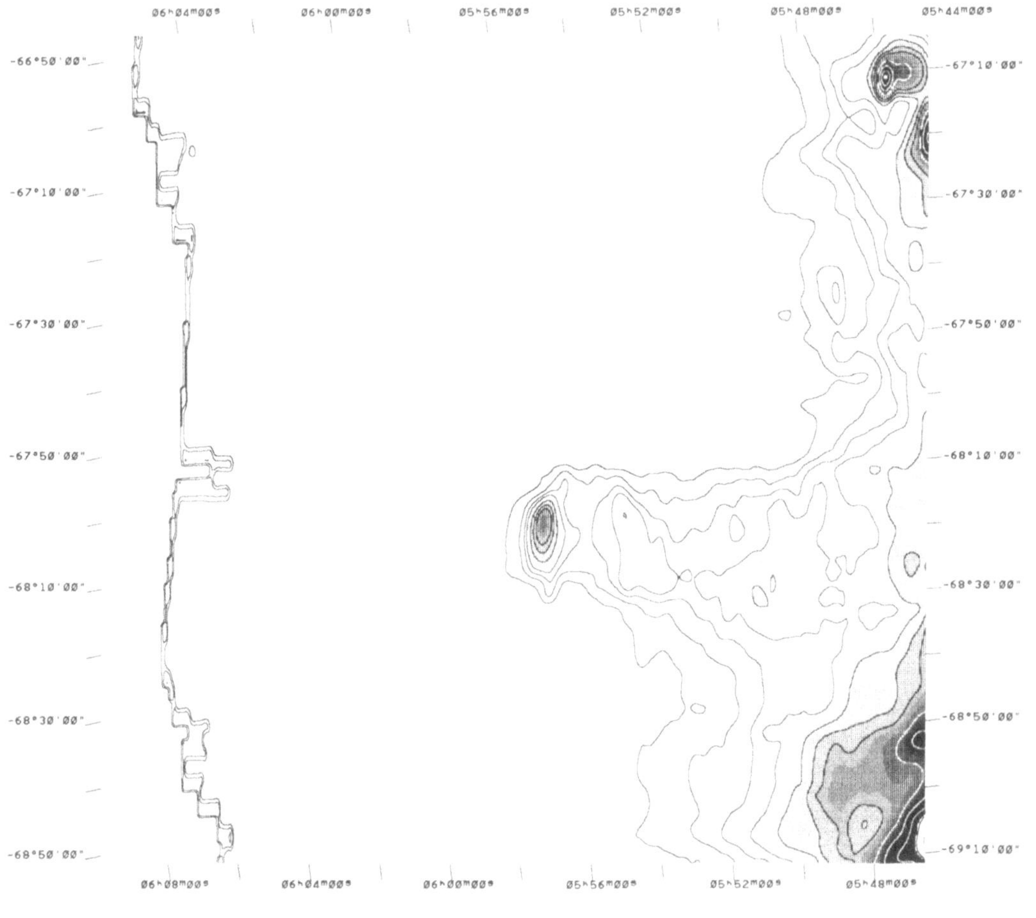

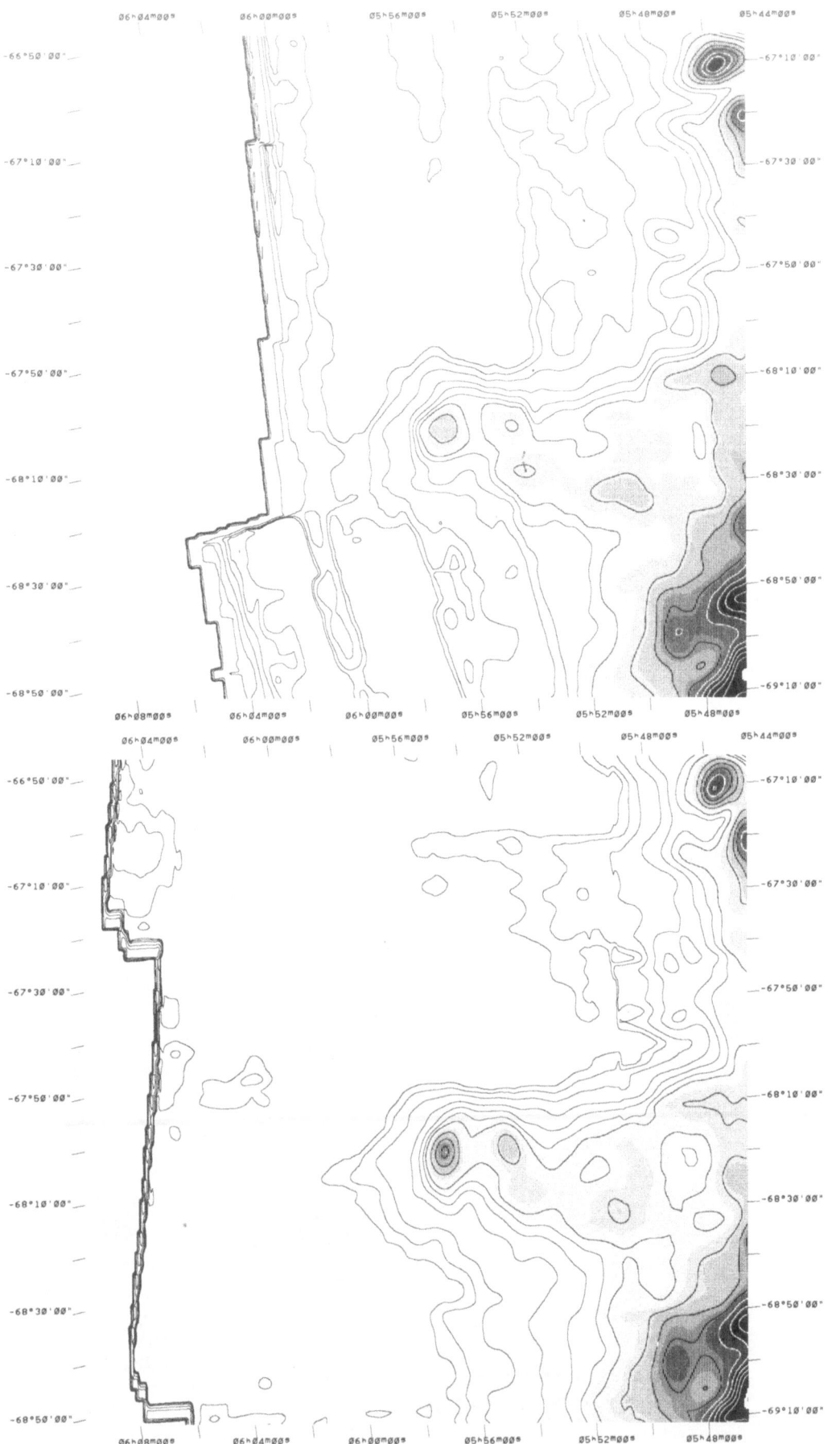

LMC
Field 5
100 μm

LMC
Field 6
12 μm

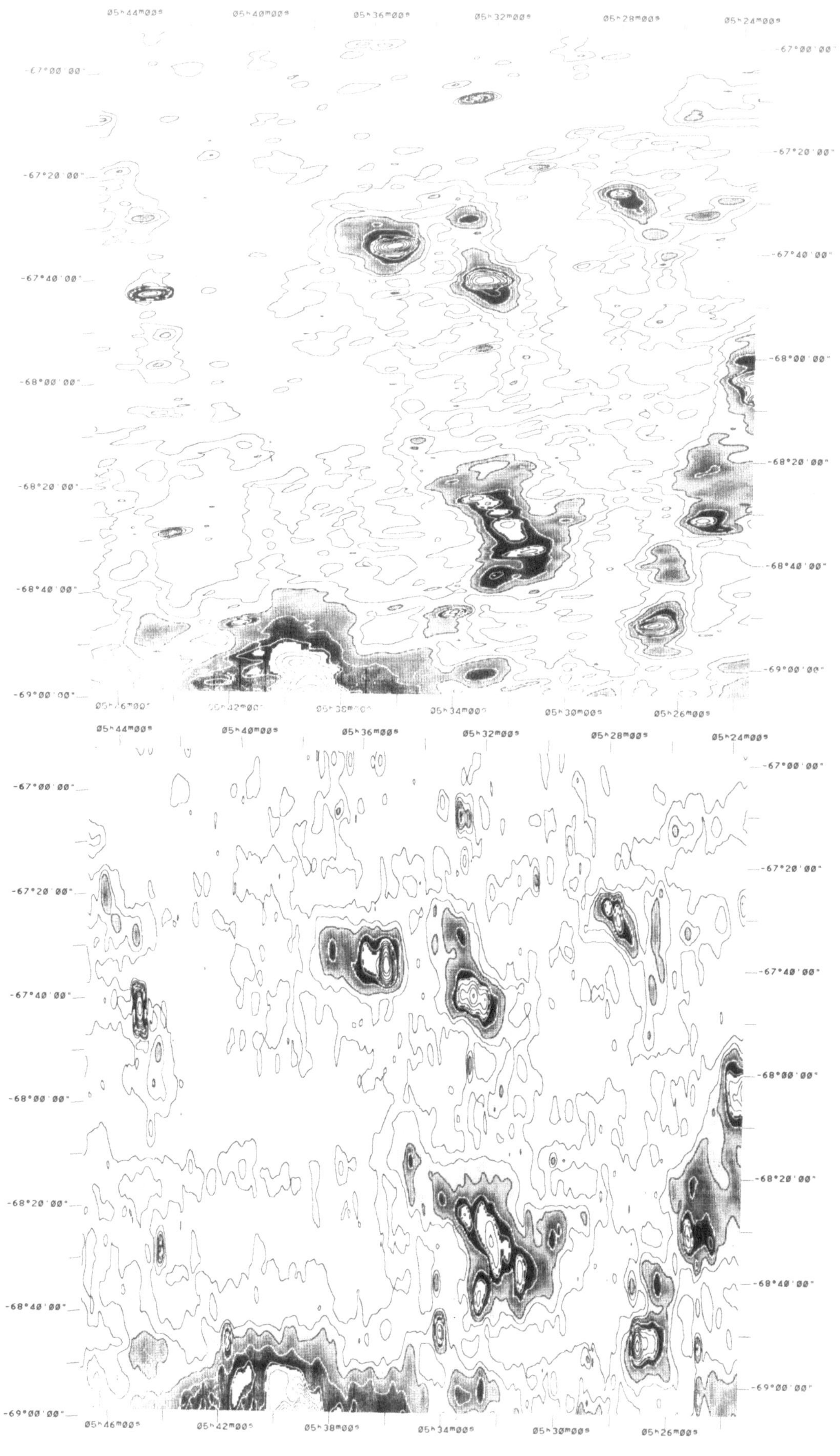

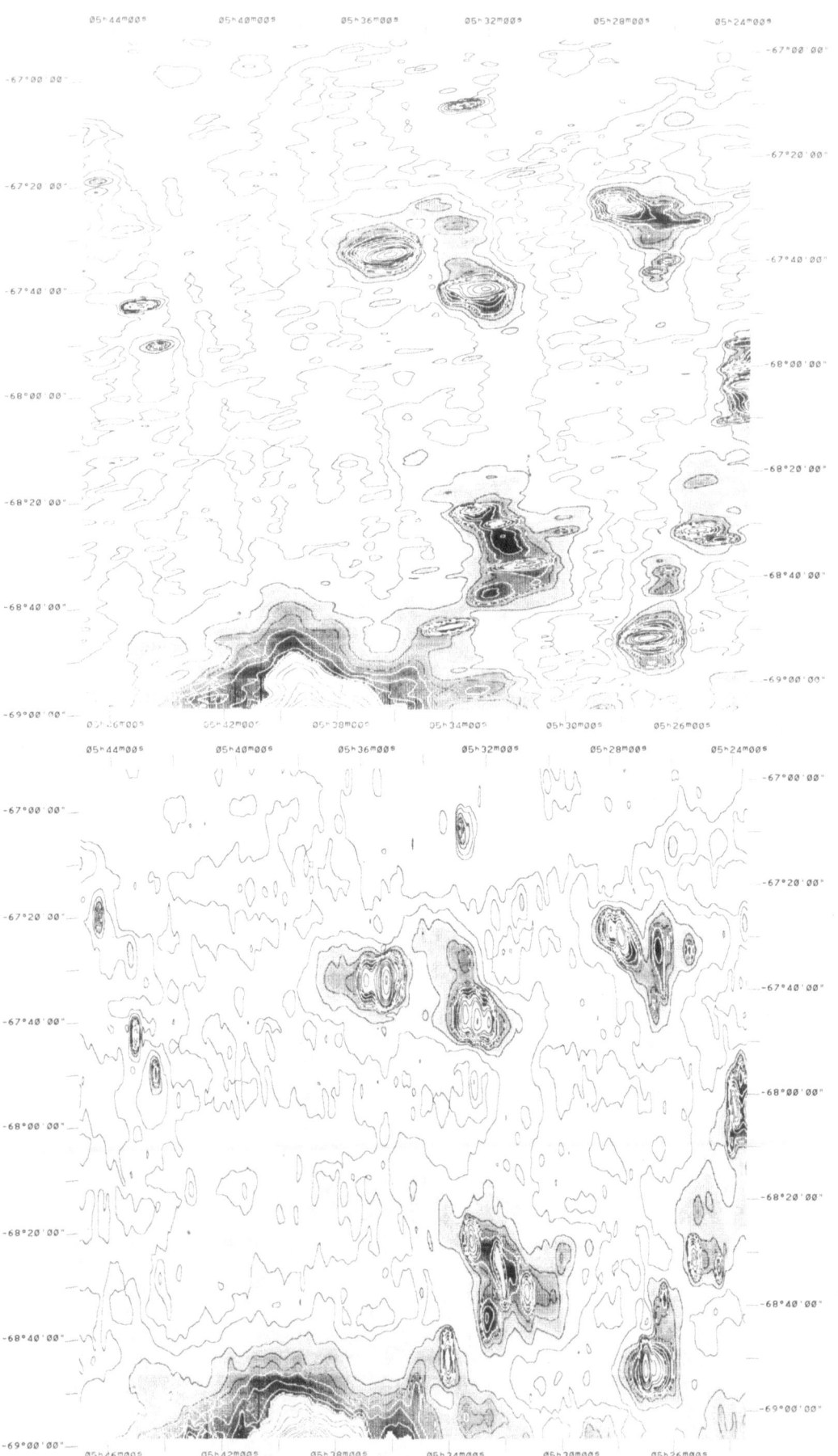

LMC
Field 6
25 μm

LMC
Field 6
60 μm

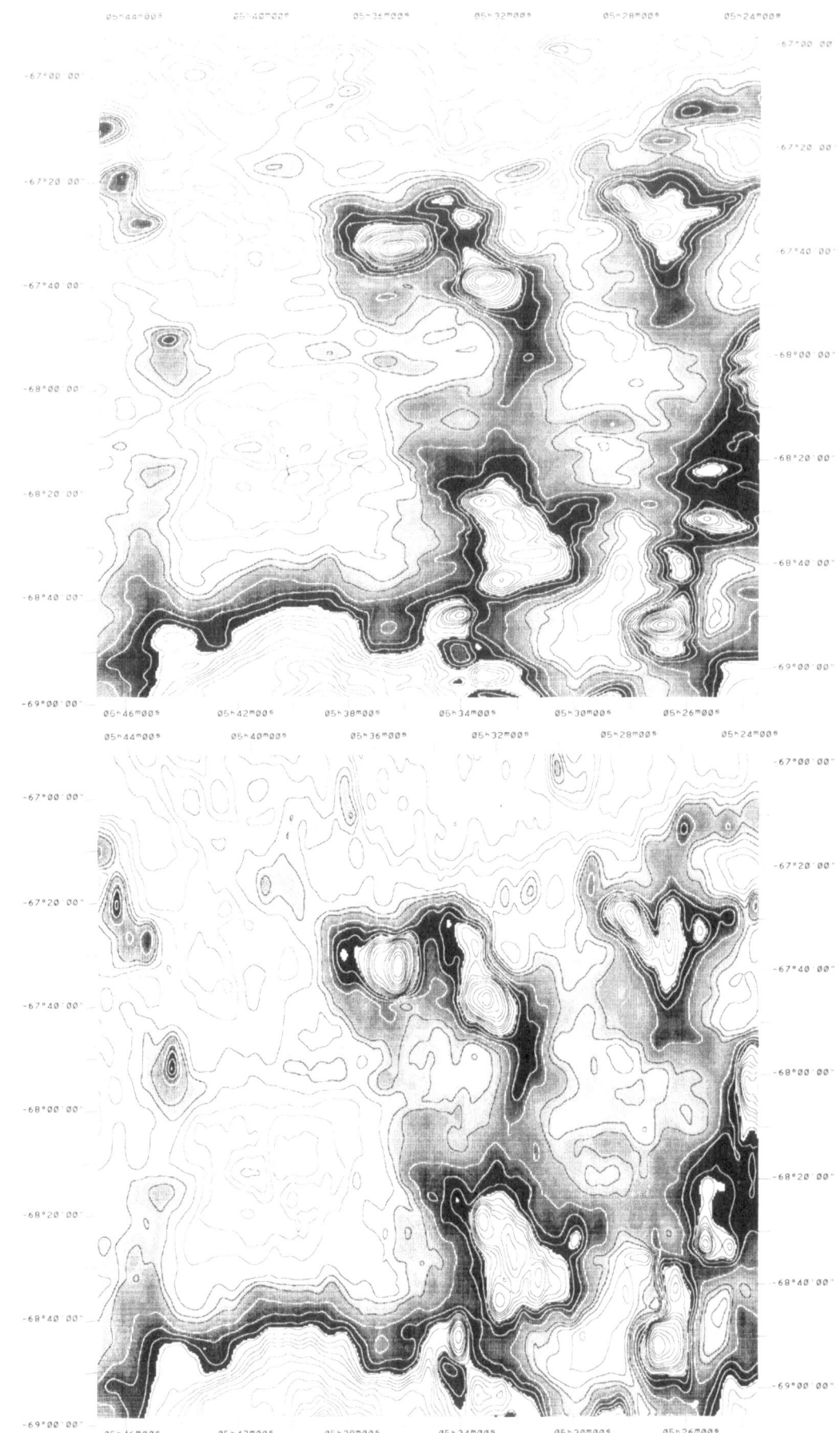

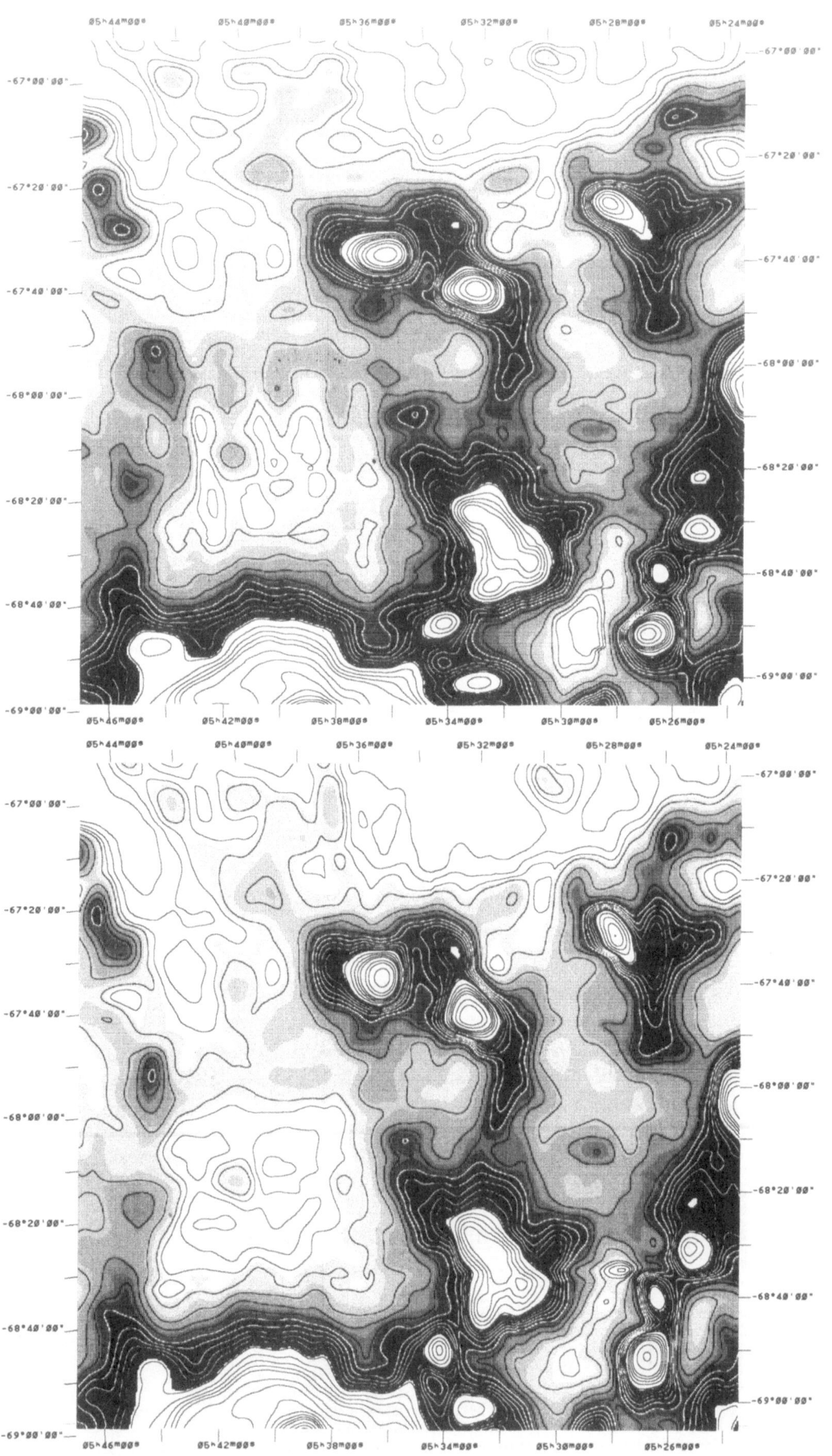

LMC
Field 6
100 µm

LMC
Field 7
12 μm

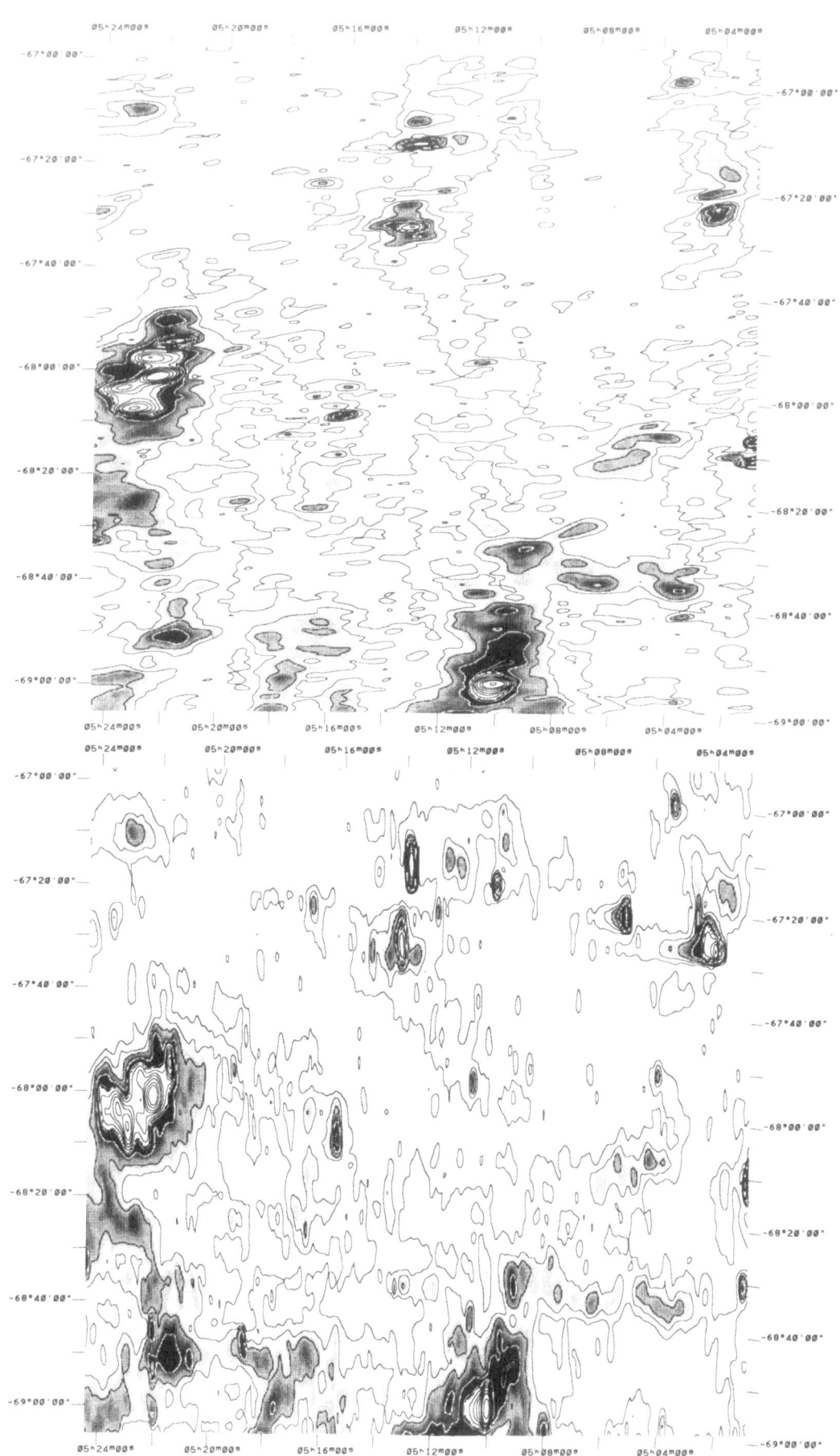

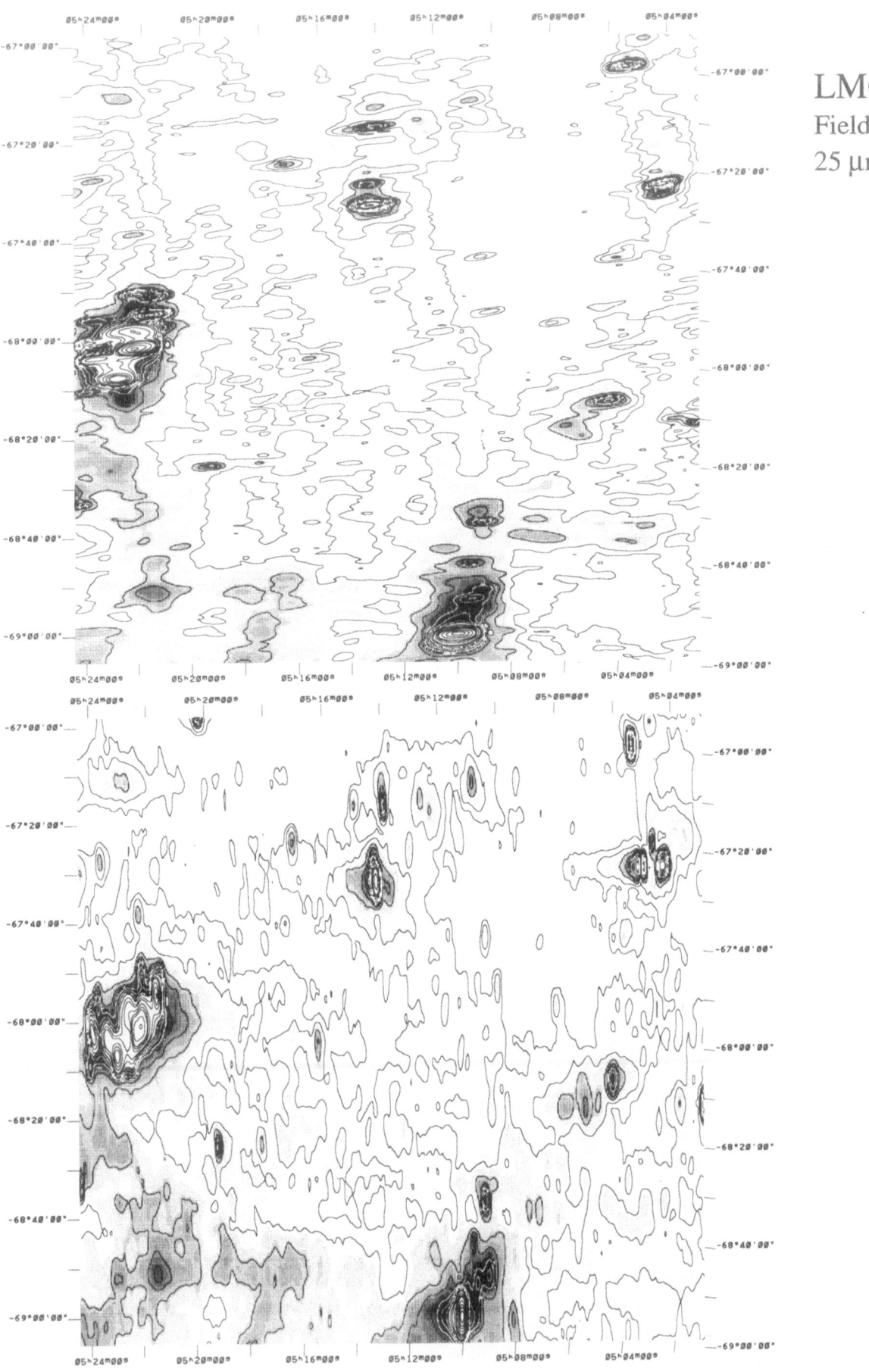

LMC
Field 7
25 μm

LMC
Field 7
60 μm

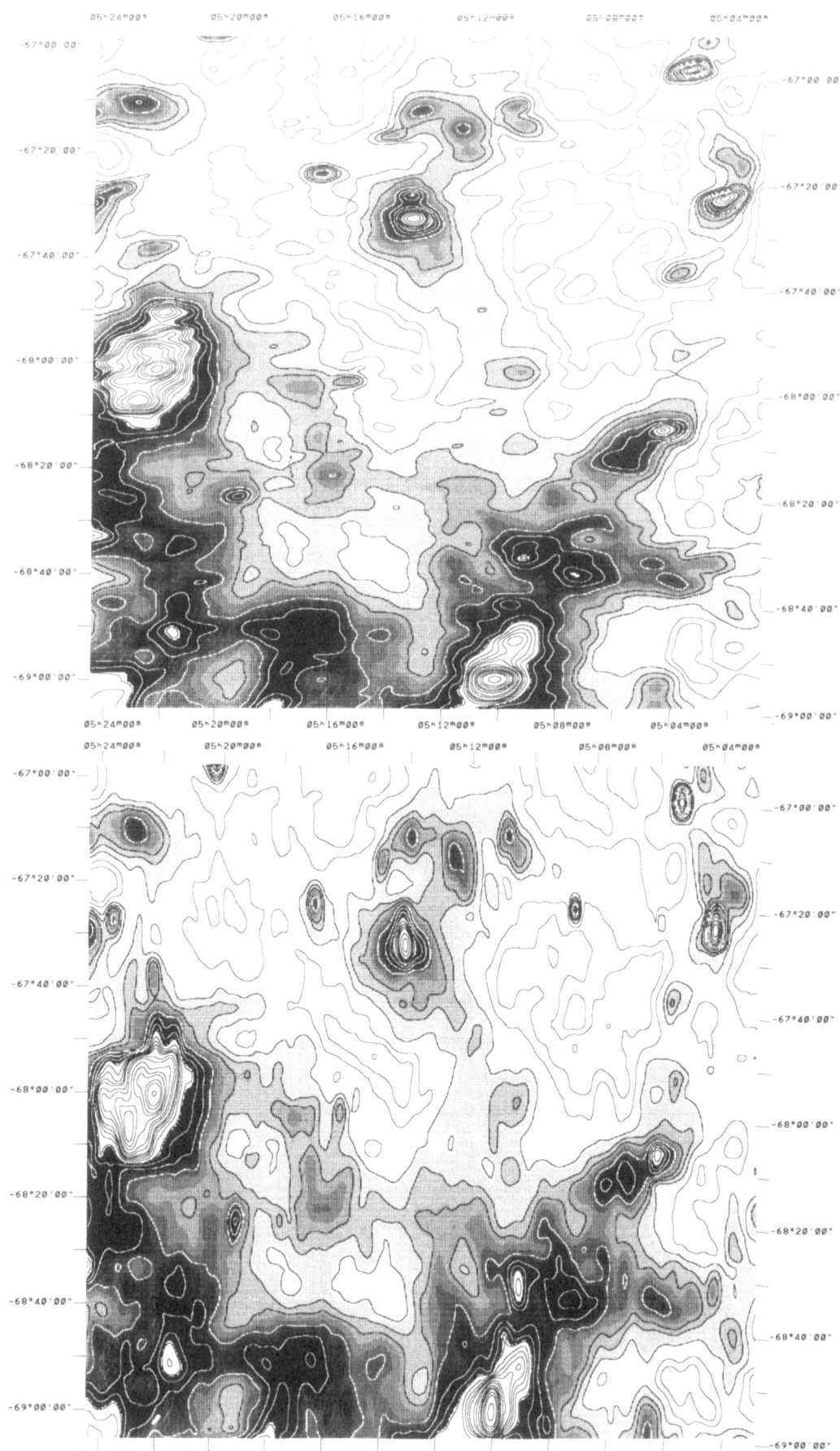

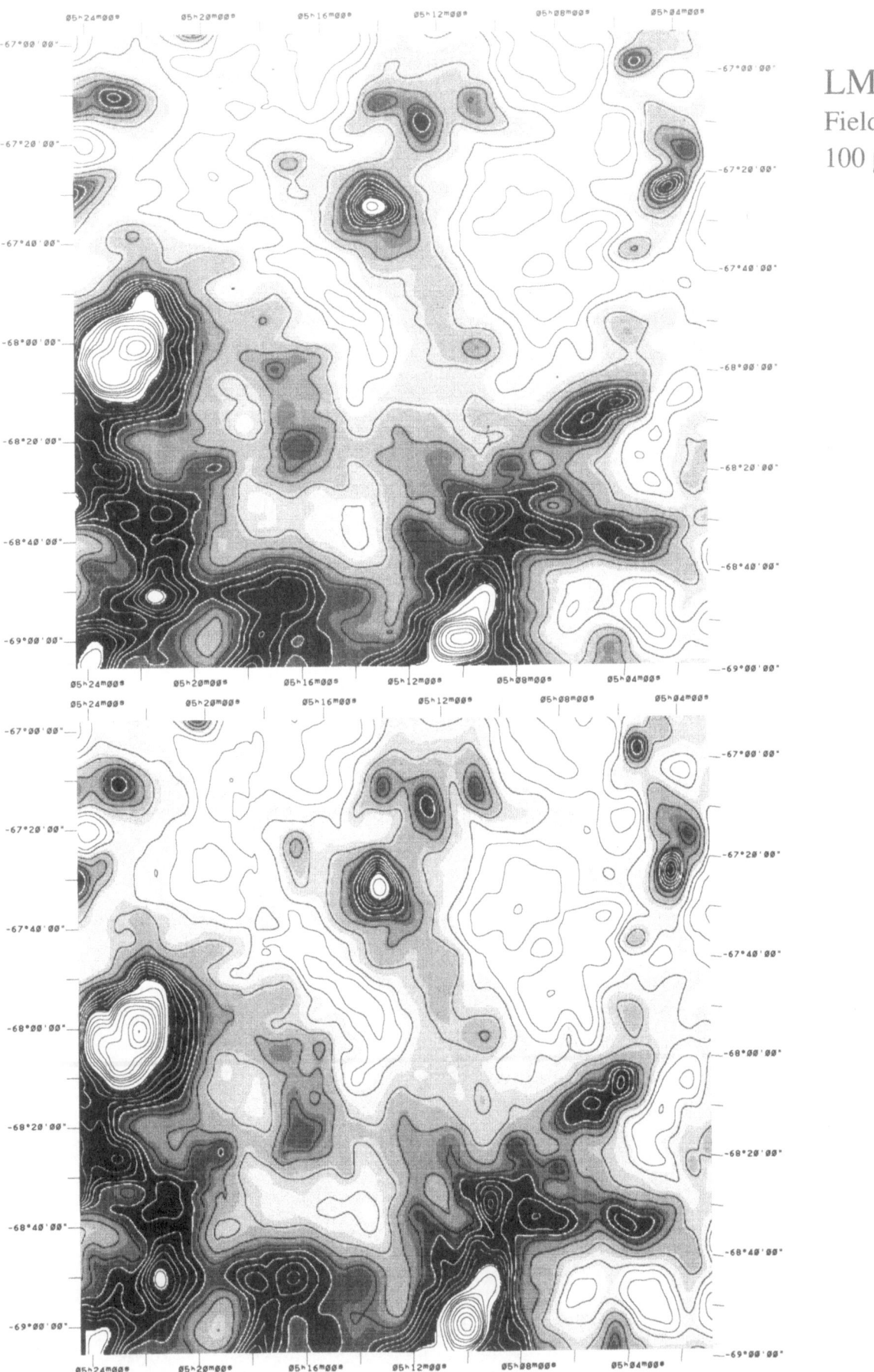

LMC
Field 7
100 μm

LMC
Field 8
12 μm

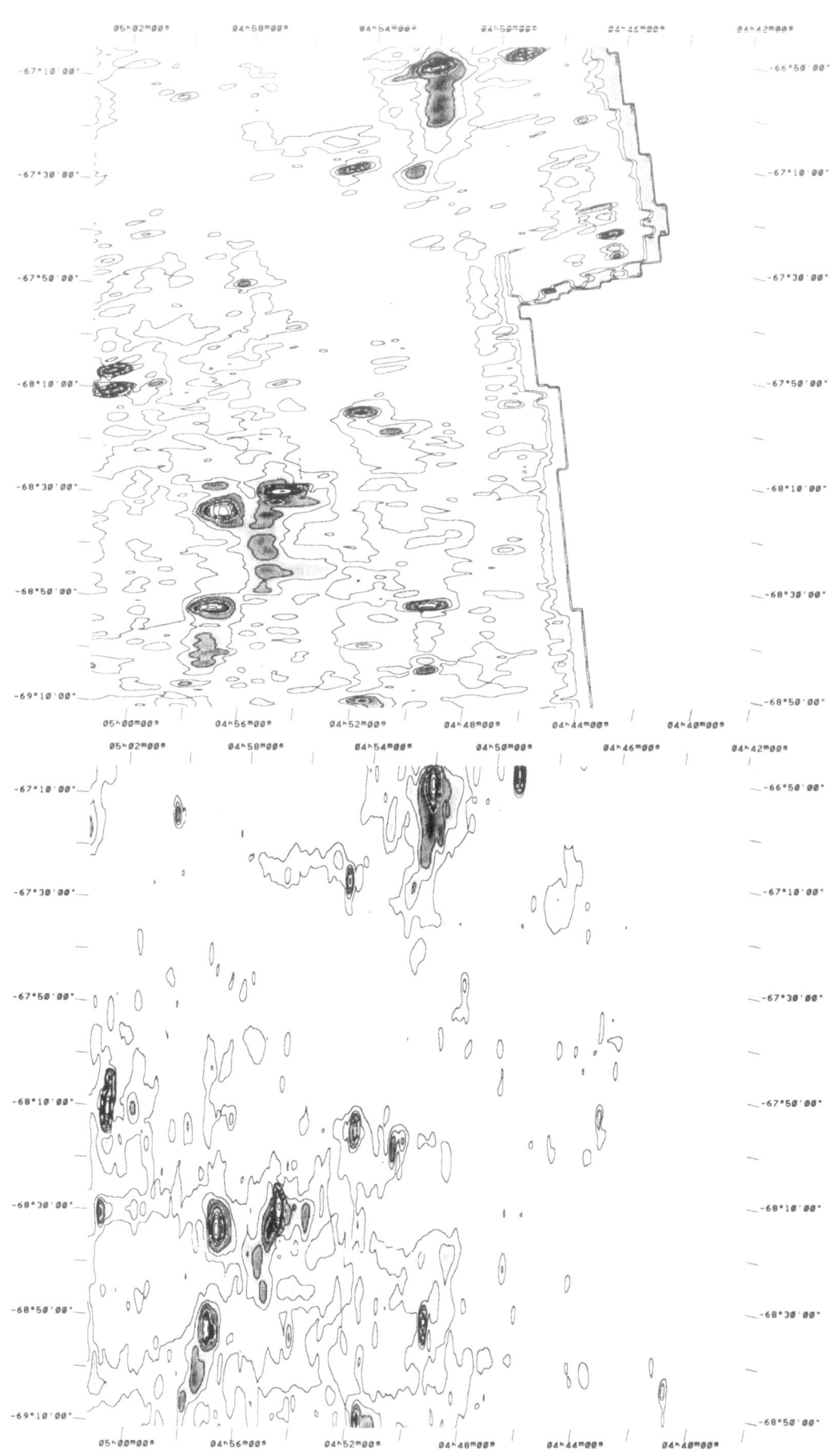

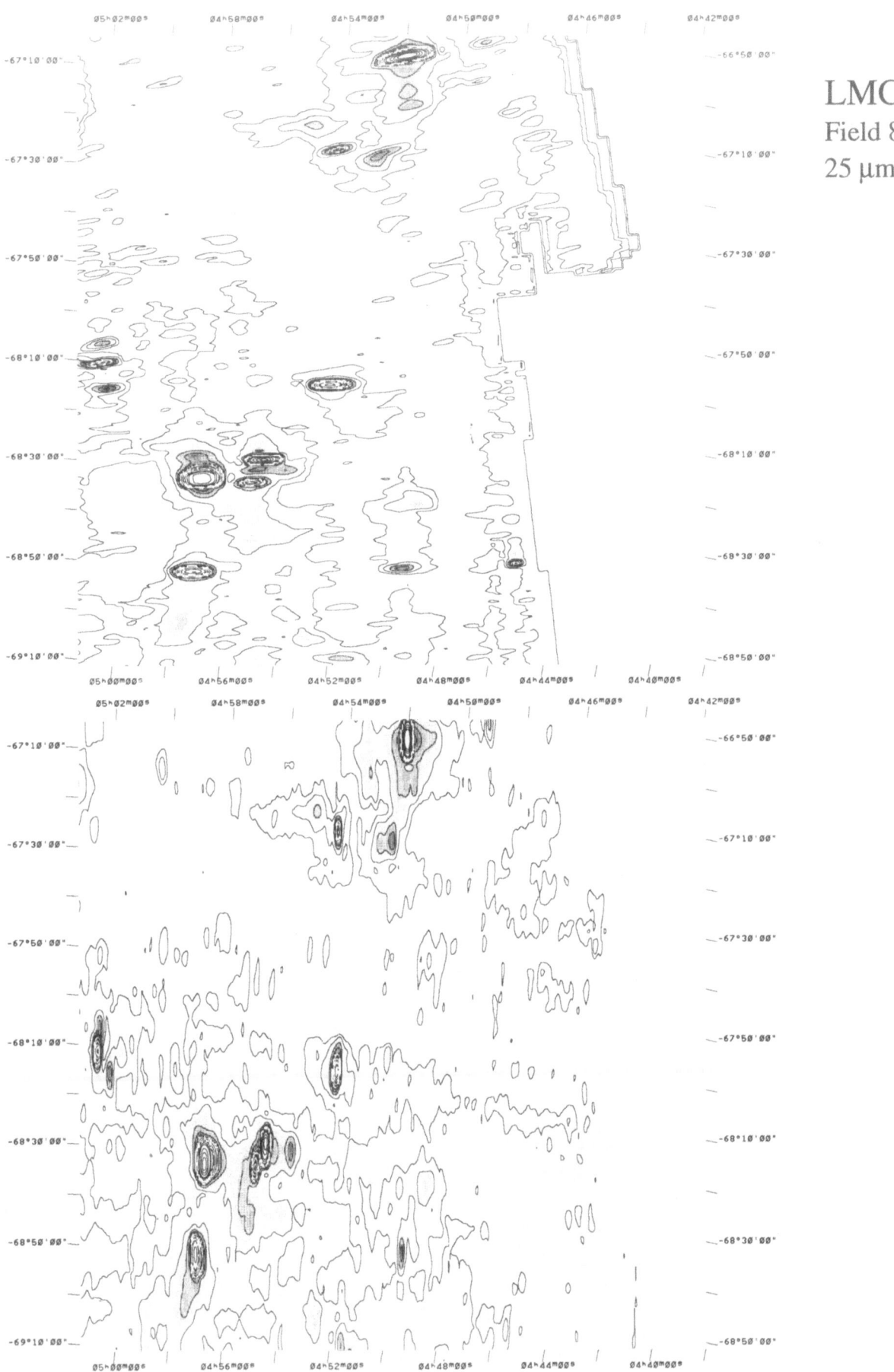

LMC
Field 8
25 μm

LMC
Field 8
60 µm

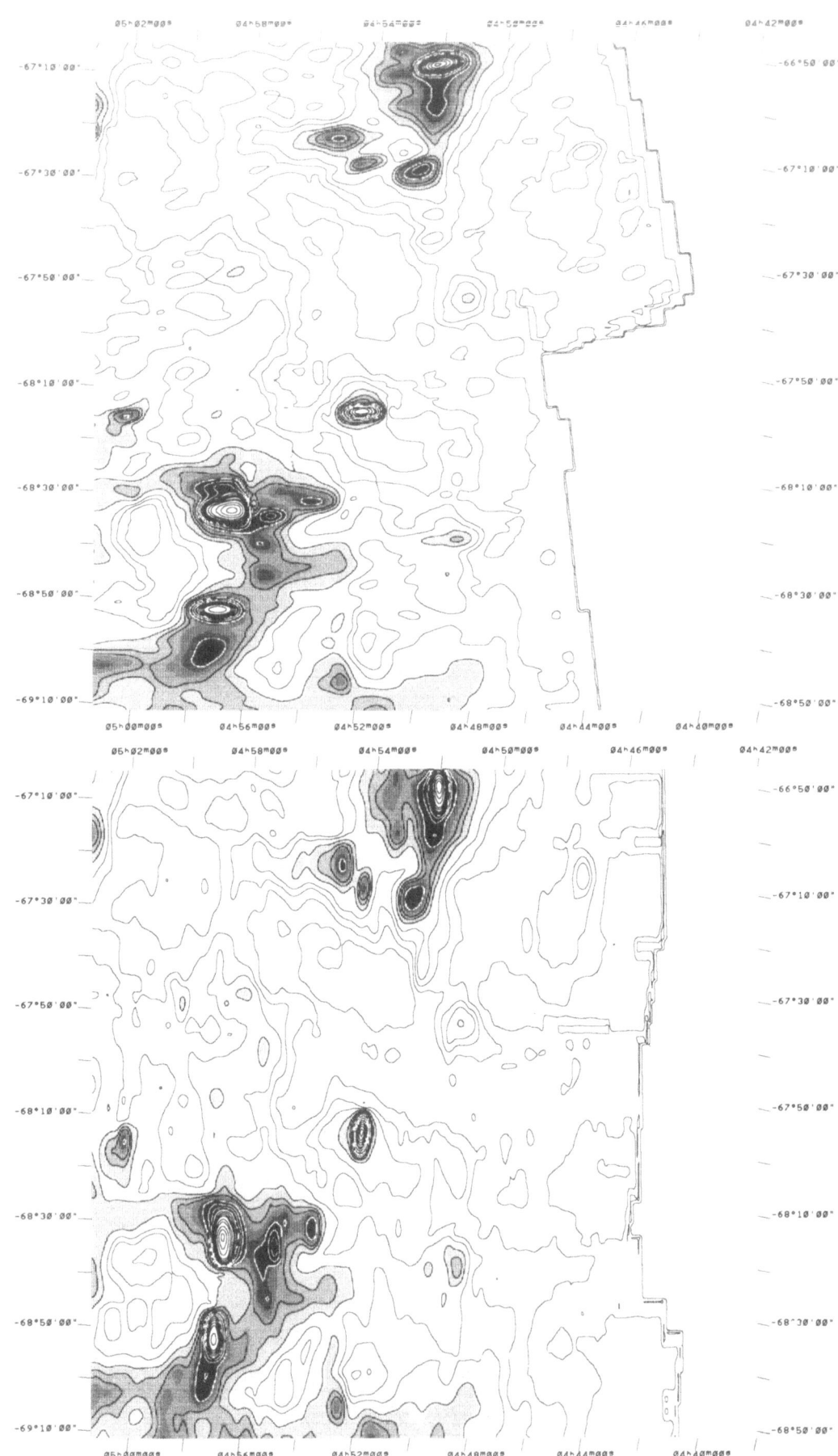

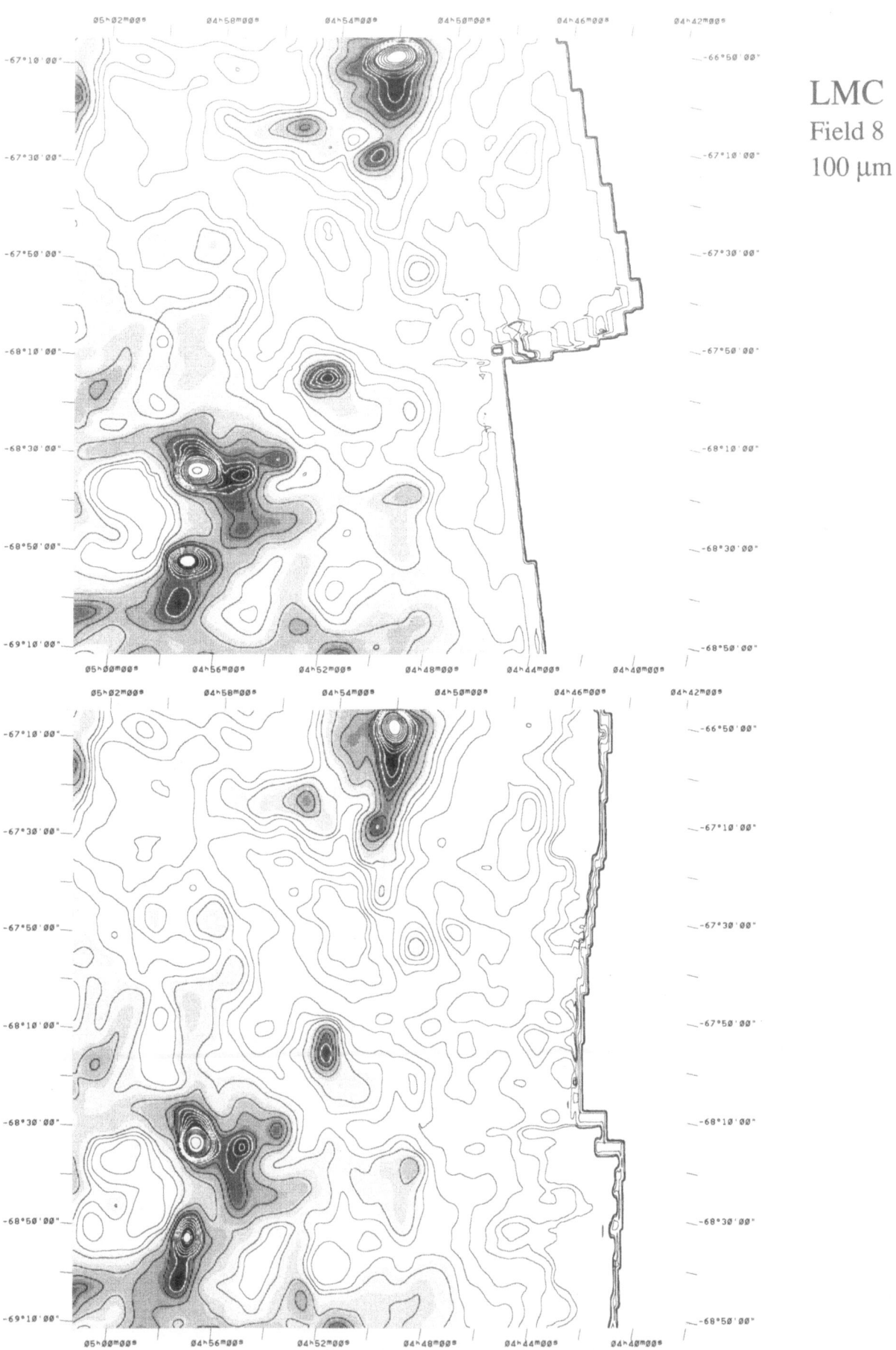

LMC
Field 9
12 μm

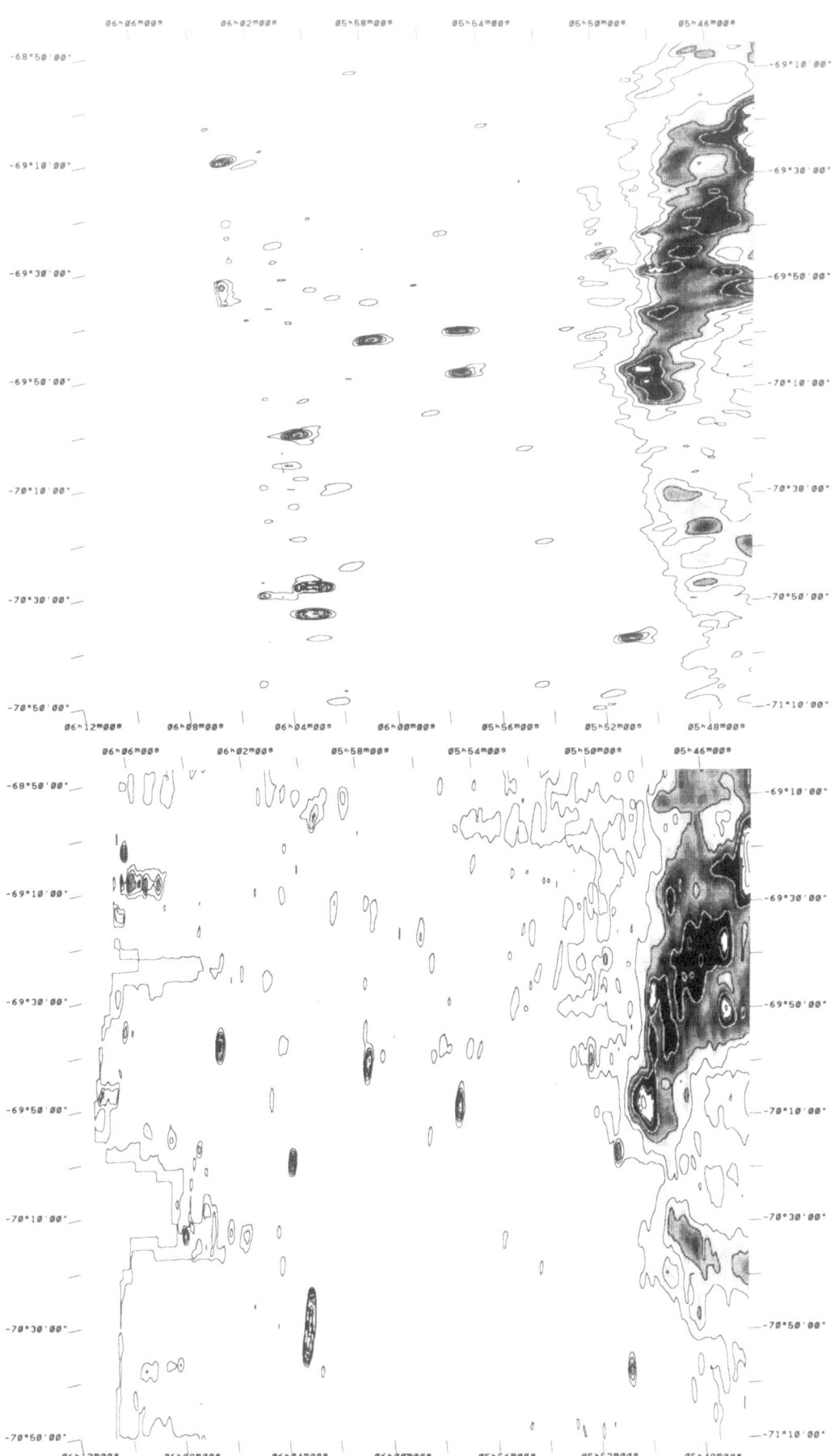

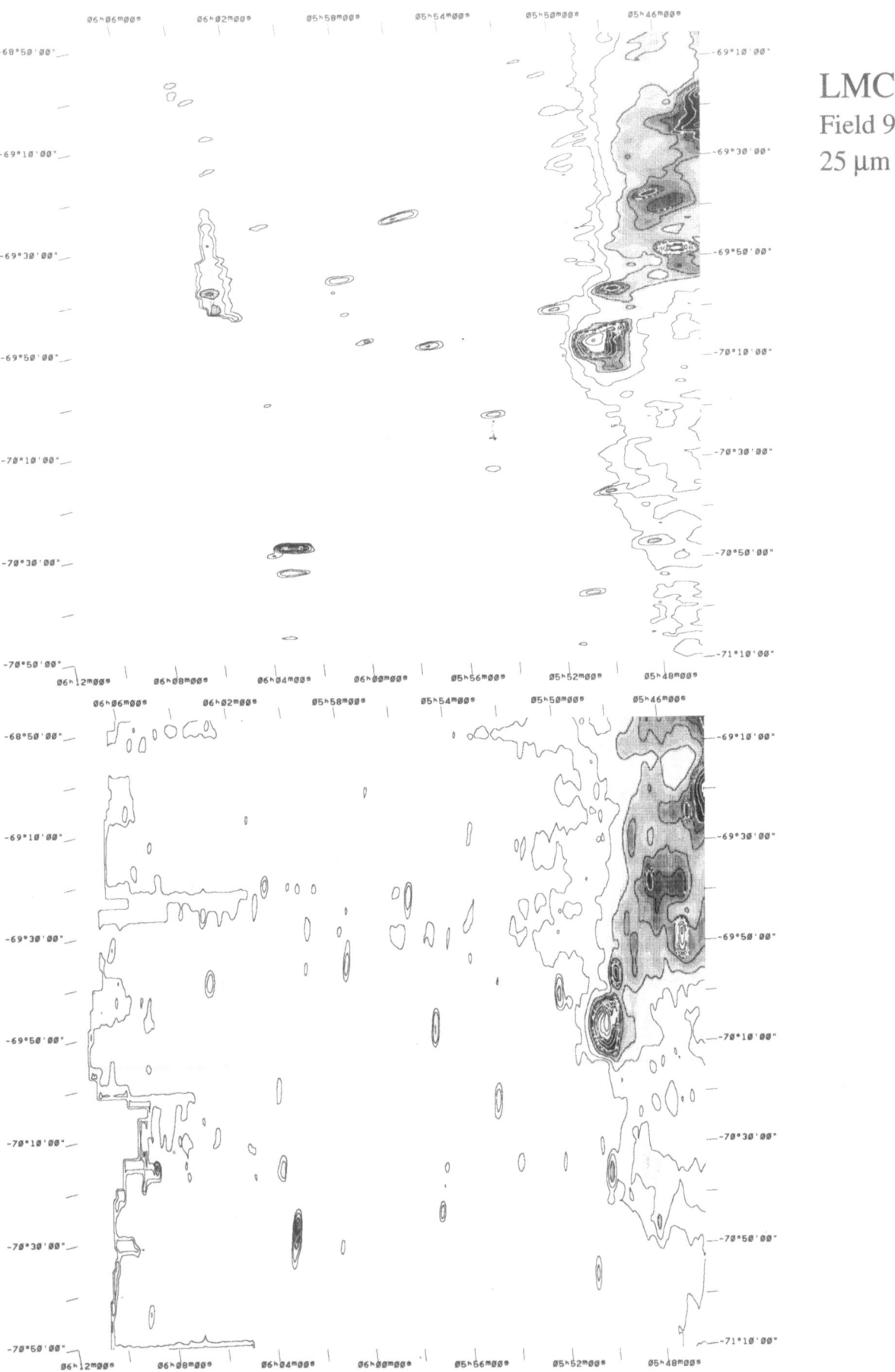

LMC
Field 9
25 μm

LMC
Field 9
60 μm

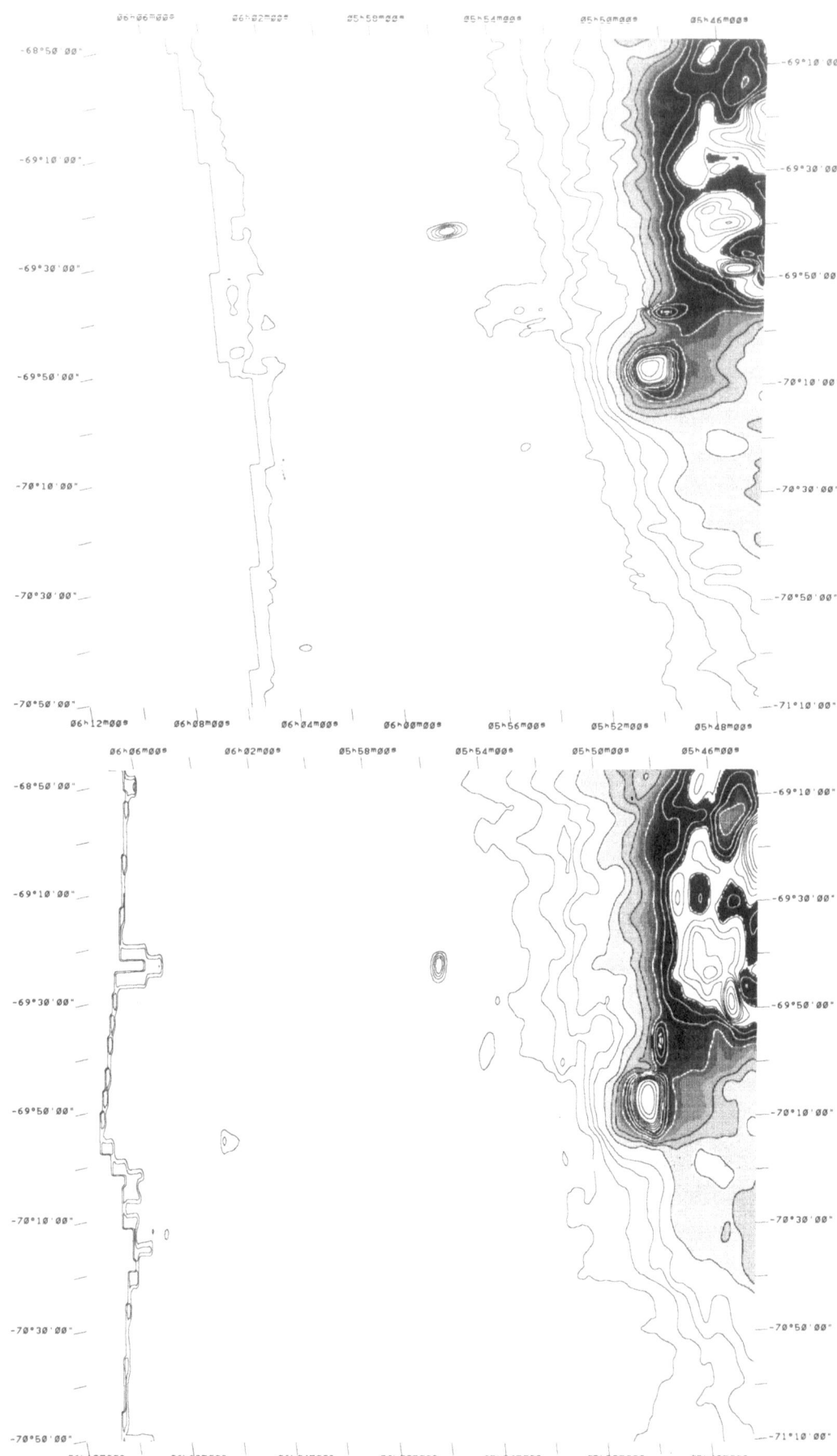

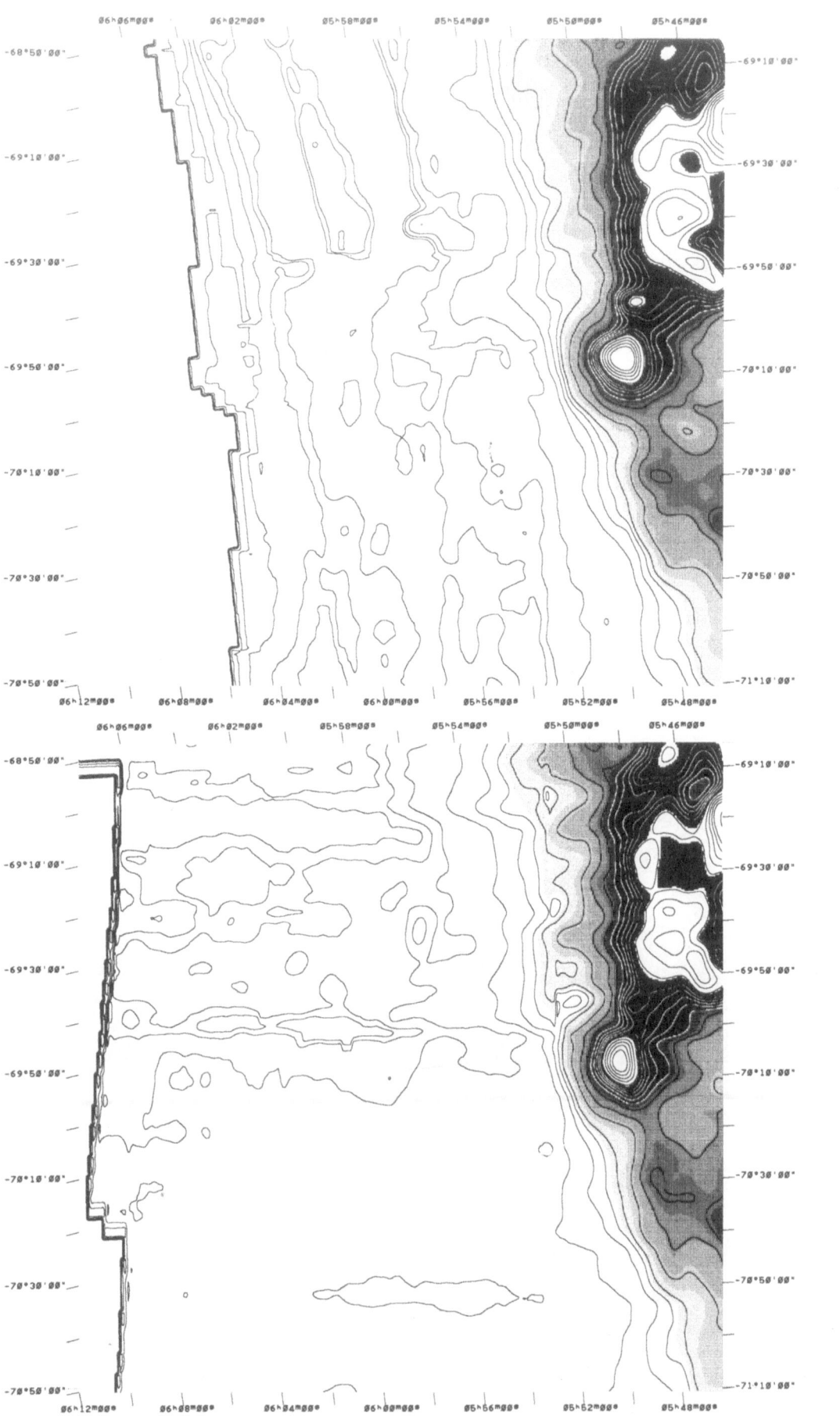

LMC
Field 9
100 μm

LMC
Field 10
12 μm

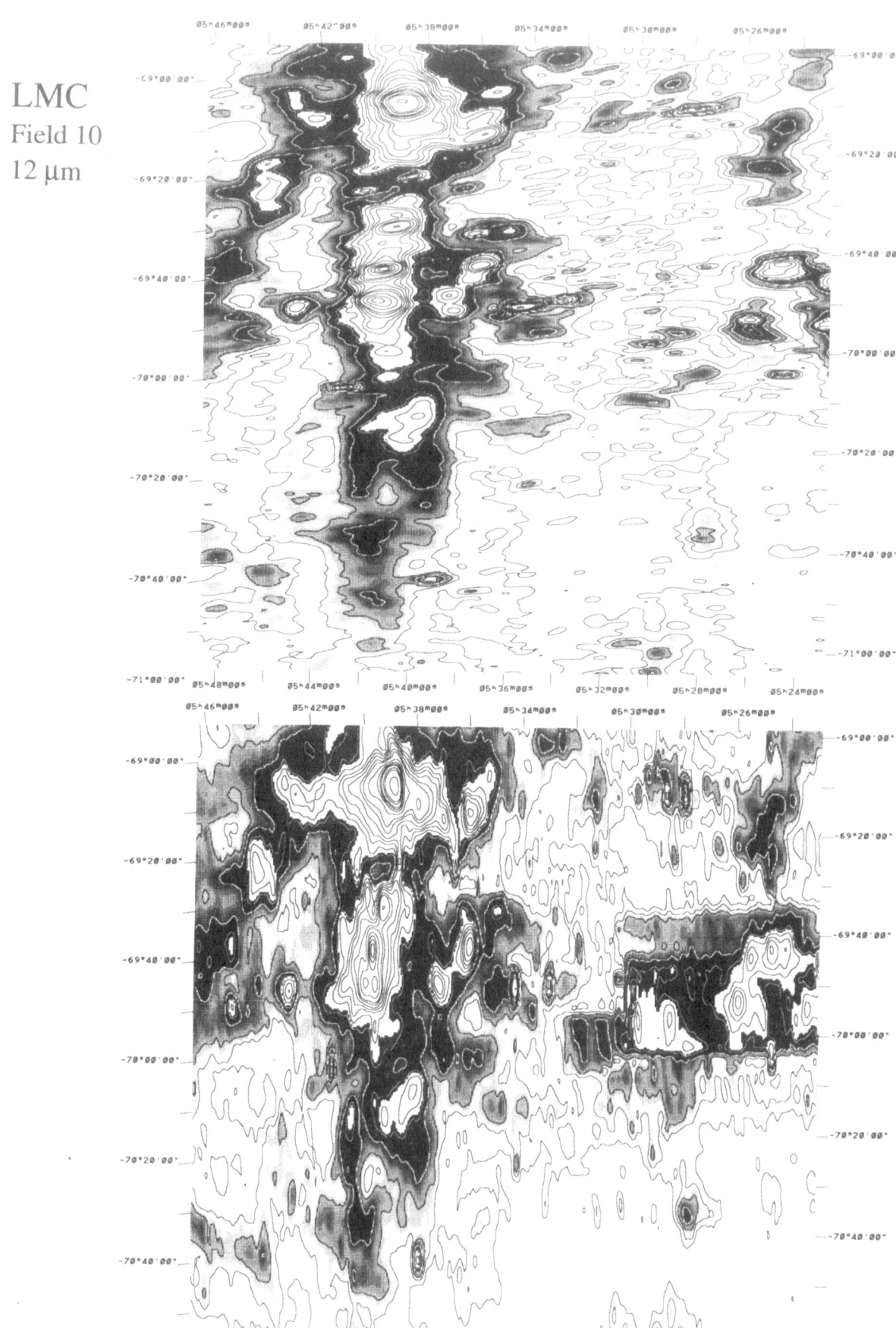

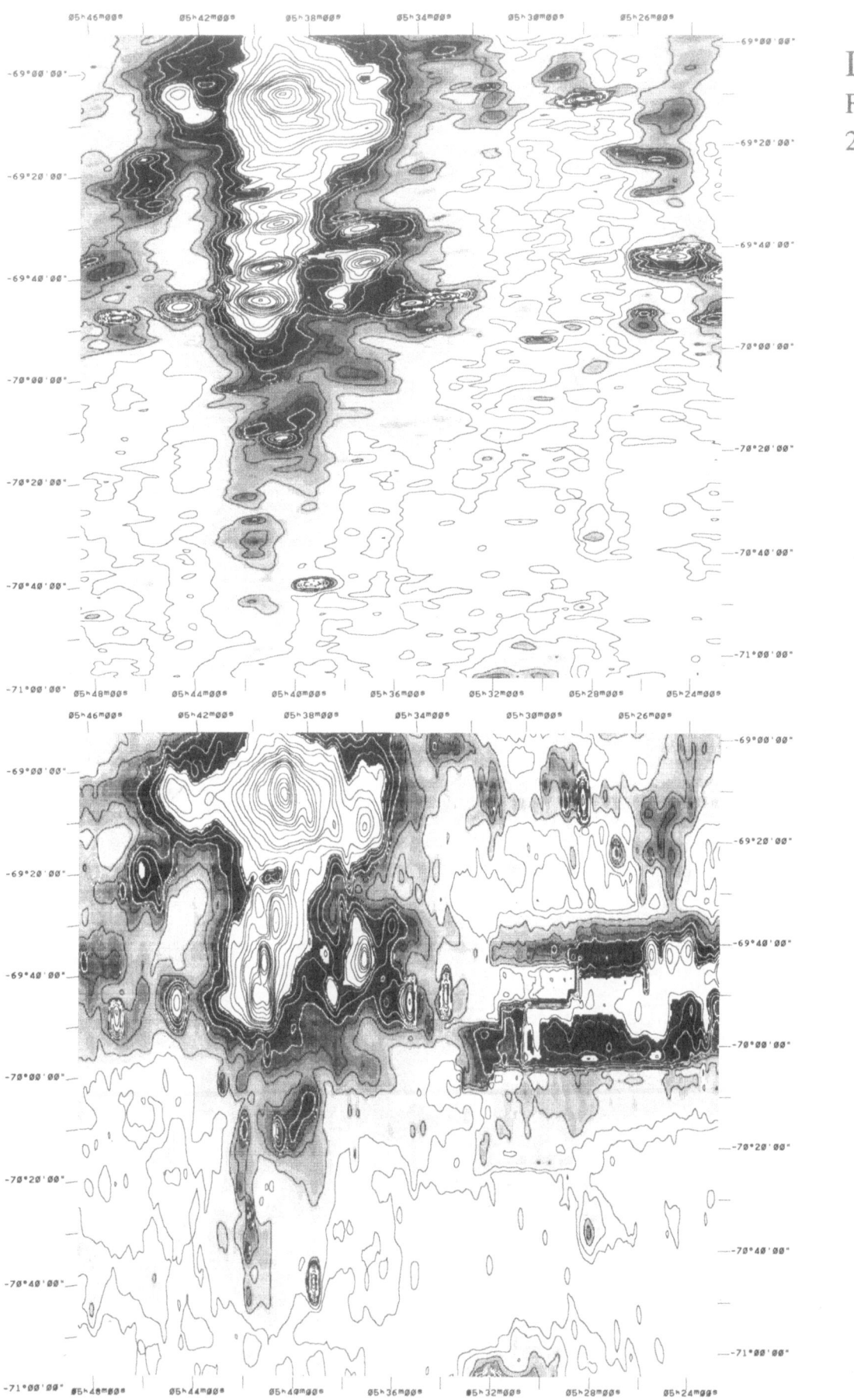

LMC
Field 10
25 µm

LMC
Field 10
60 μm

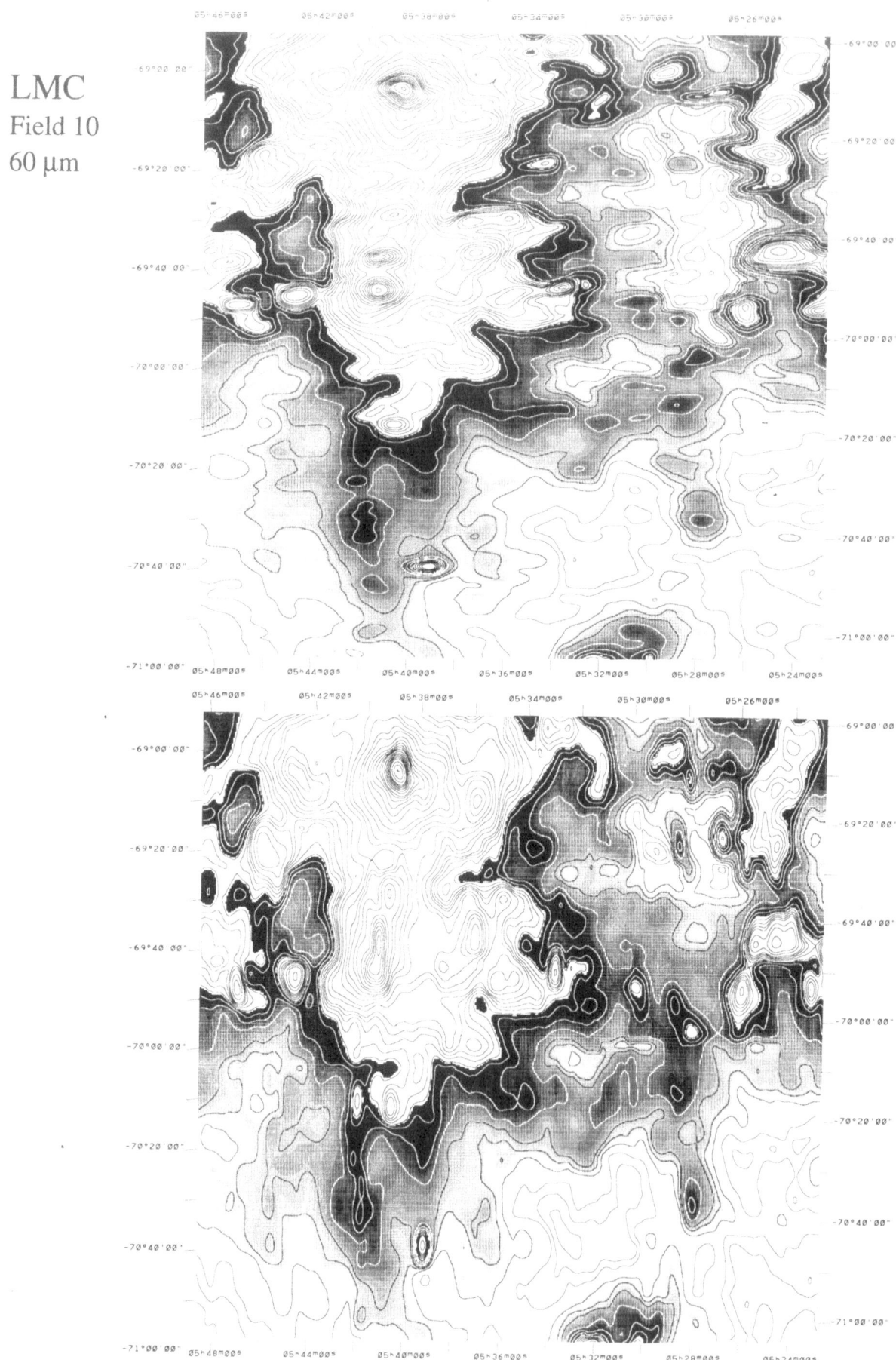

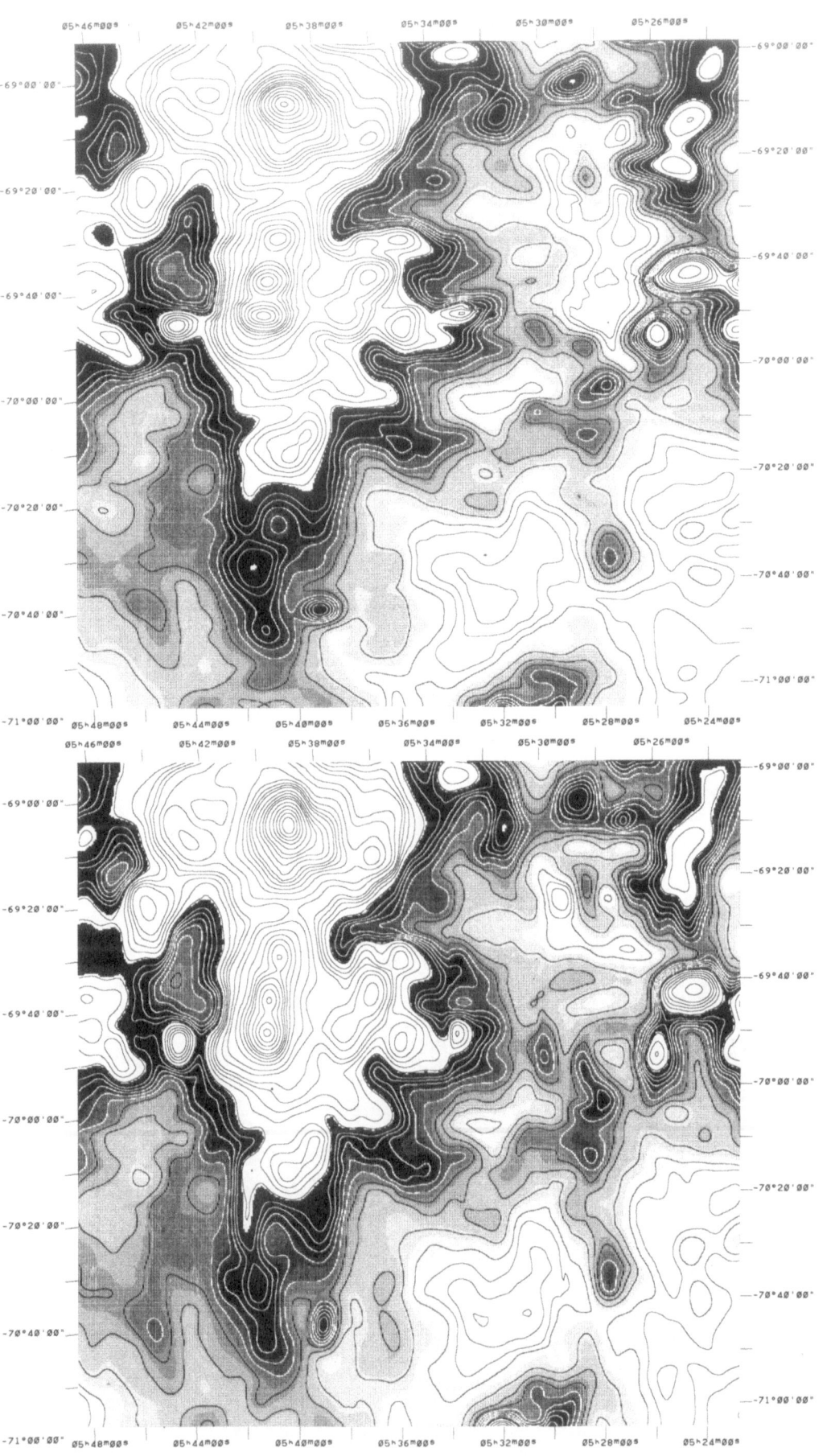

LMC
Field 10
100 μm

LMC
Field 11
12 μm

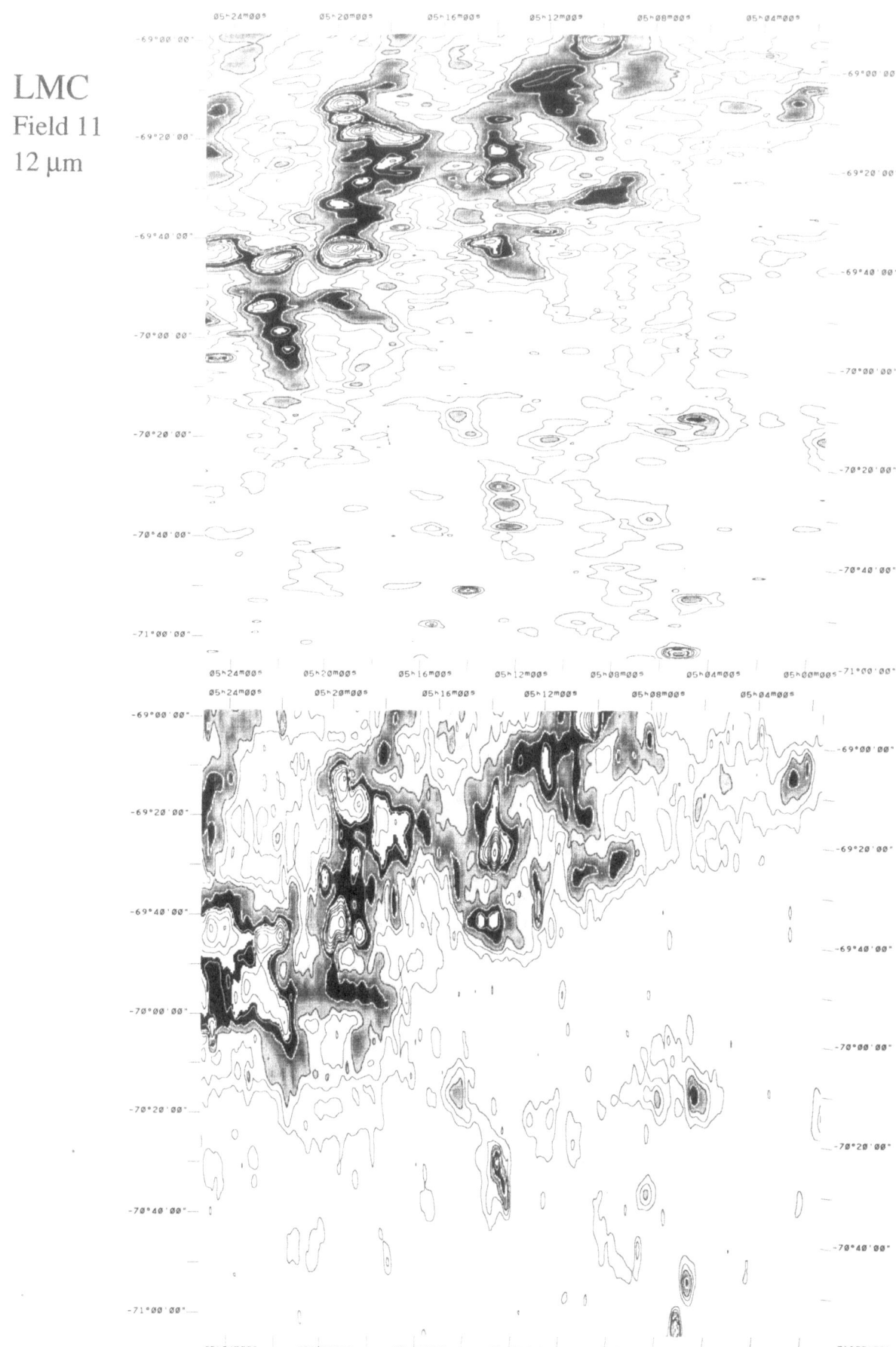

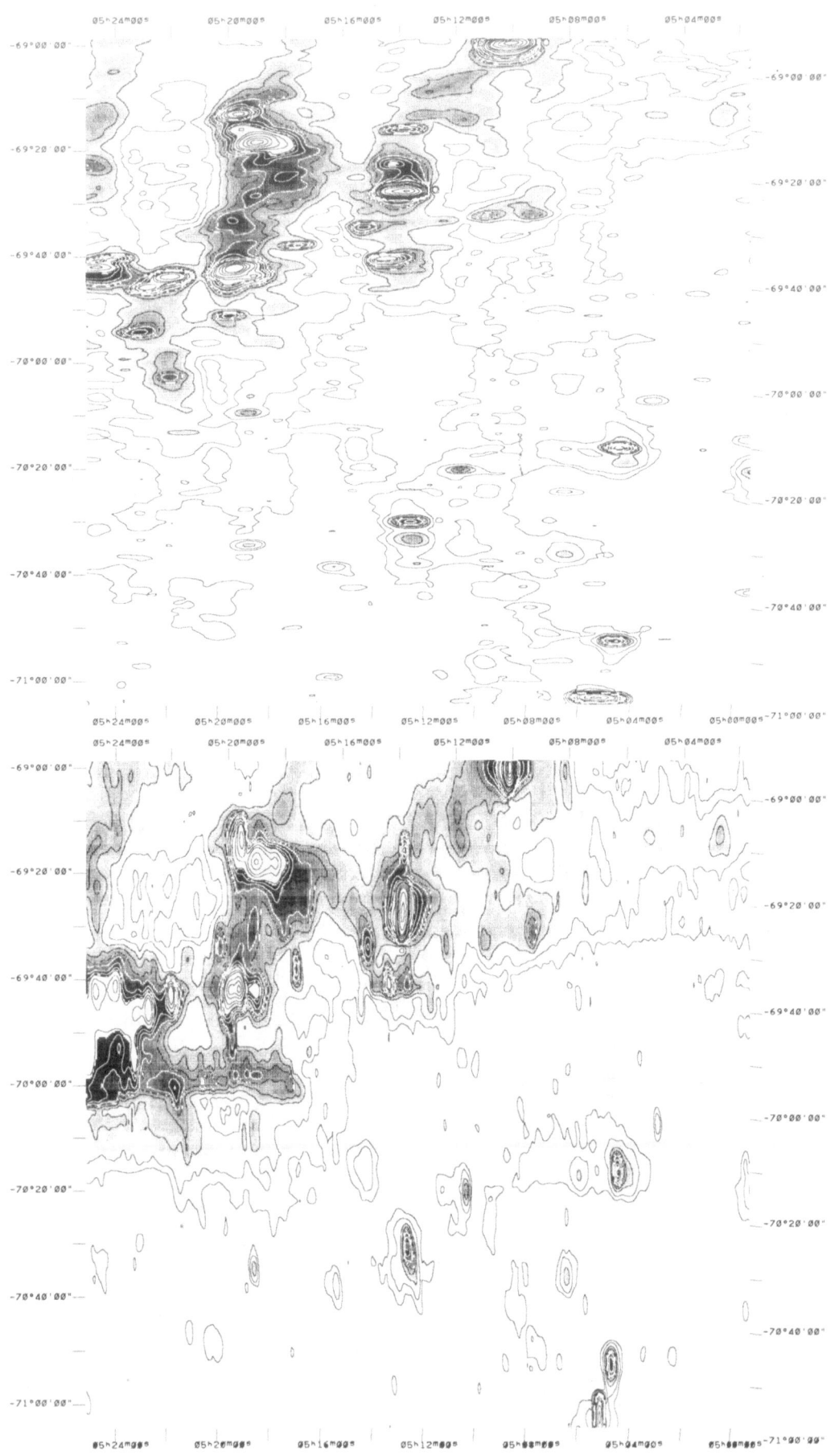

LMC
Field 11
25 μm

LMC
Field 11
60 μm

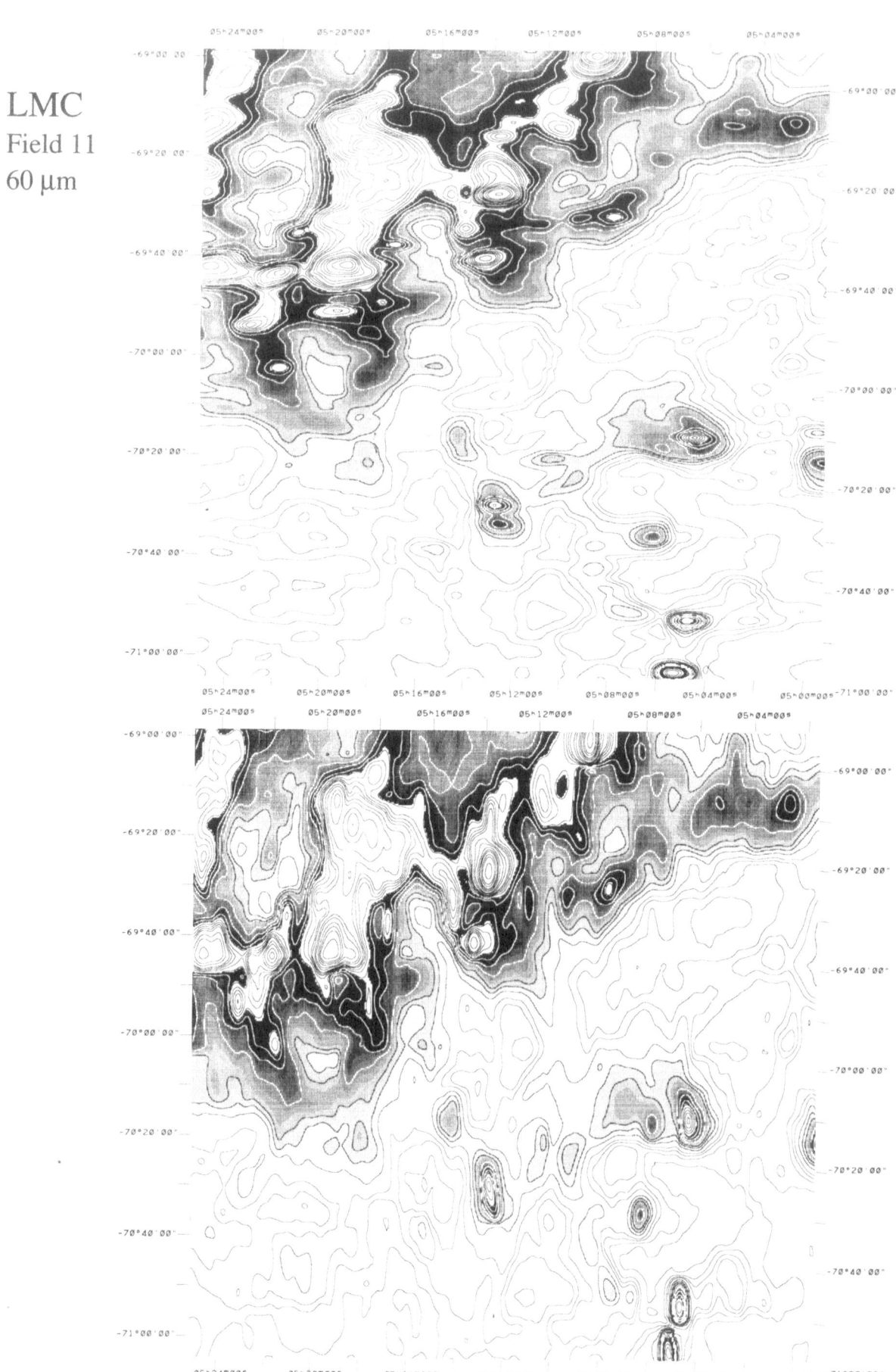

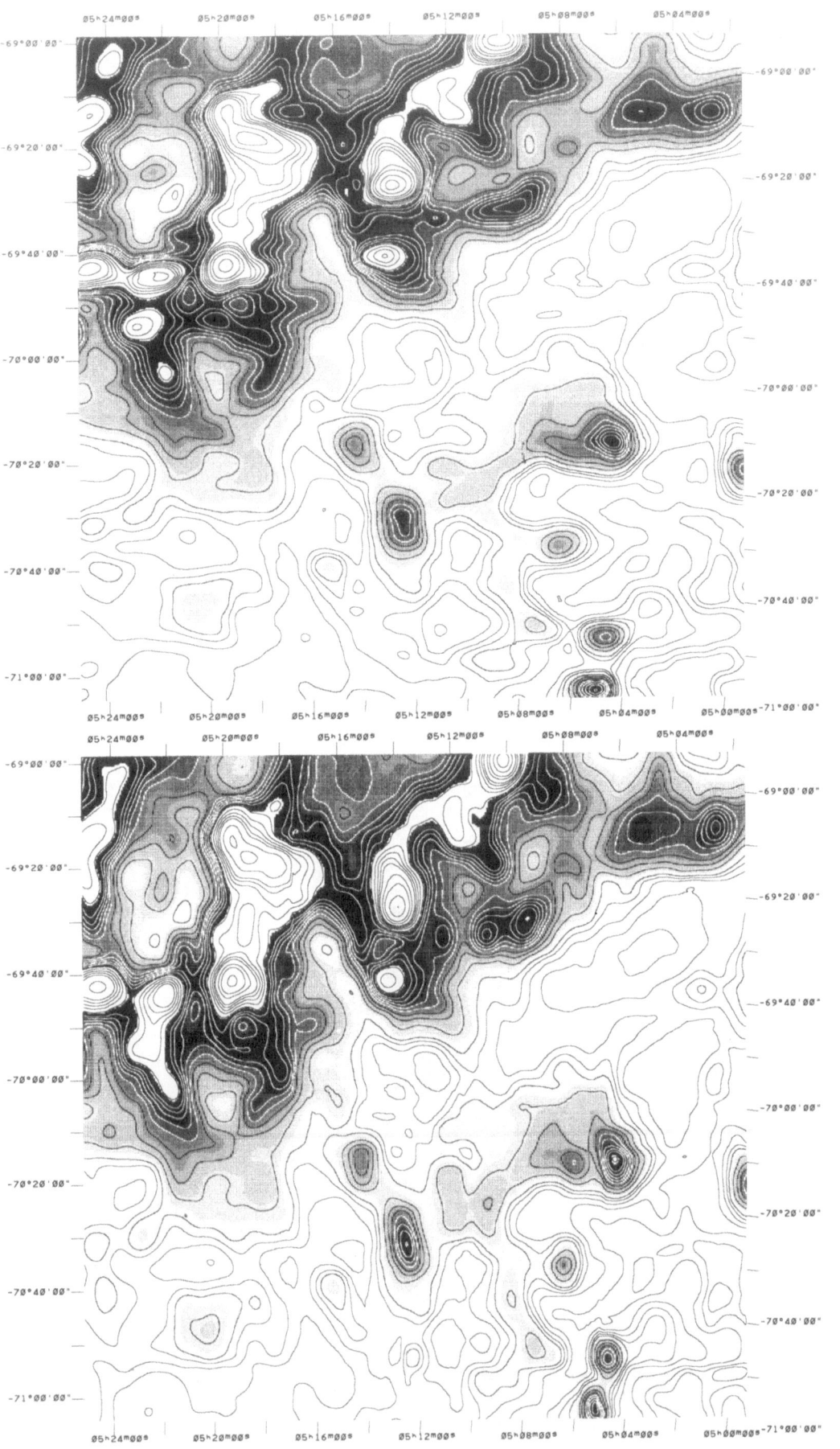

LMC
Field 11
100 µm

LMC
Field 12
12 μm

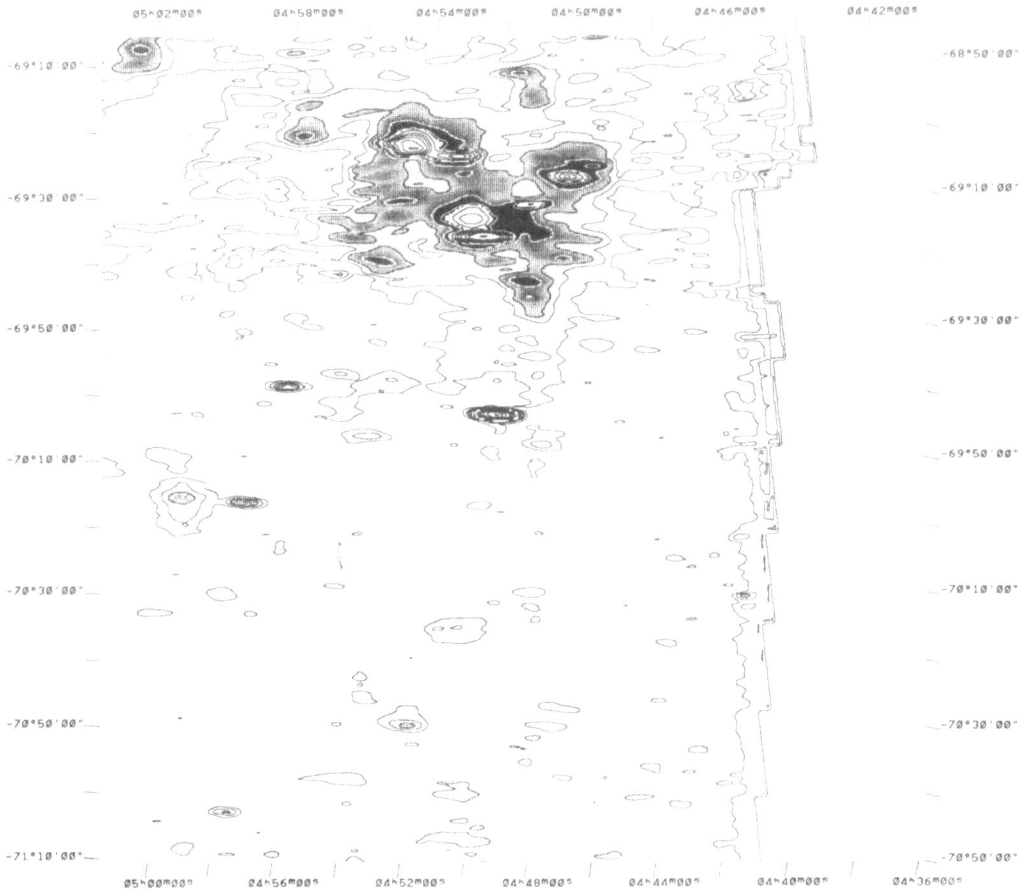

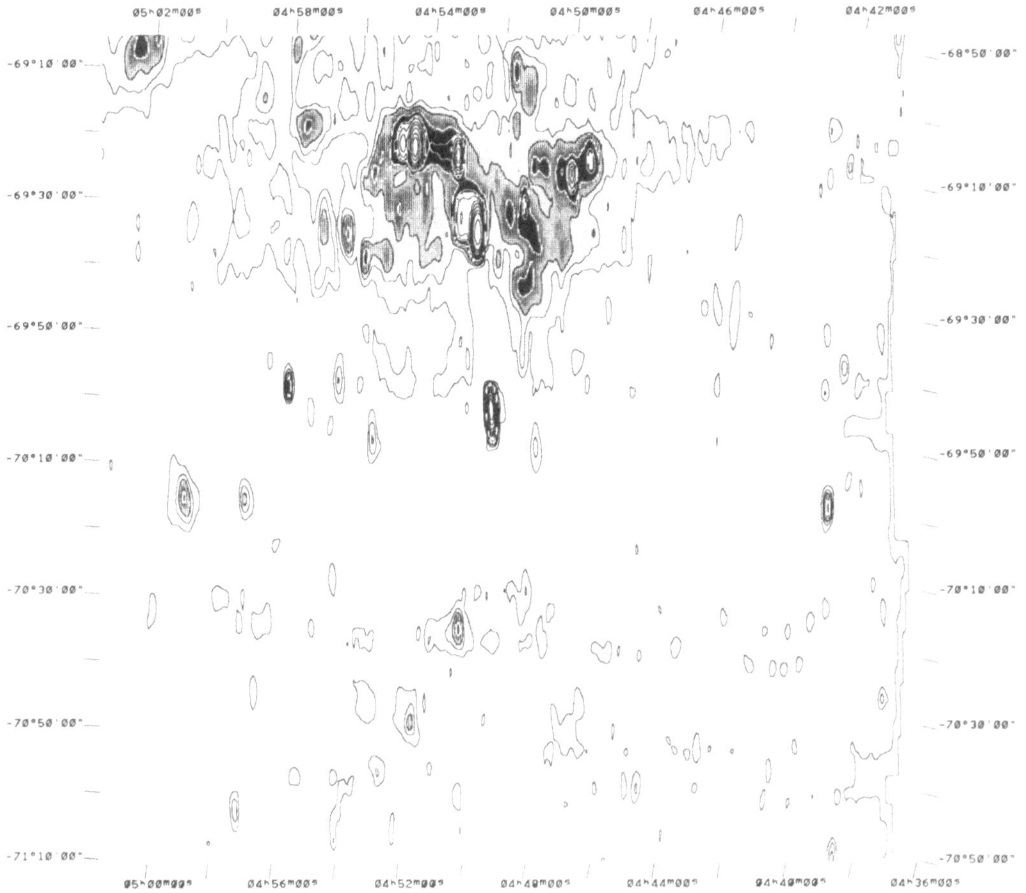

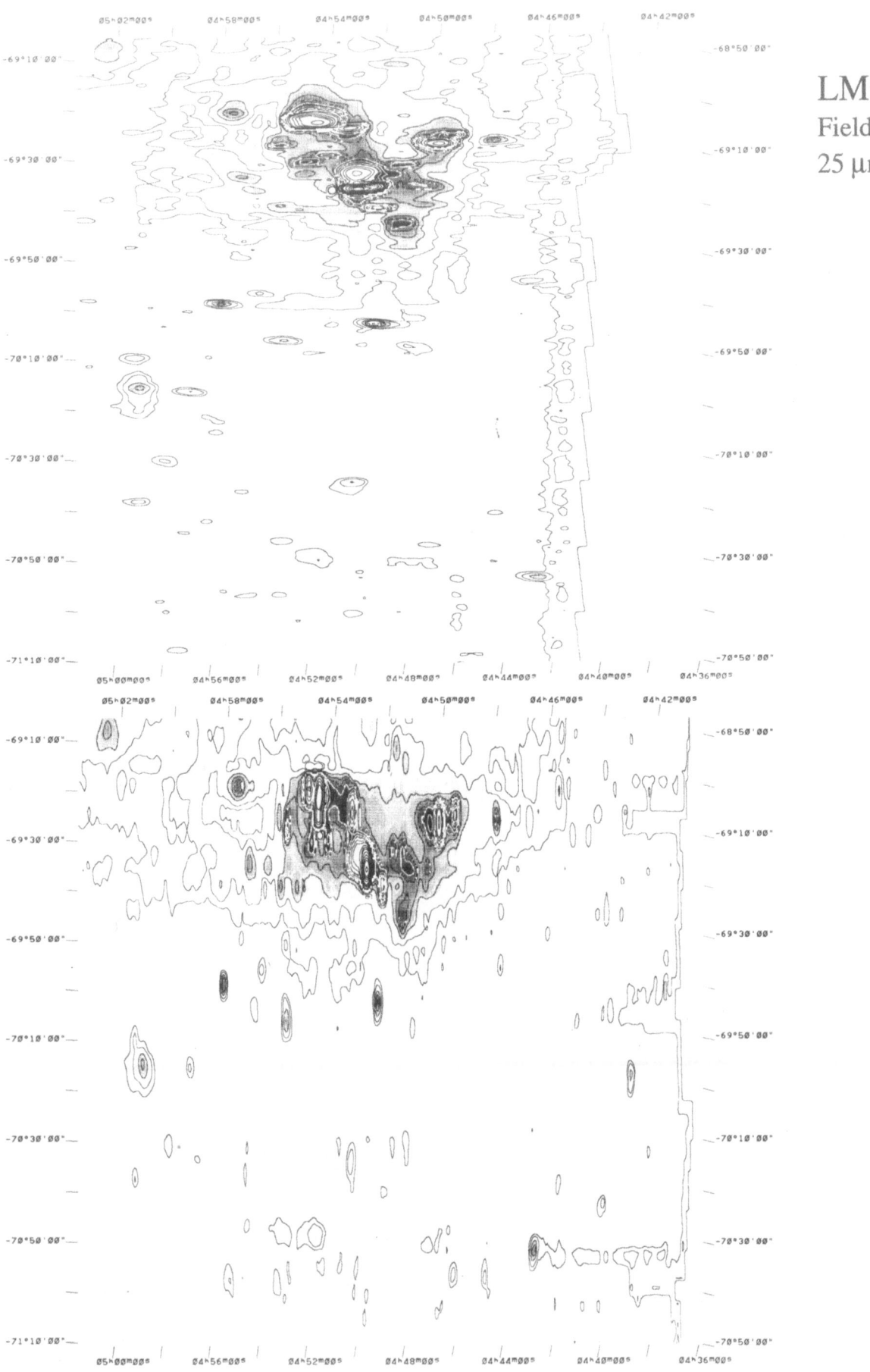

LMC
Field 12
25 µm

LMC
Field 12
60 μm

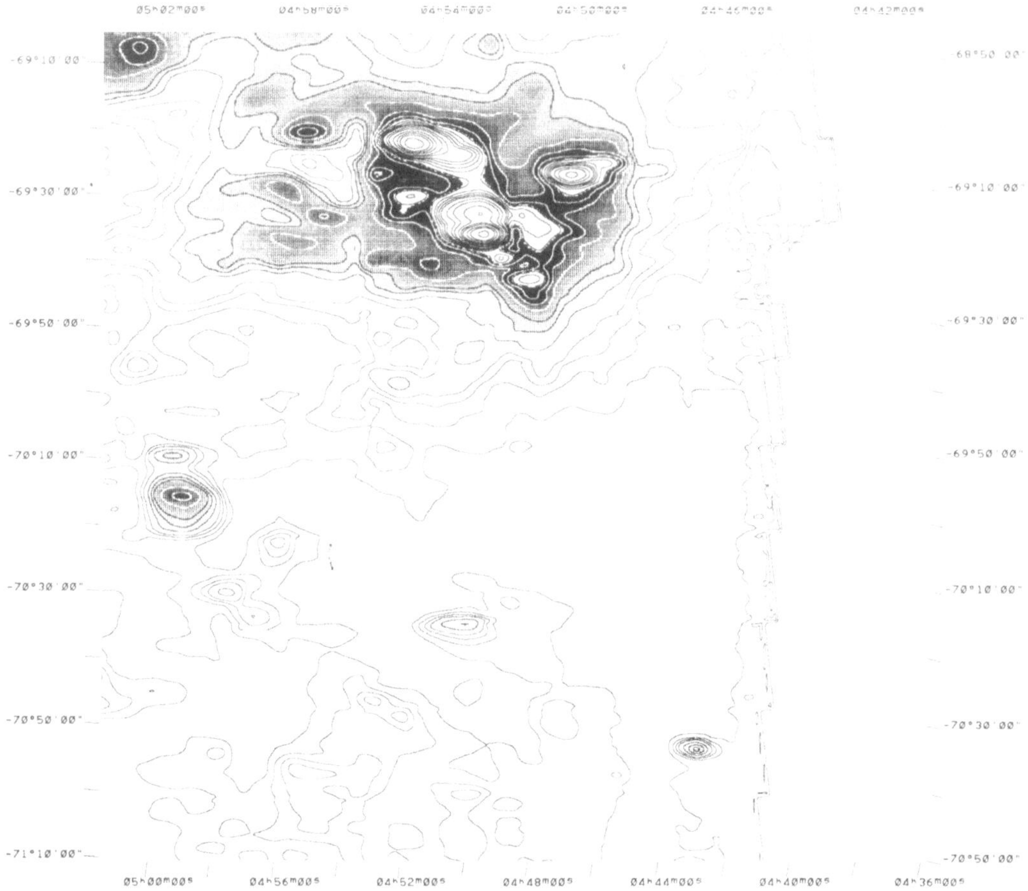

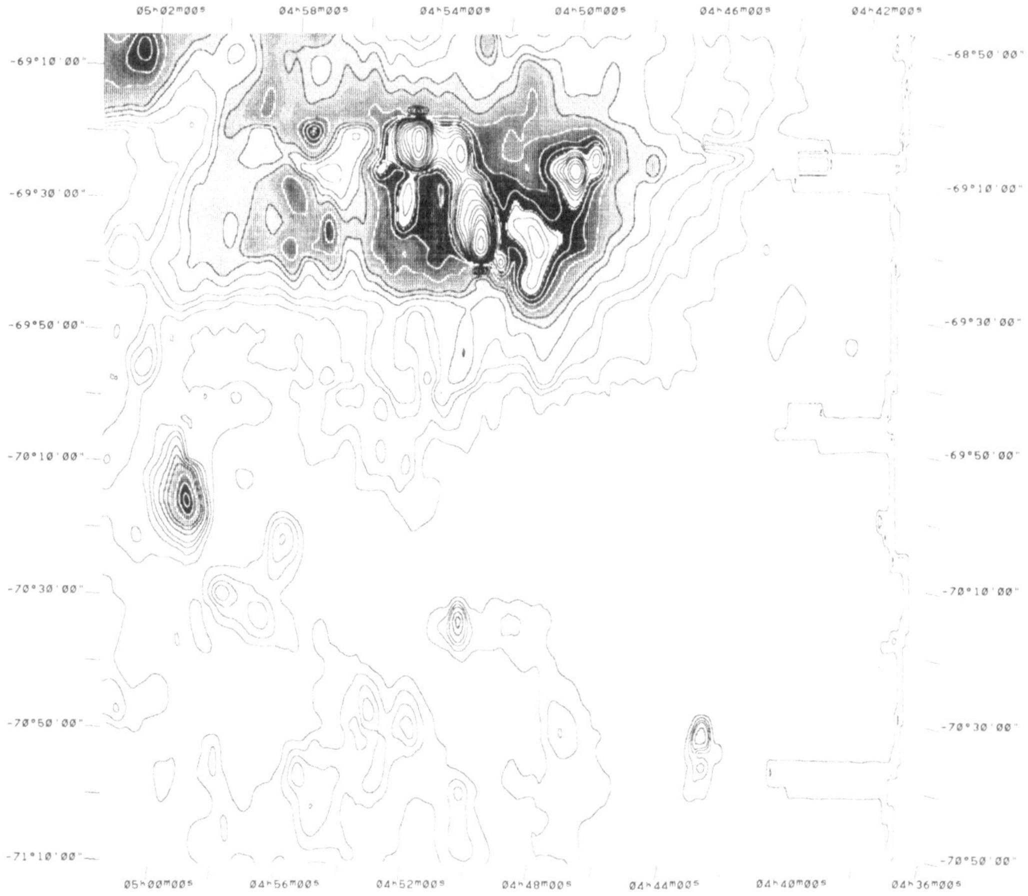

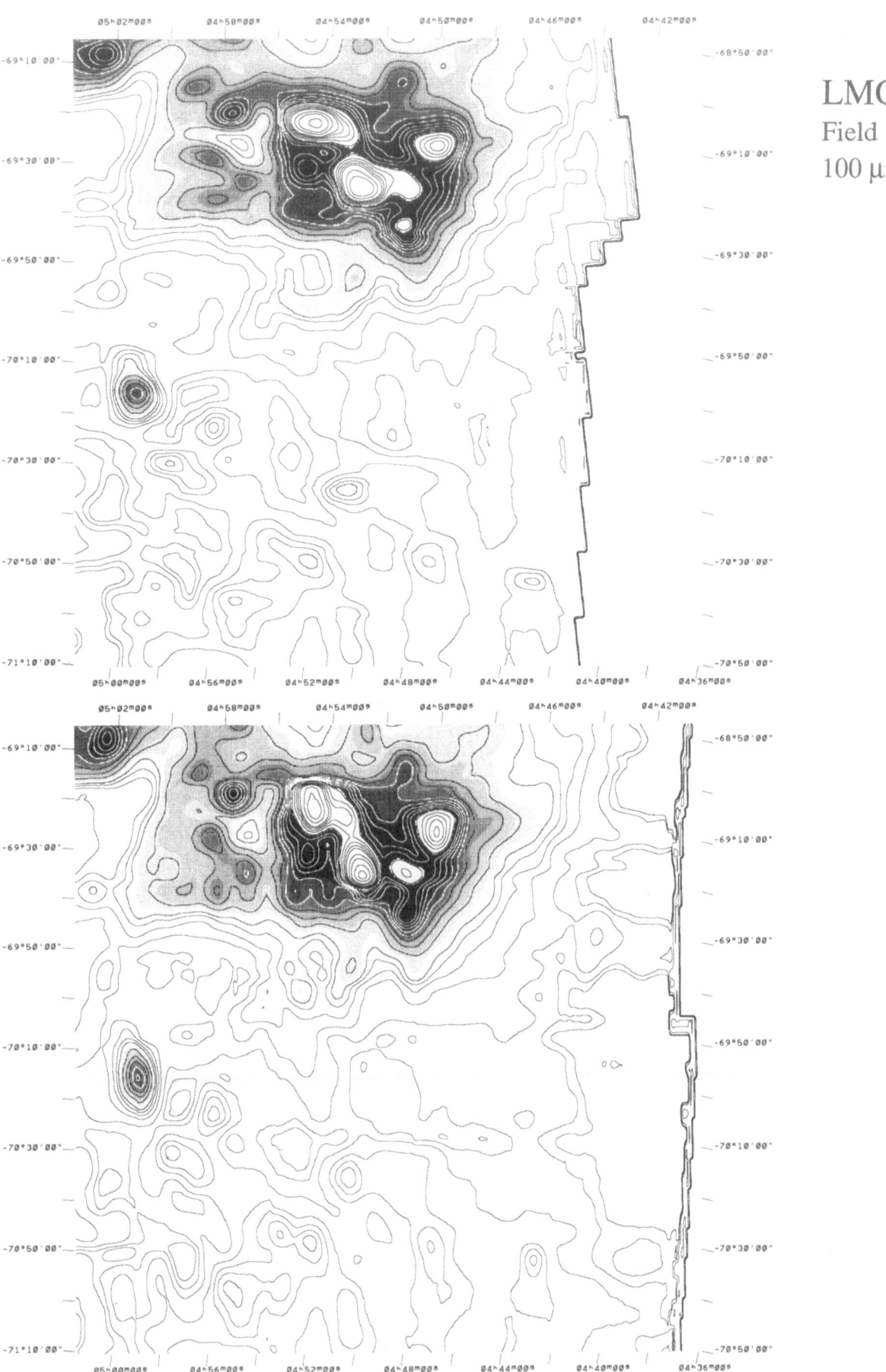

LMC
Field 12
100 μm

LMC
Field 13
12 μm

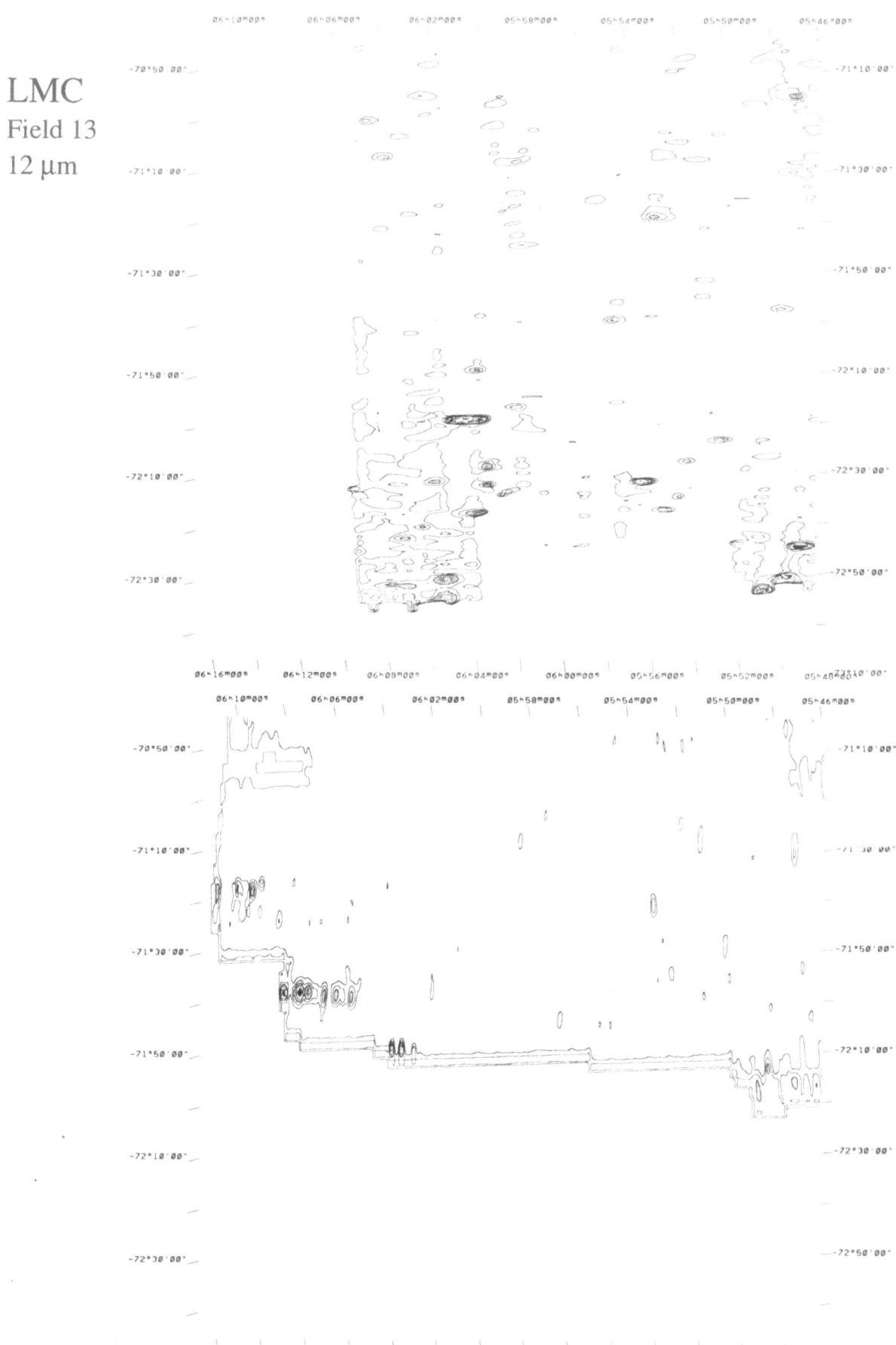

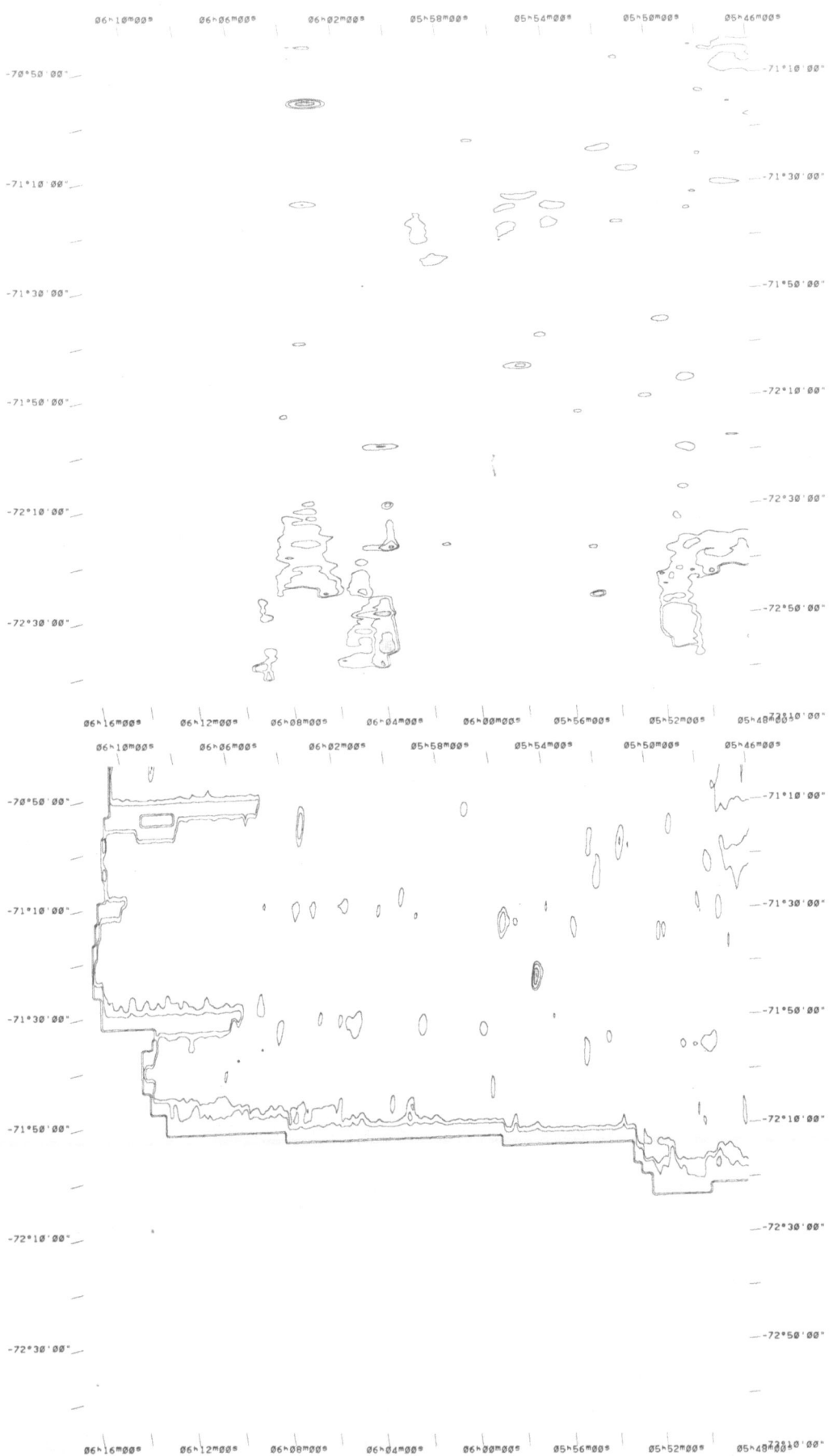

LMC
Field 13
25 μm

LMC
Field 13
60 μm

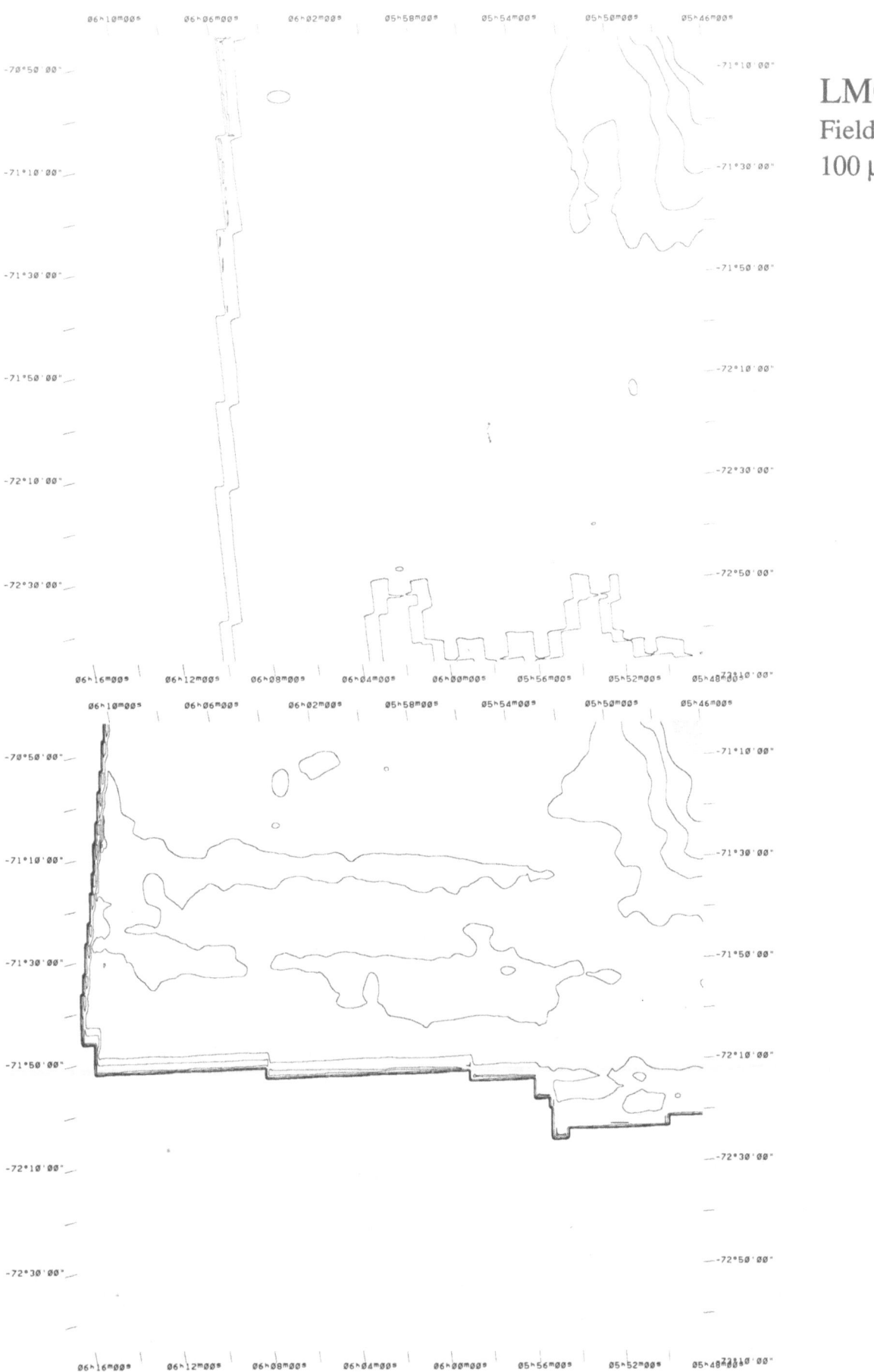

LMC
Field 13
100 μm

LMC
Field 14
12 μm

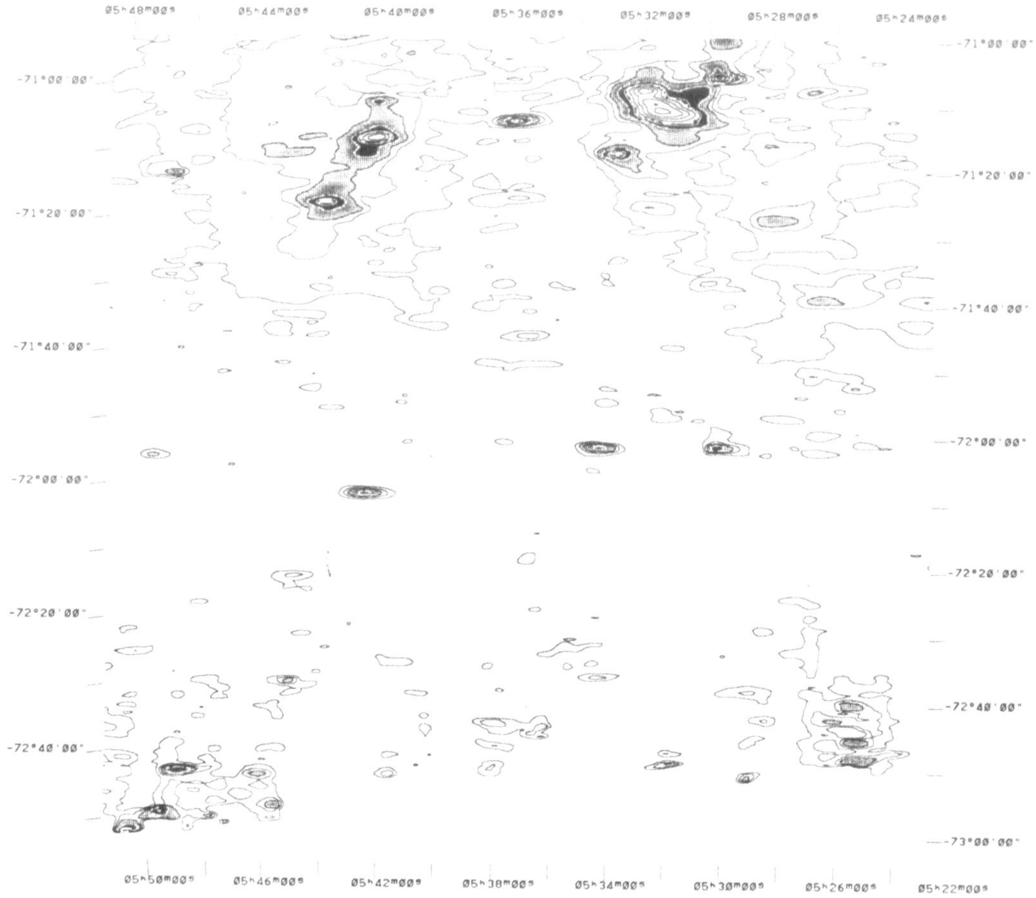

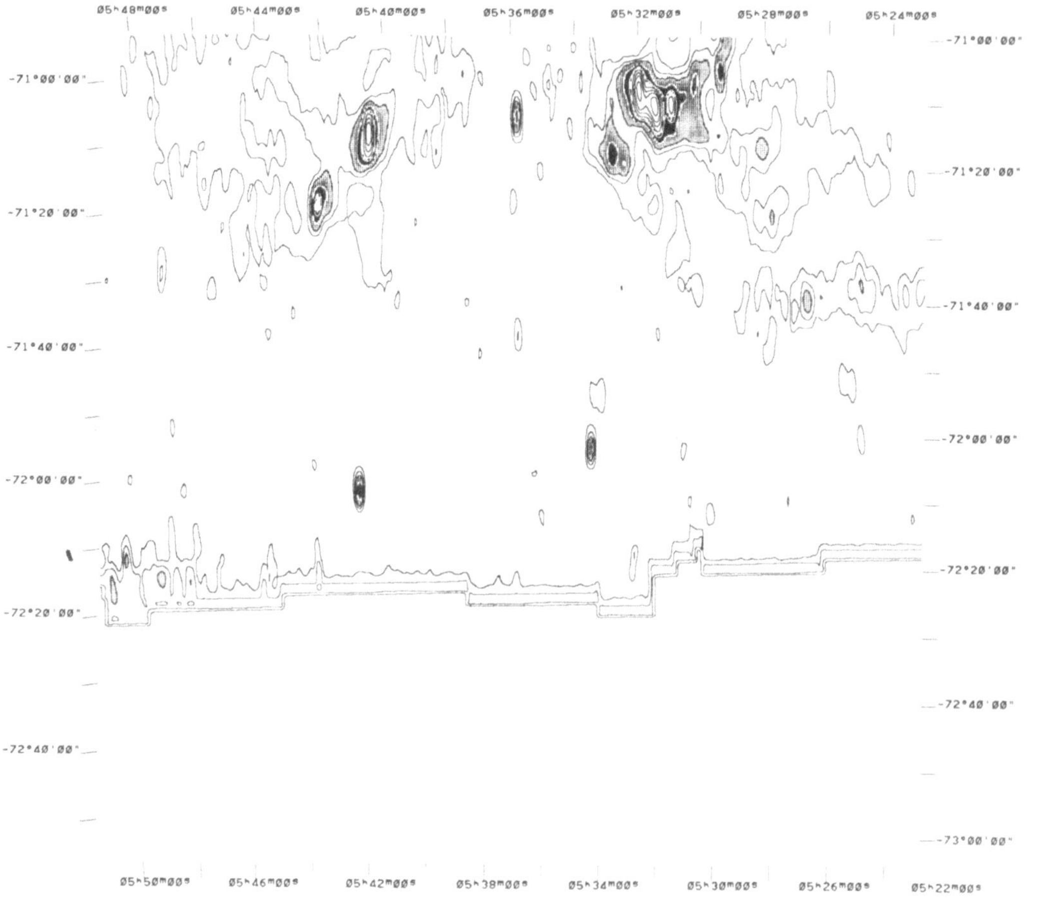

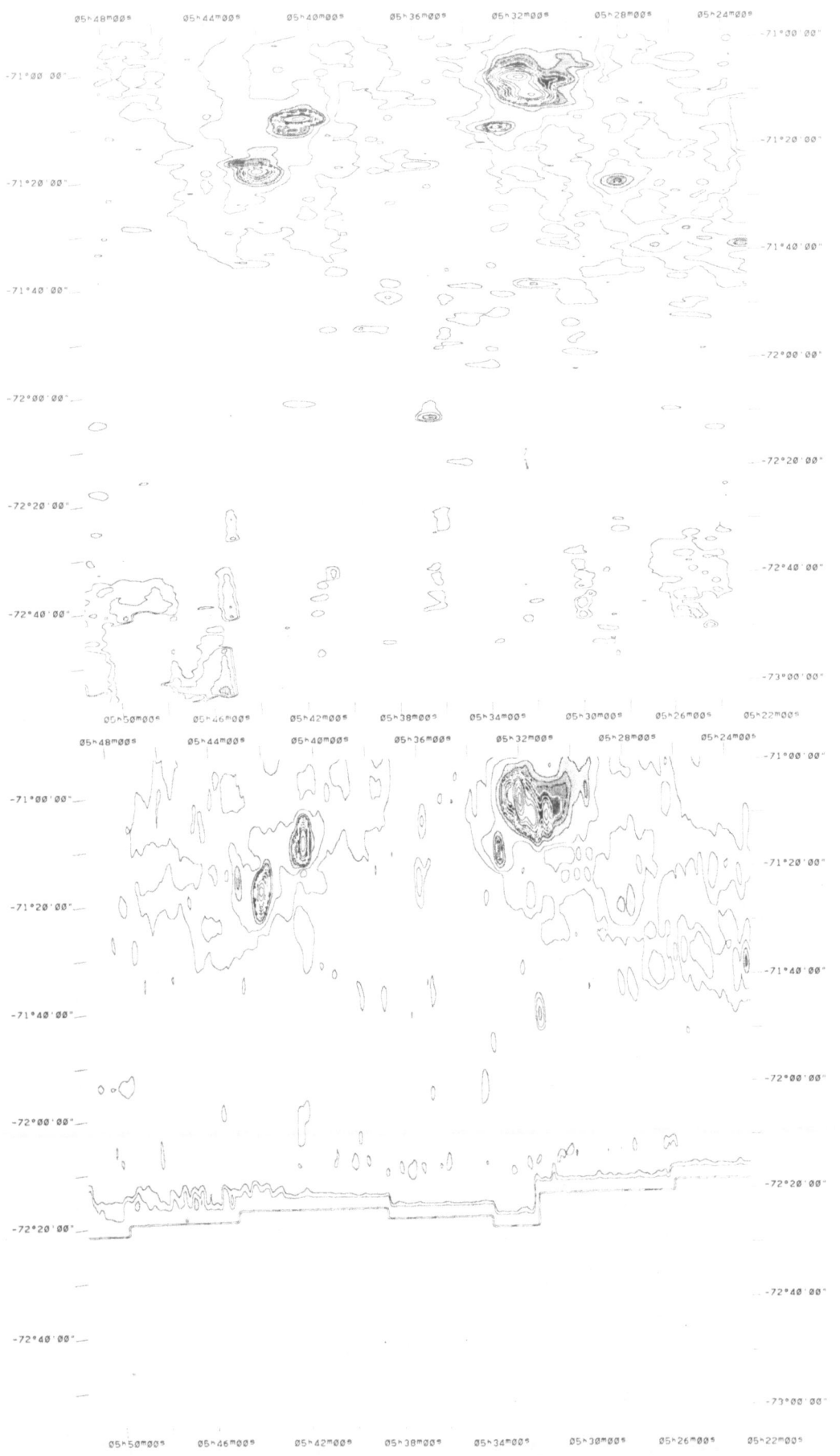

LMC
Field 14
25 μm

LMC
Field 14
60 μm

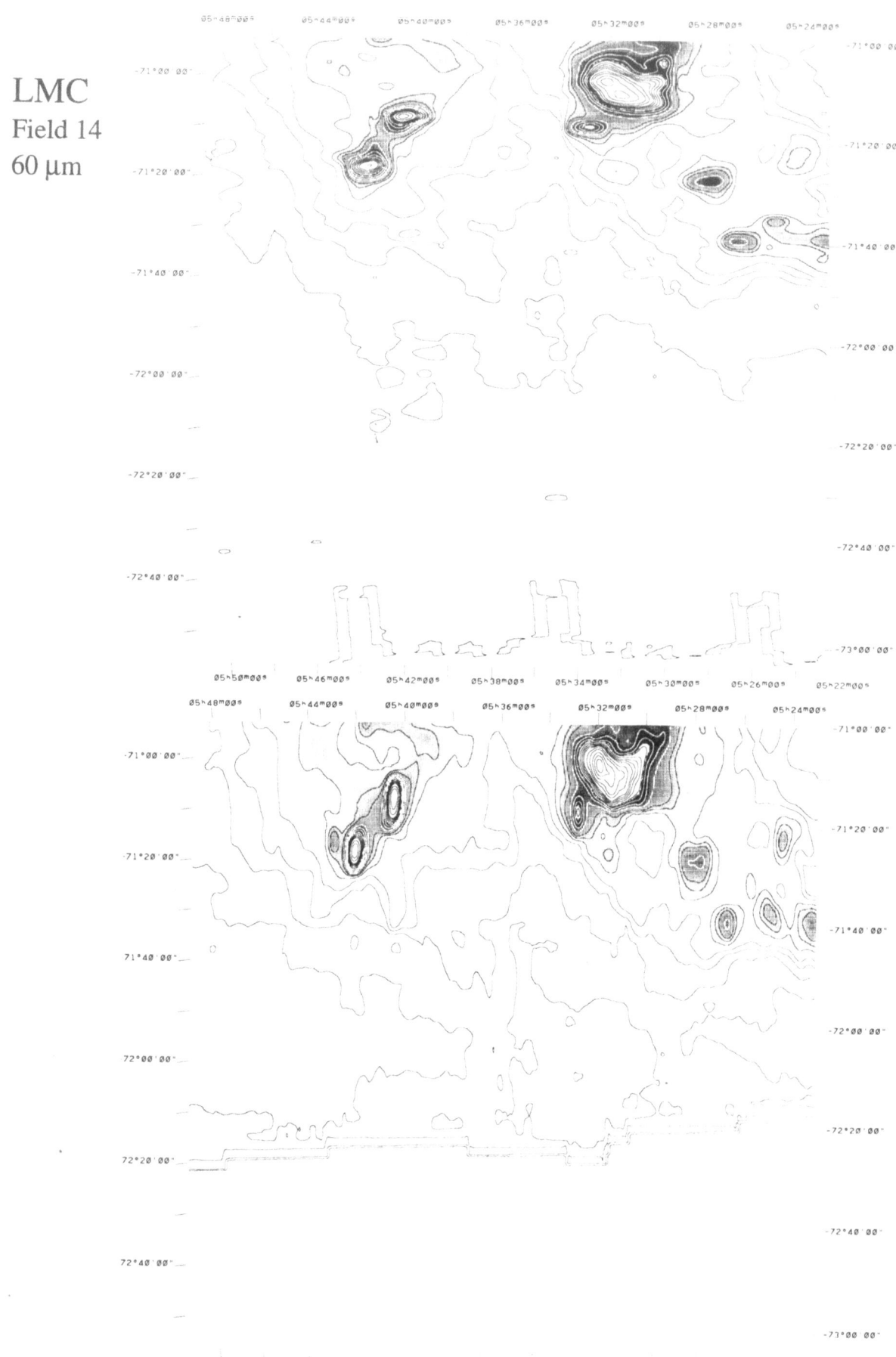

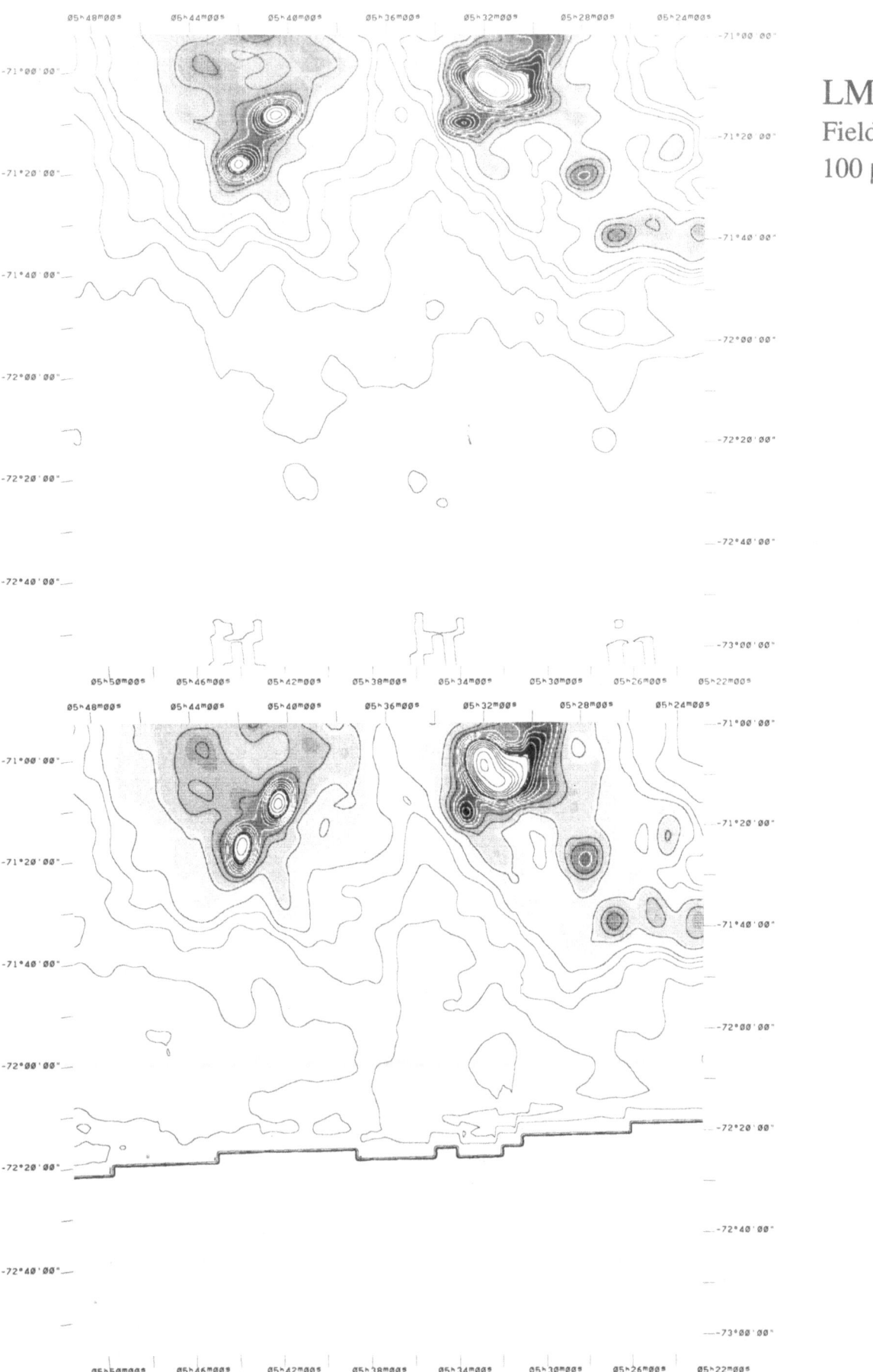

LMC
Field 14
100 μm

LMC
Field 15
12 μm

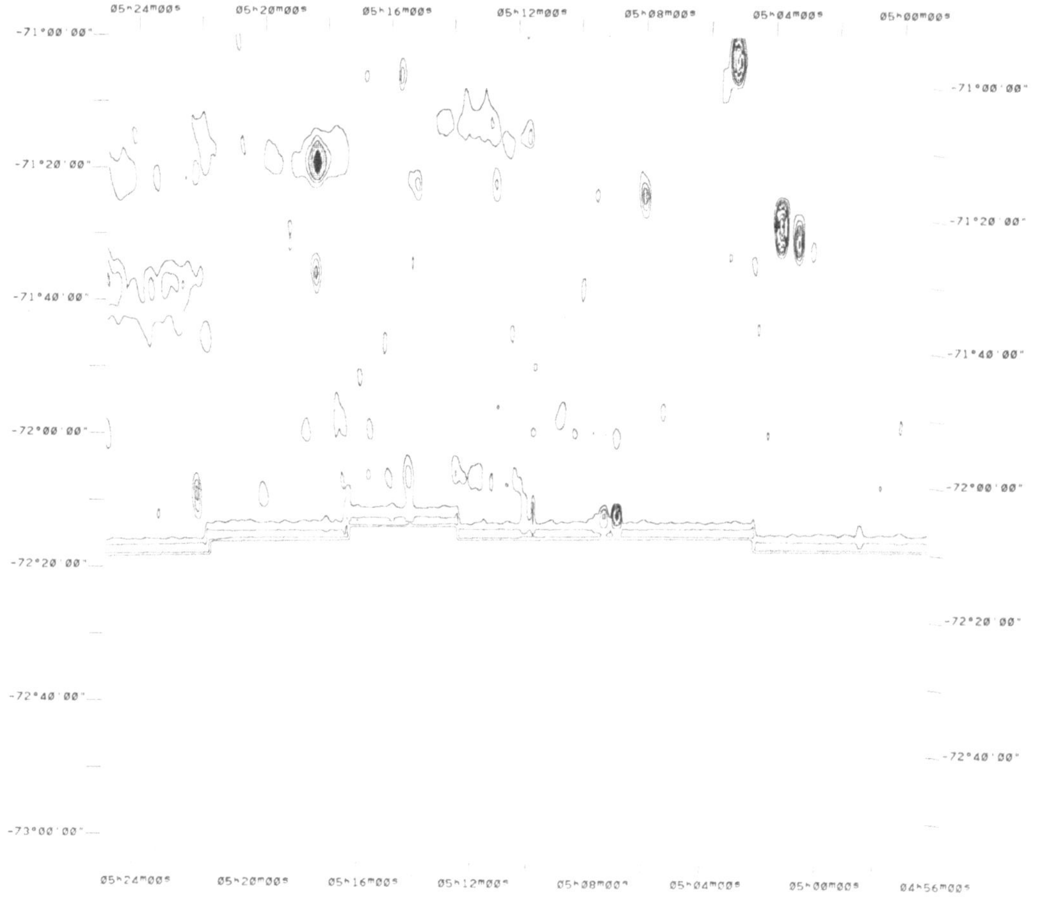

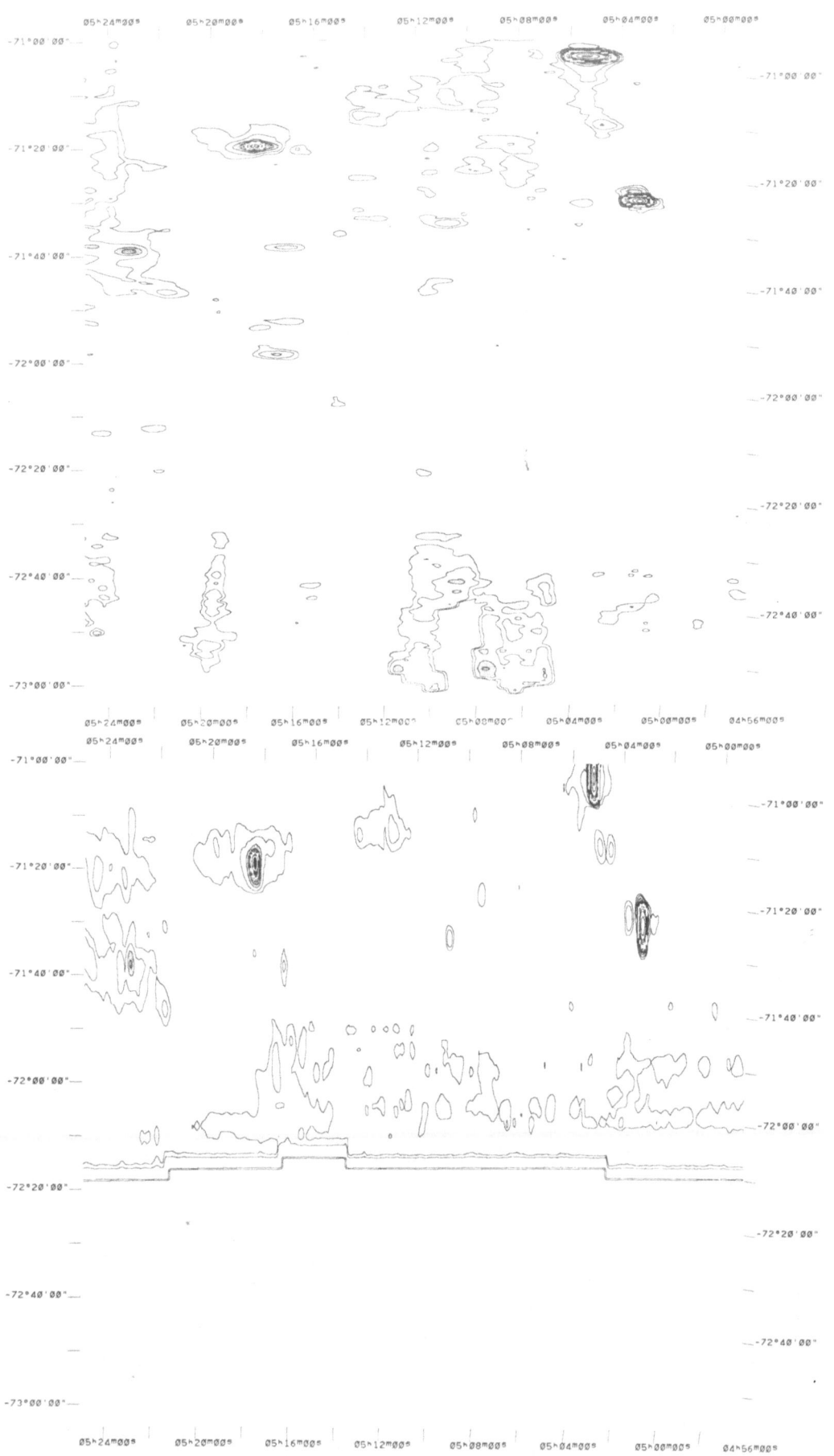

LMC
Field 15
25 μm

LMC
Field 15
60 μm

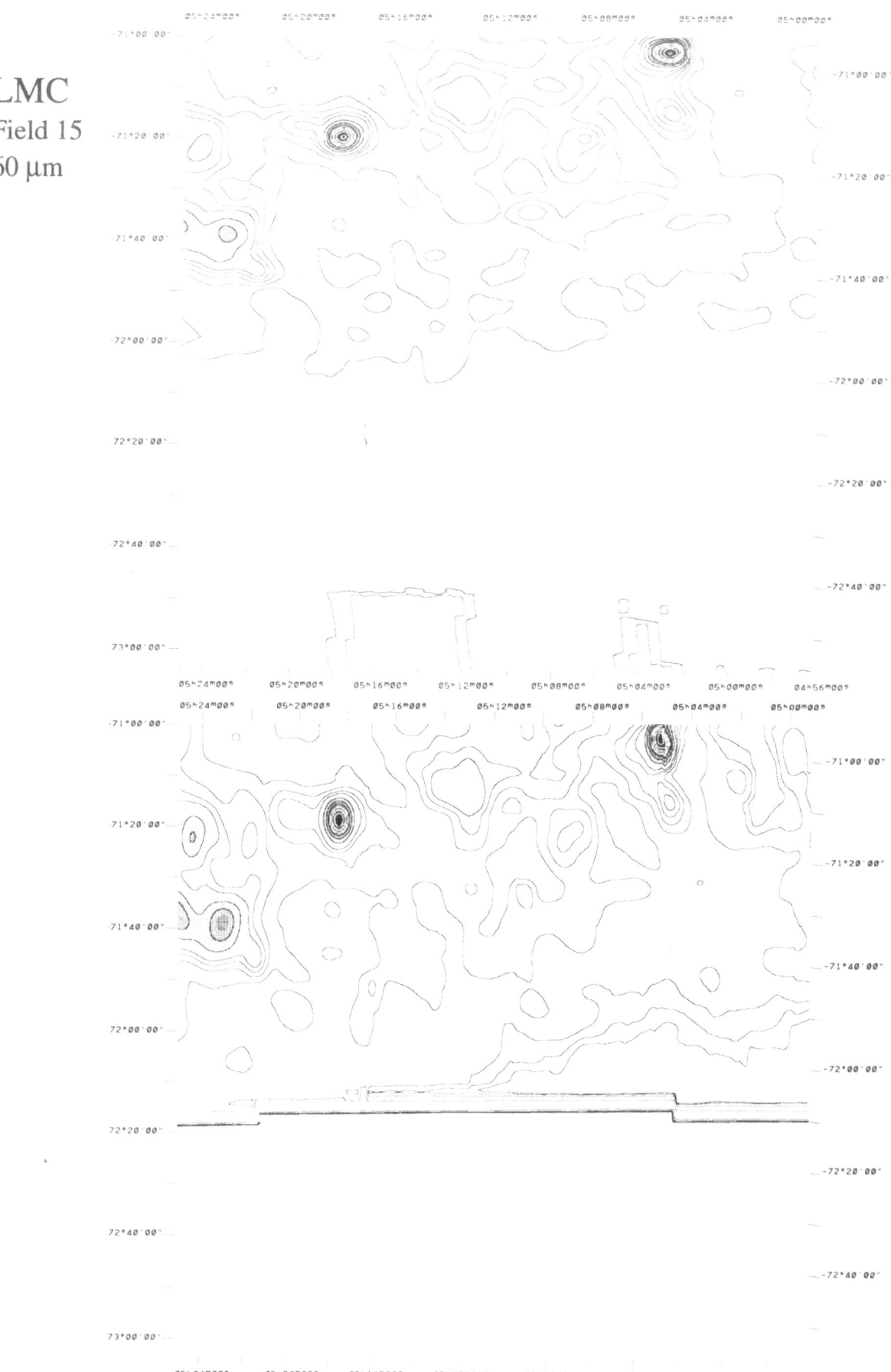

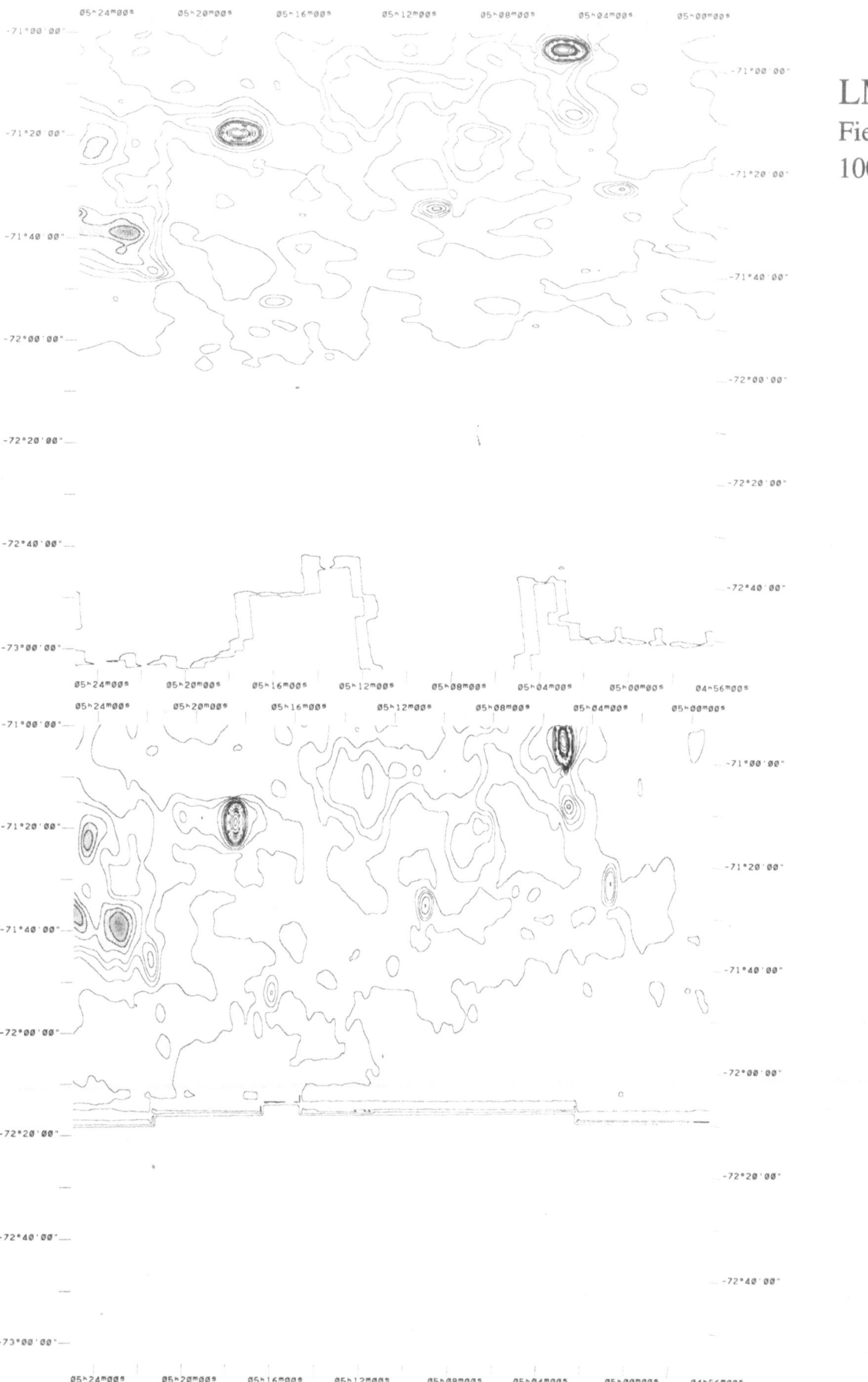

LMC
Field 15
100 μm

LMC
Field 16
12 μm

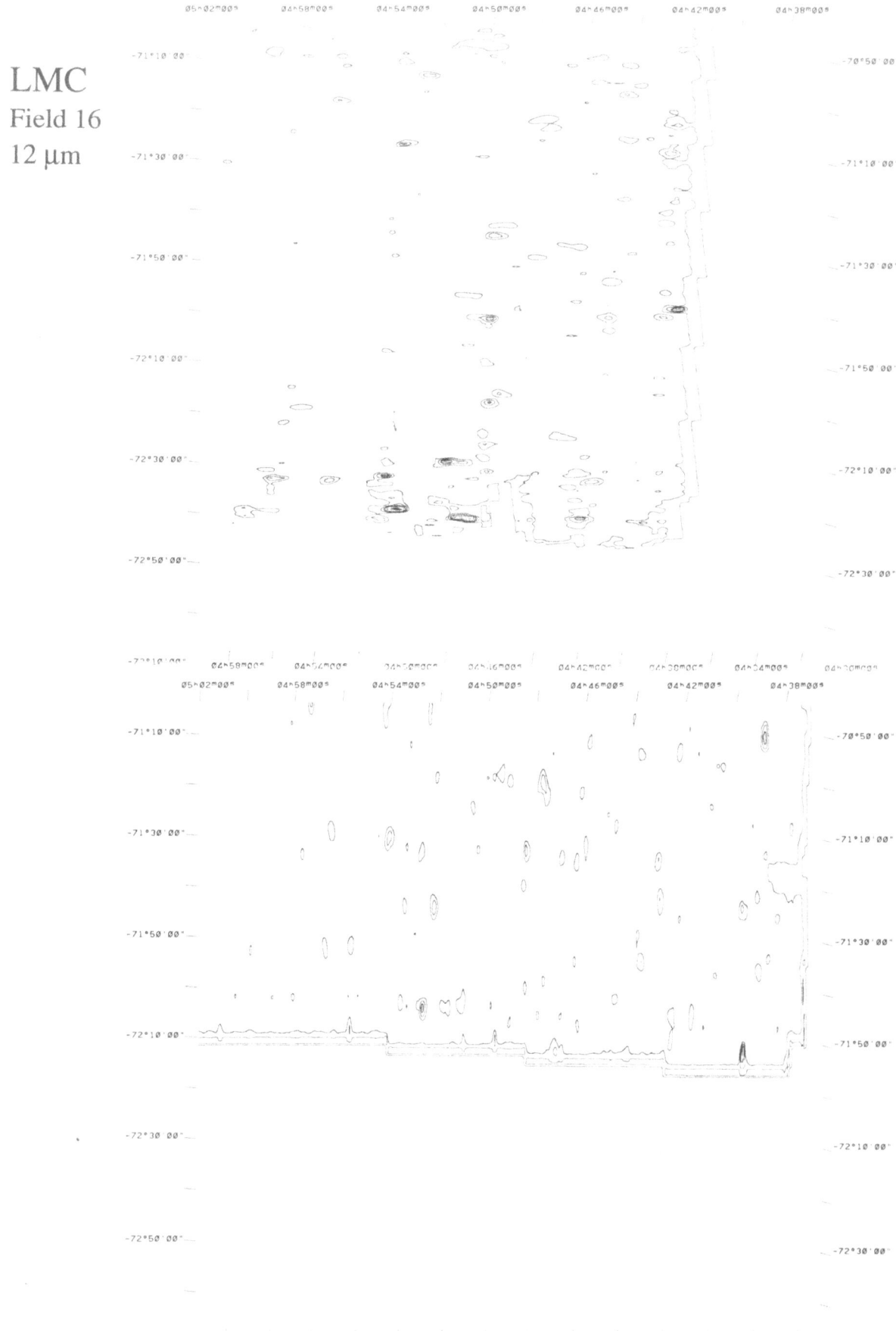

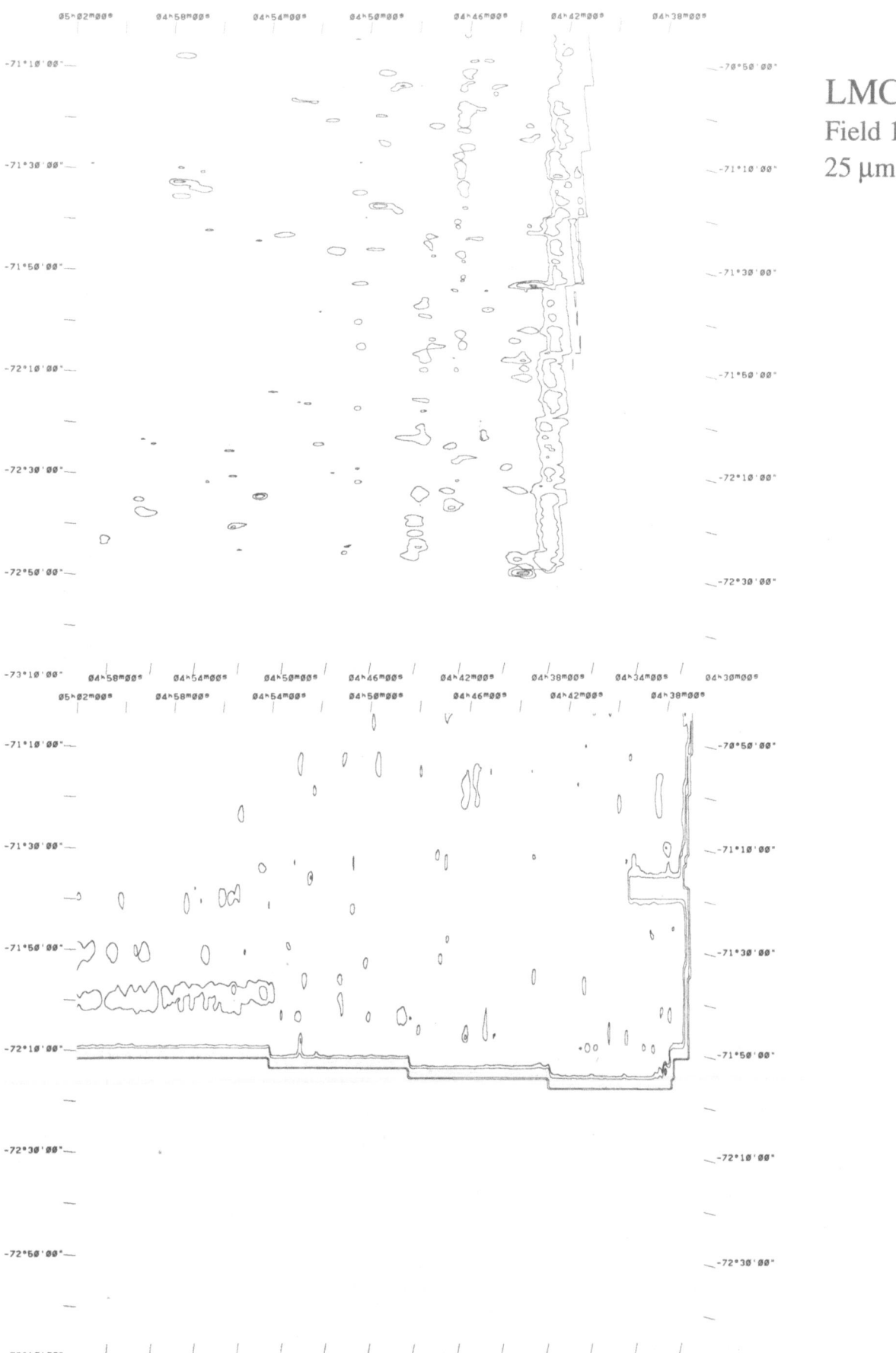

LMC
Field 16
25 μm

LMC
Field 16
60 μm

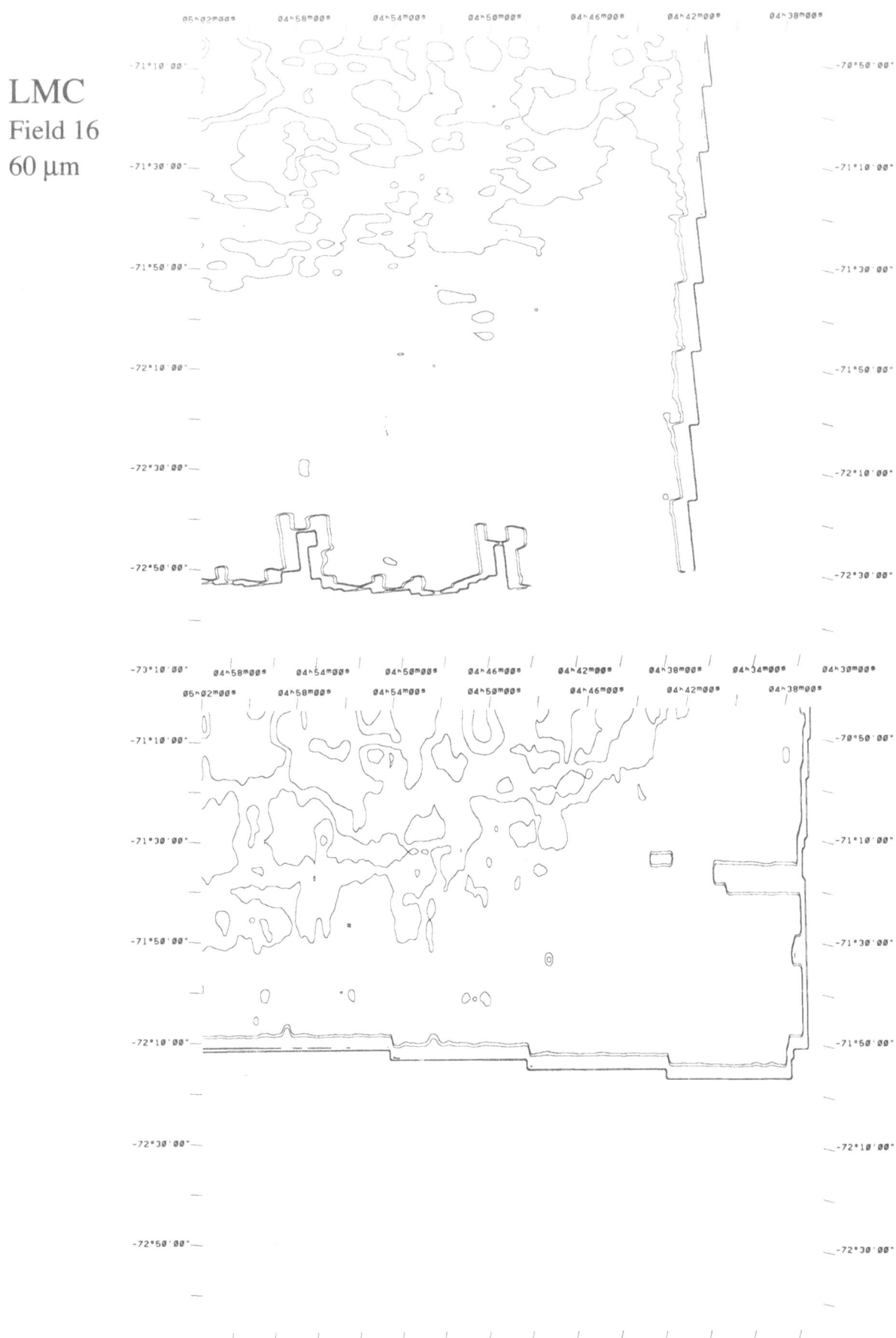

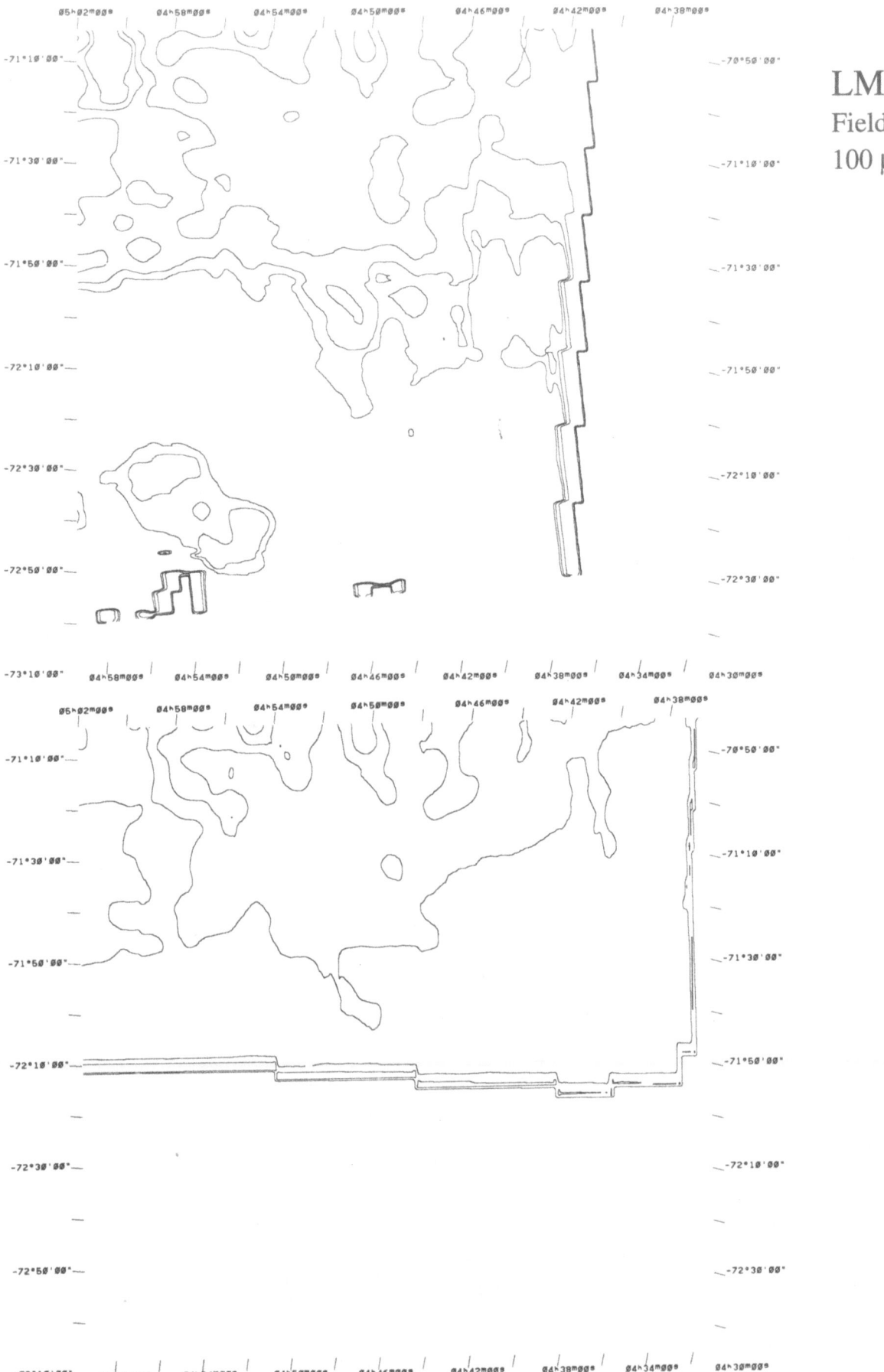

LMC
Field 16
100 μm

LMC
Field 17
12 μm

LMC
Field 17
25 μm

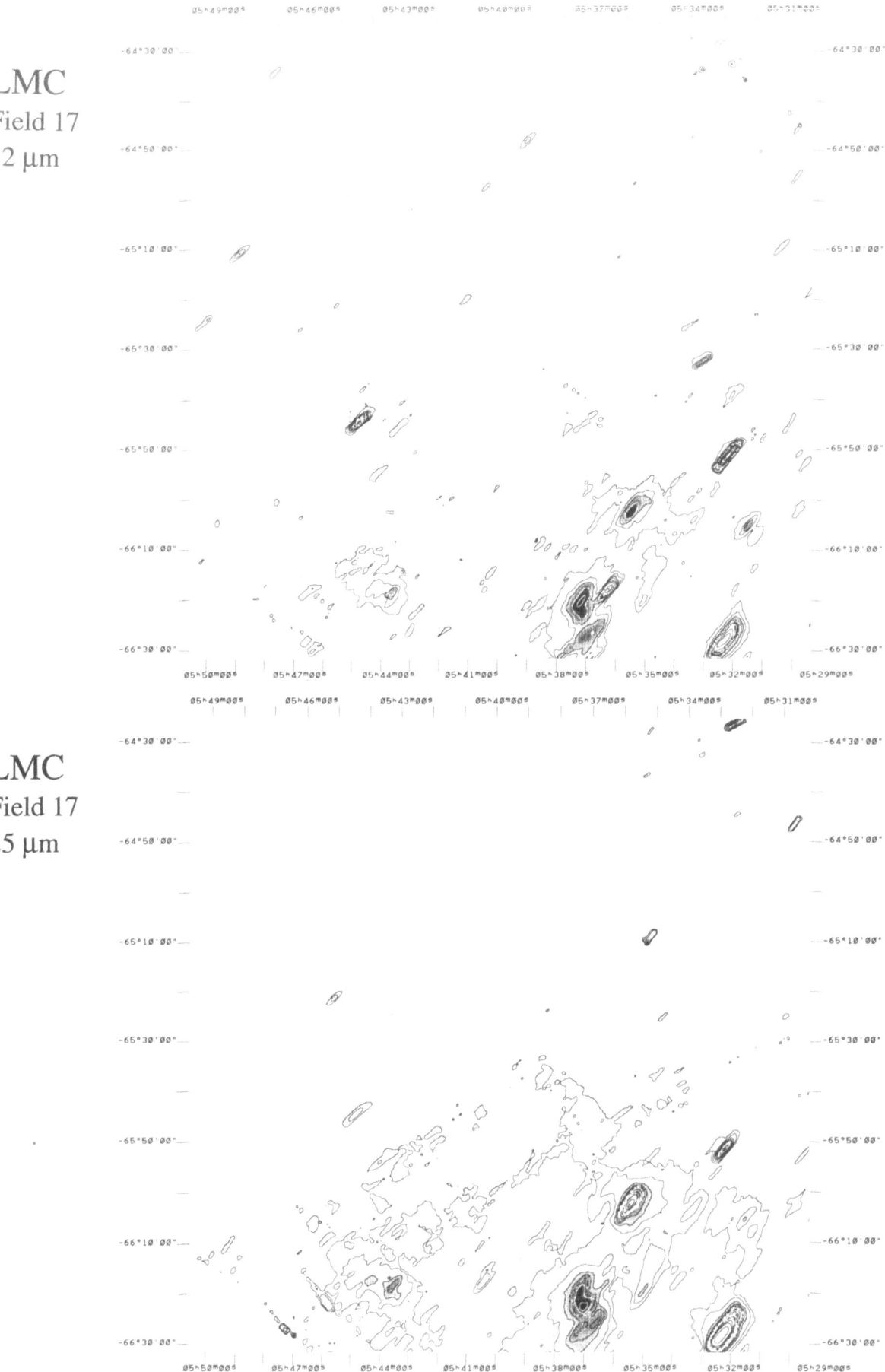

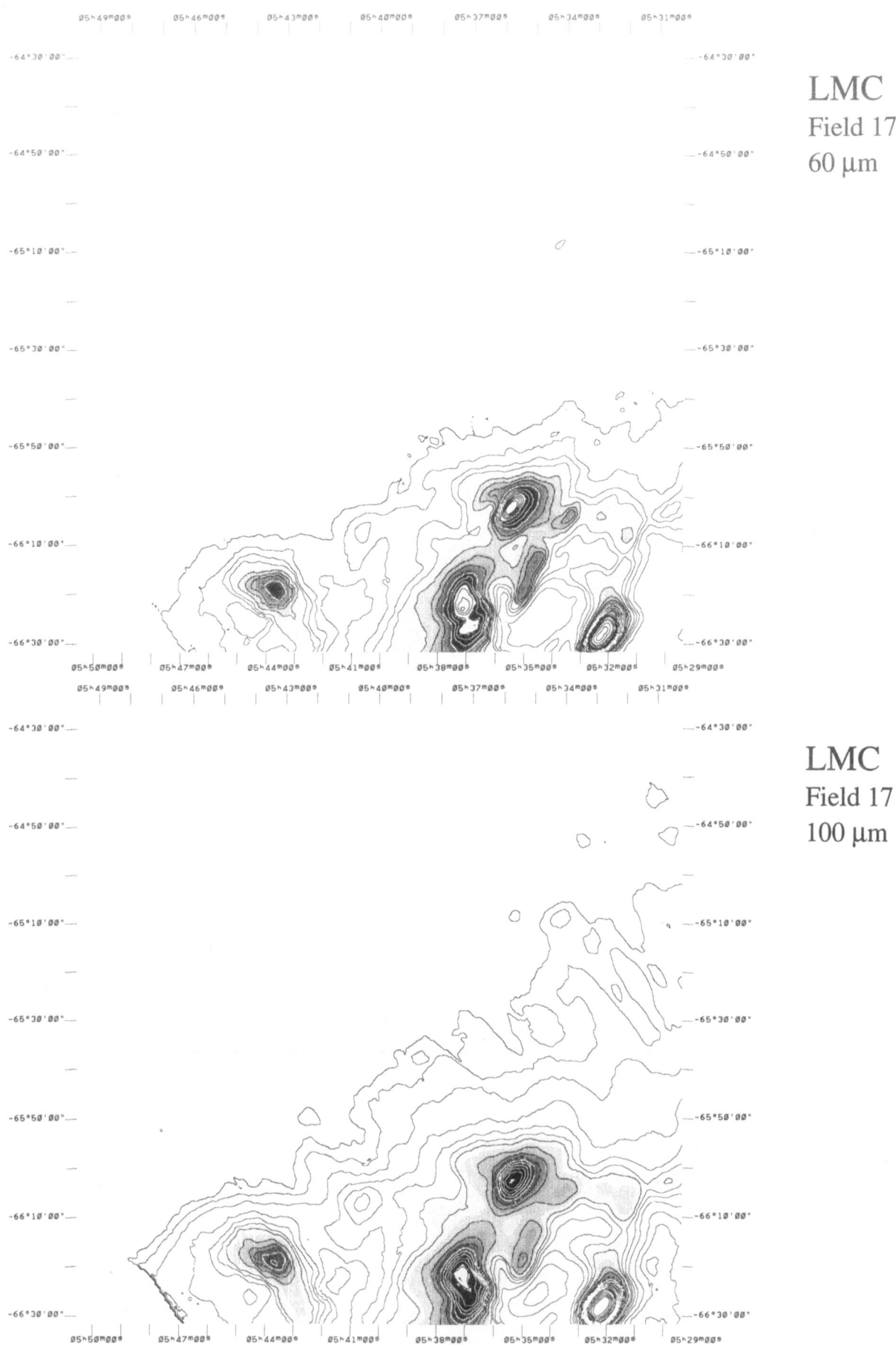

LMC
Field 17
60 μm

LMC
Field 17
100 μm

LMC
Field 18
12 μm

LMC
Field 18
25 μm

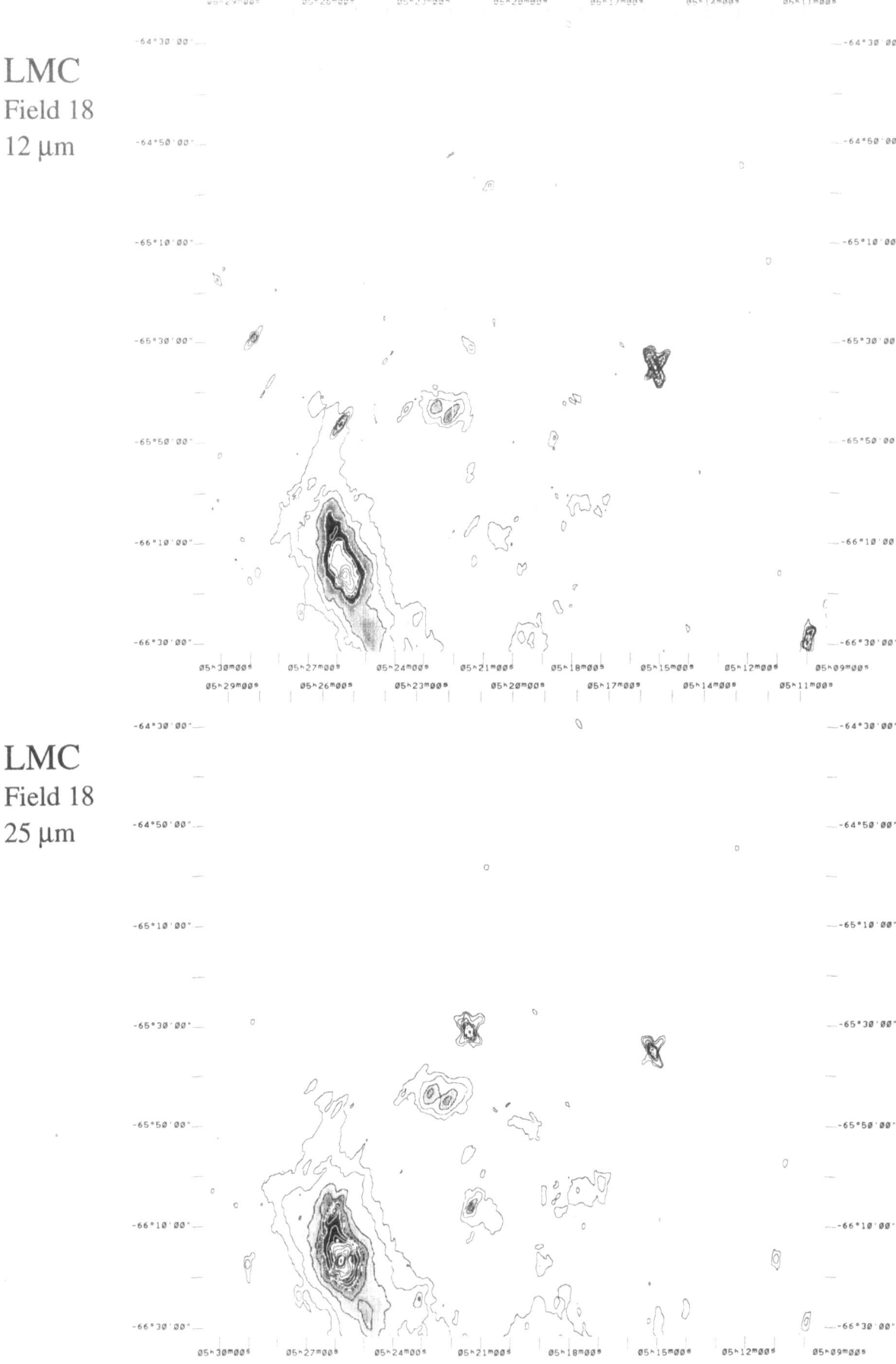

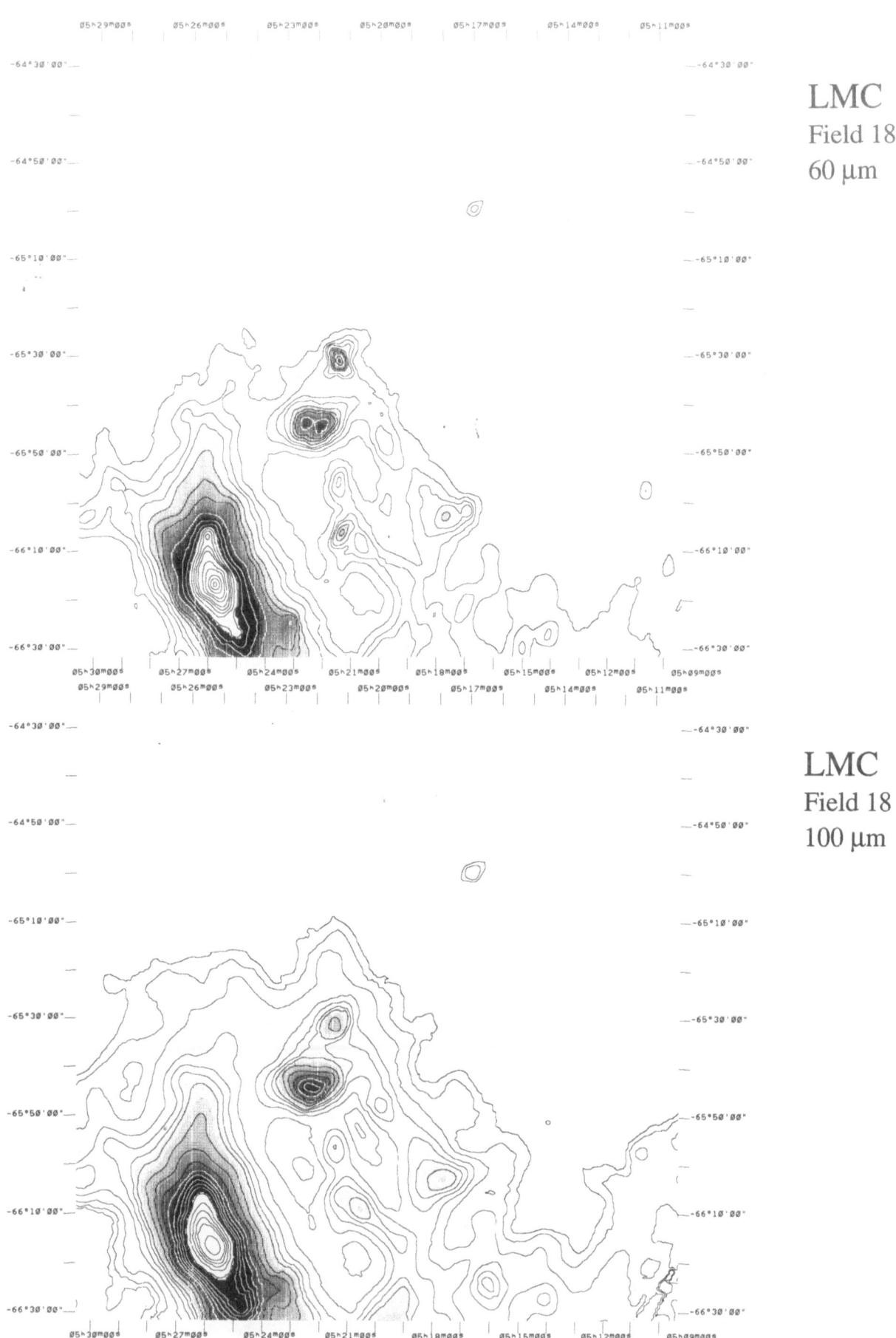

LMC
Field 18
60 µm

LMC
Field 18
100 µm

LMC
Field 19
12 μm

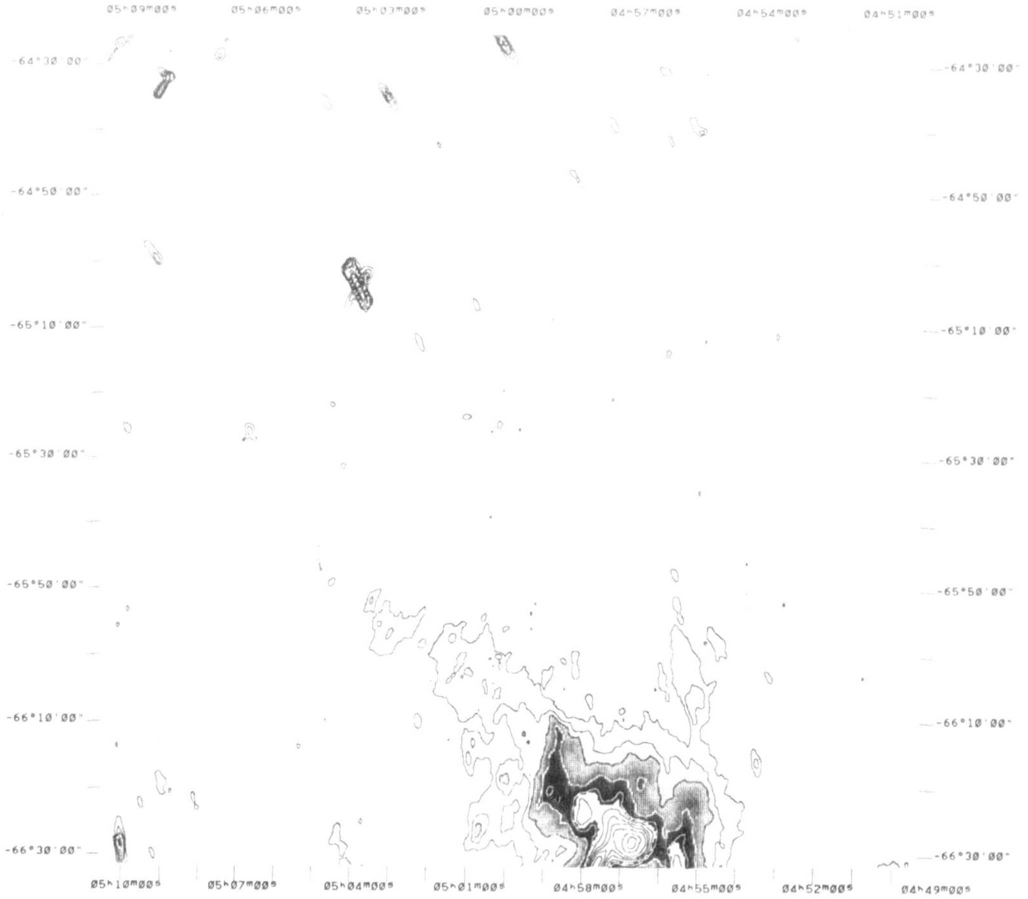

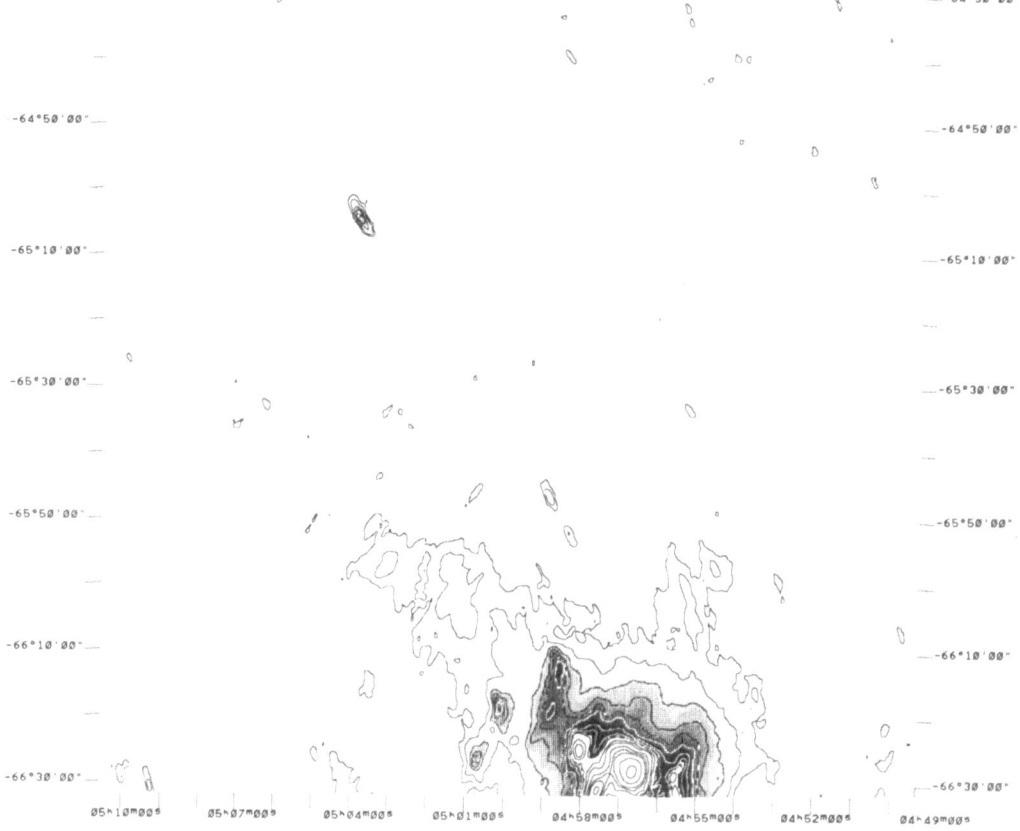

LMC
Field 19
25 μm

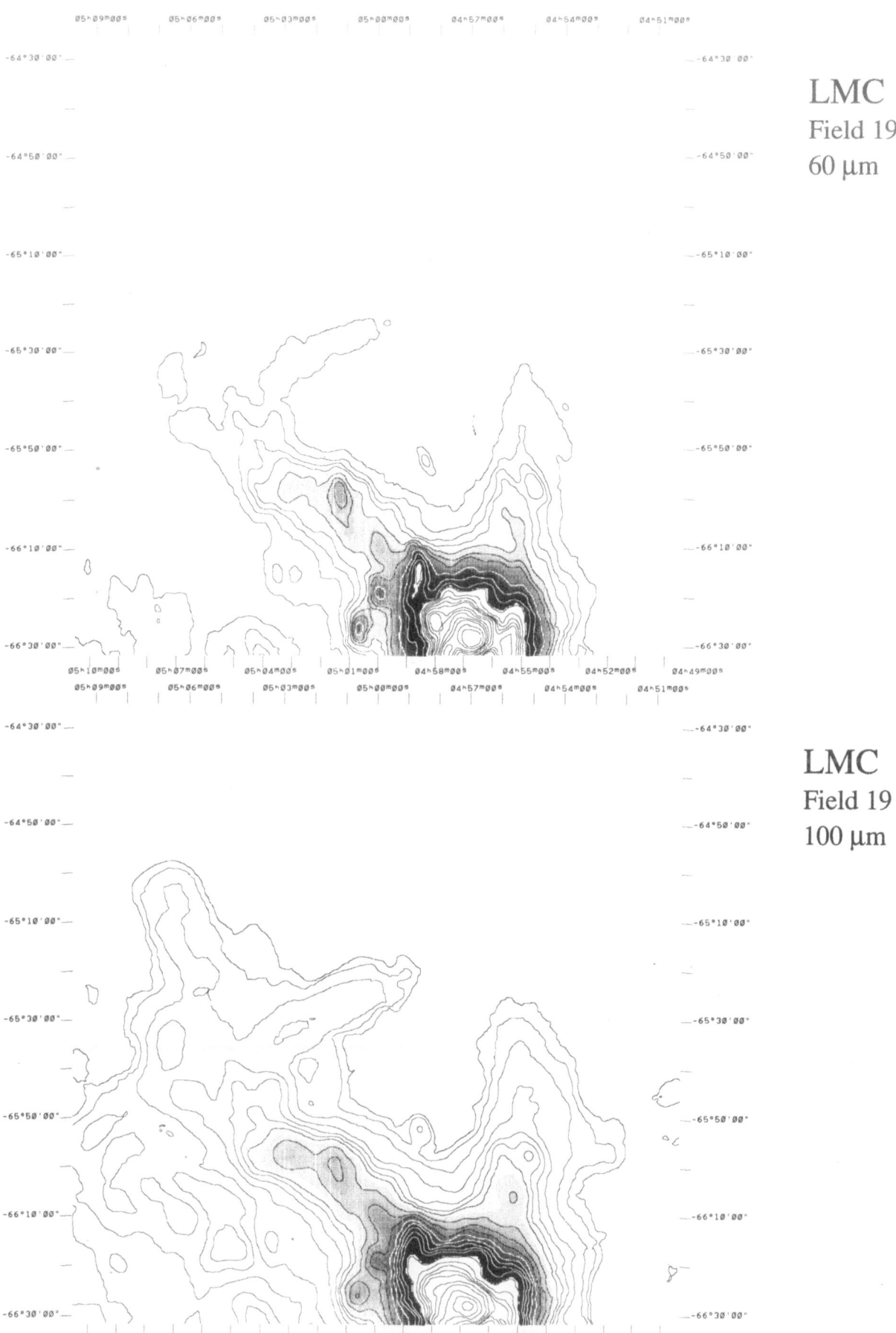

LMC
Field 19
60 μm

LMC
Field 19
100 μm

LMC
Field 20
12 μm

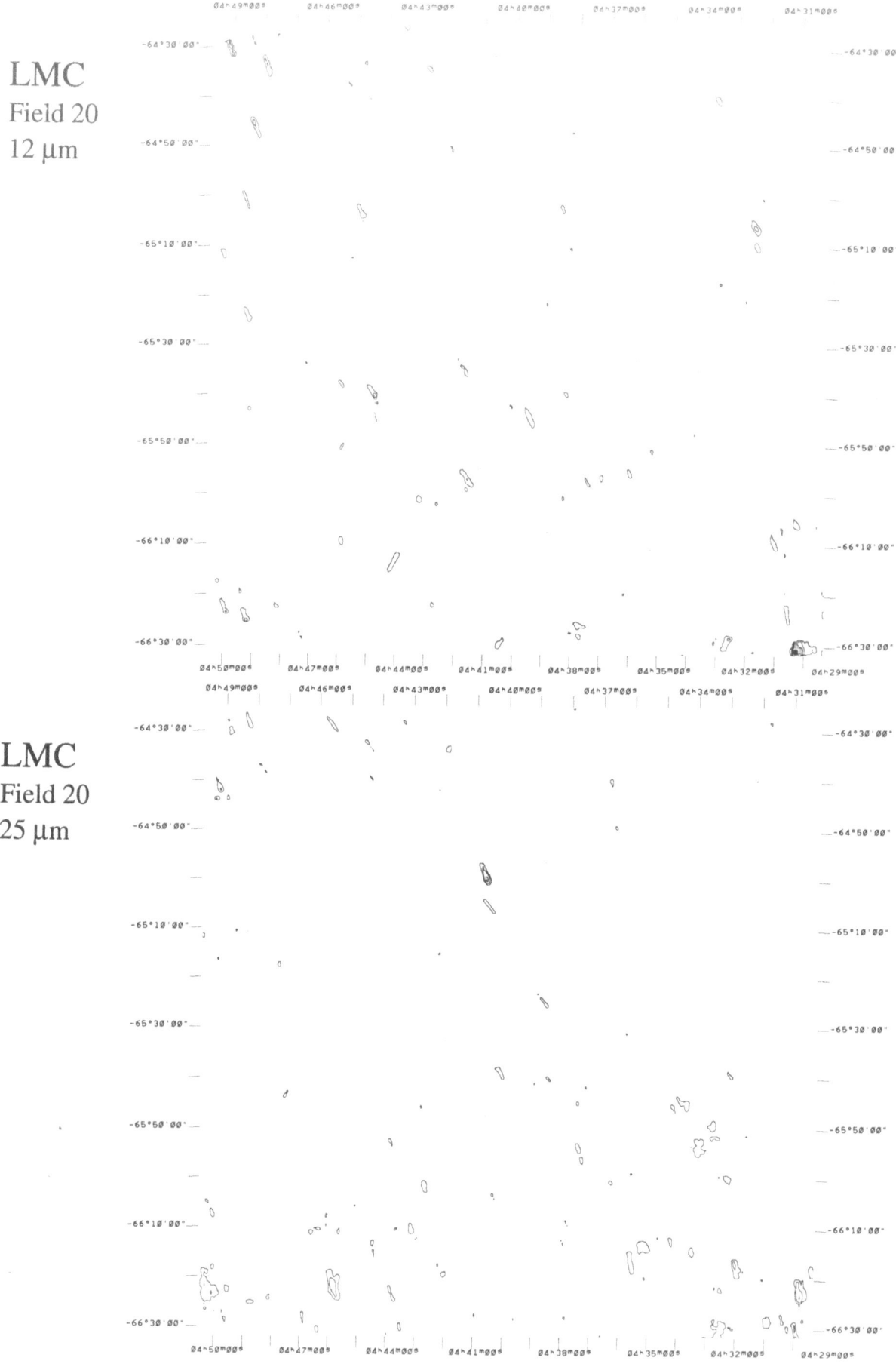

LMC
Field 20
25 μm

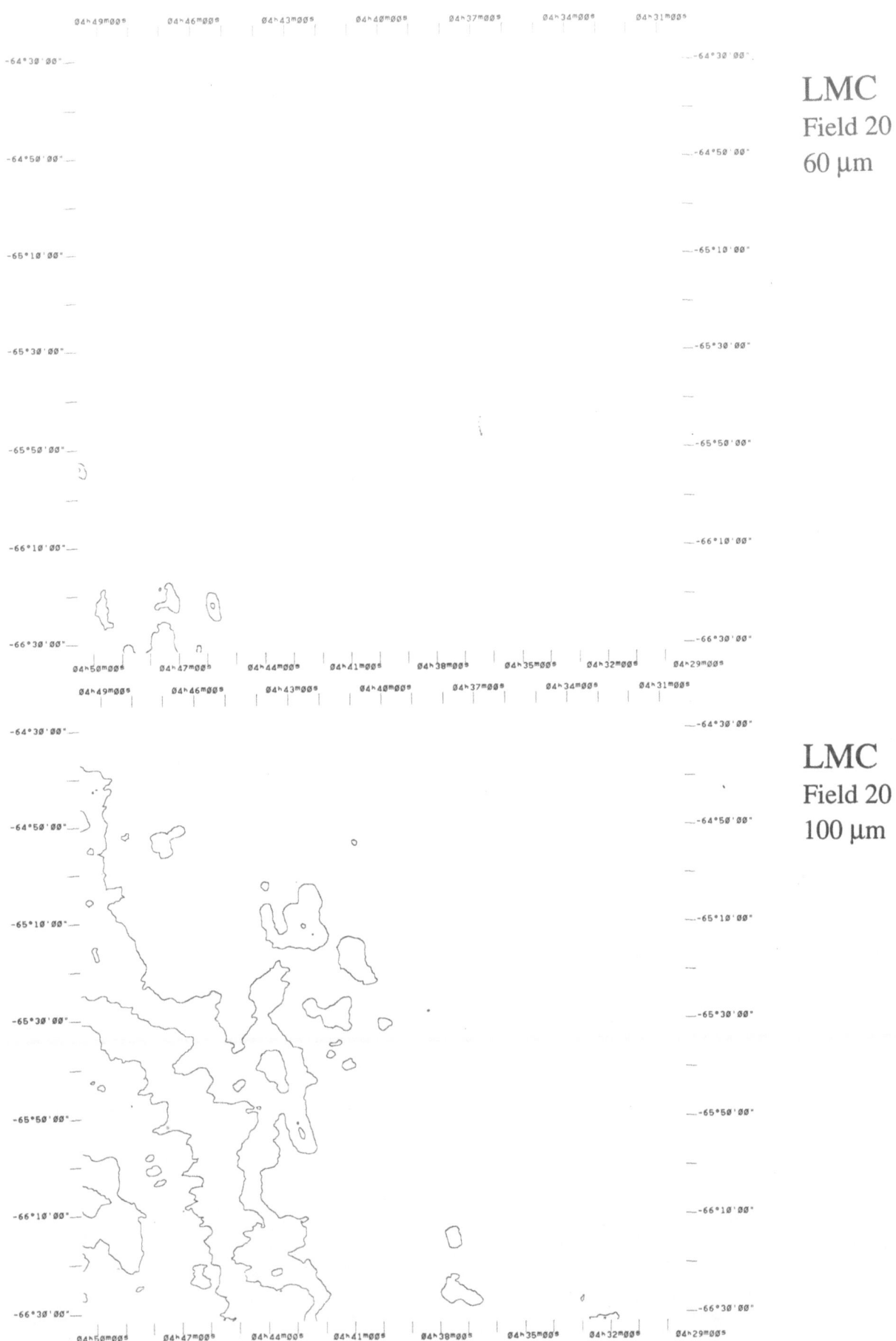

LMC
Field 20
60 µm

LMC
Field 20
100 µm

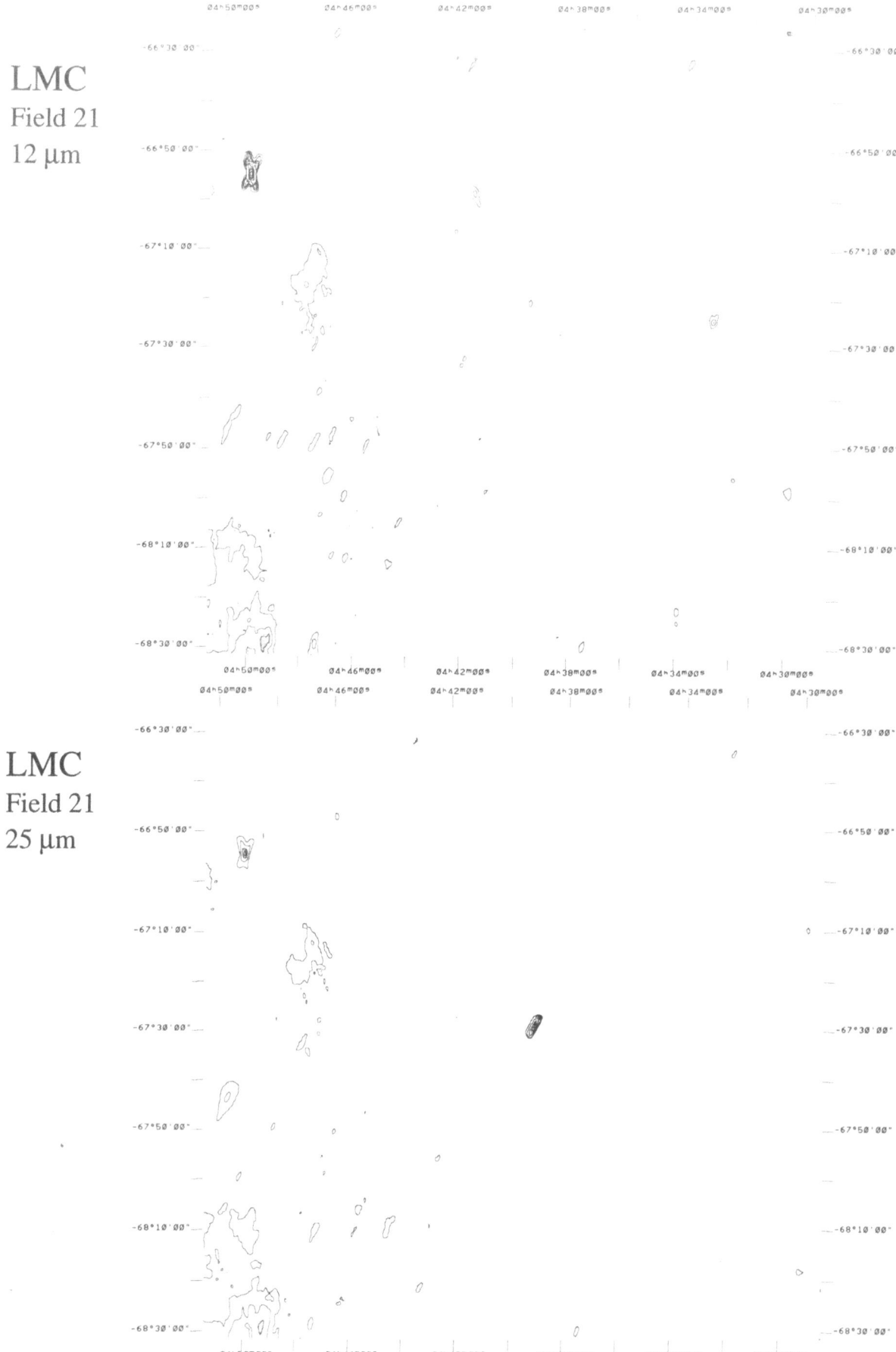

LMC
Field 21
12 μm

LMC
Field 21
25 μm

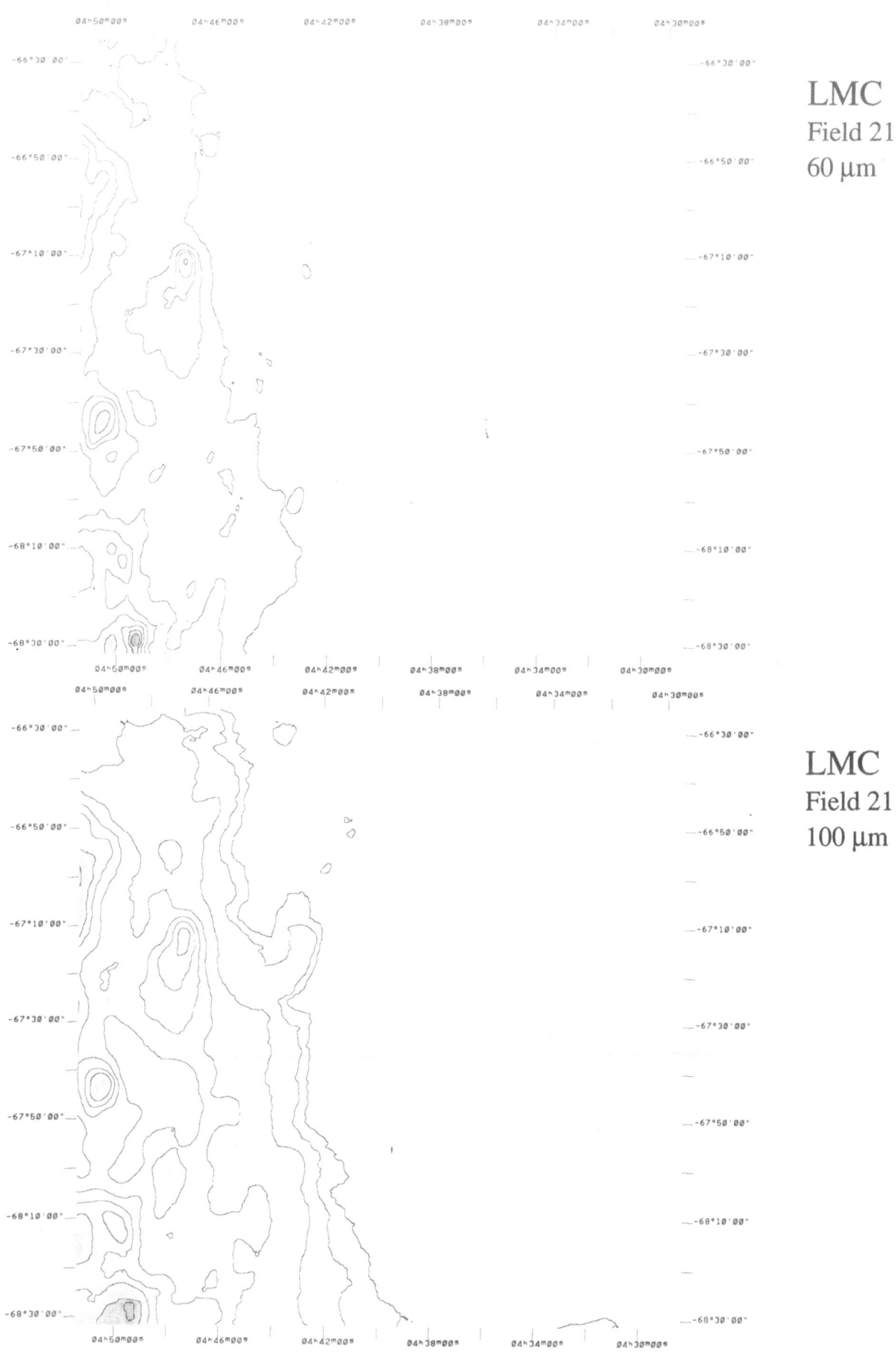

LMC
Field 21
60 μm

LMC
Field 21
100 μm

LMC
Field 22
12 μm

LMC
Field 22
25 μm

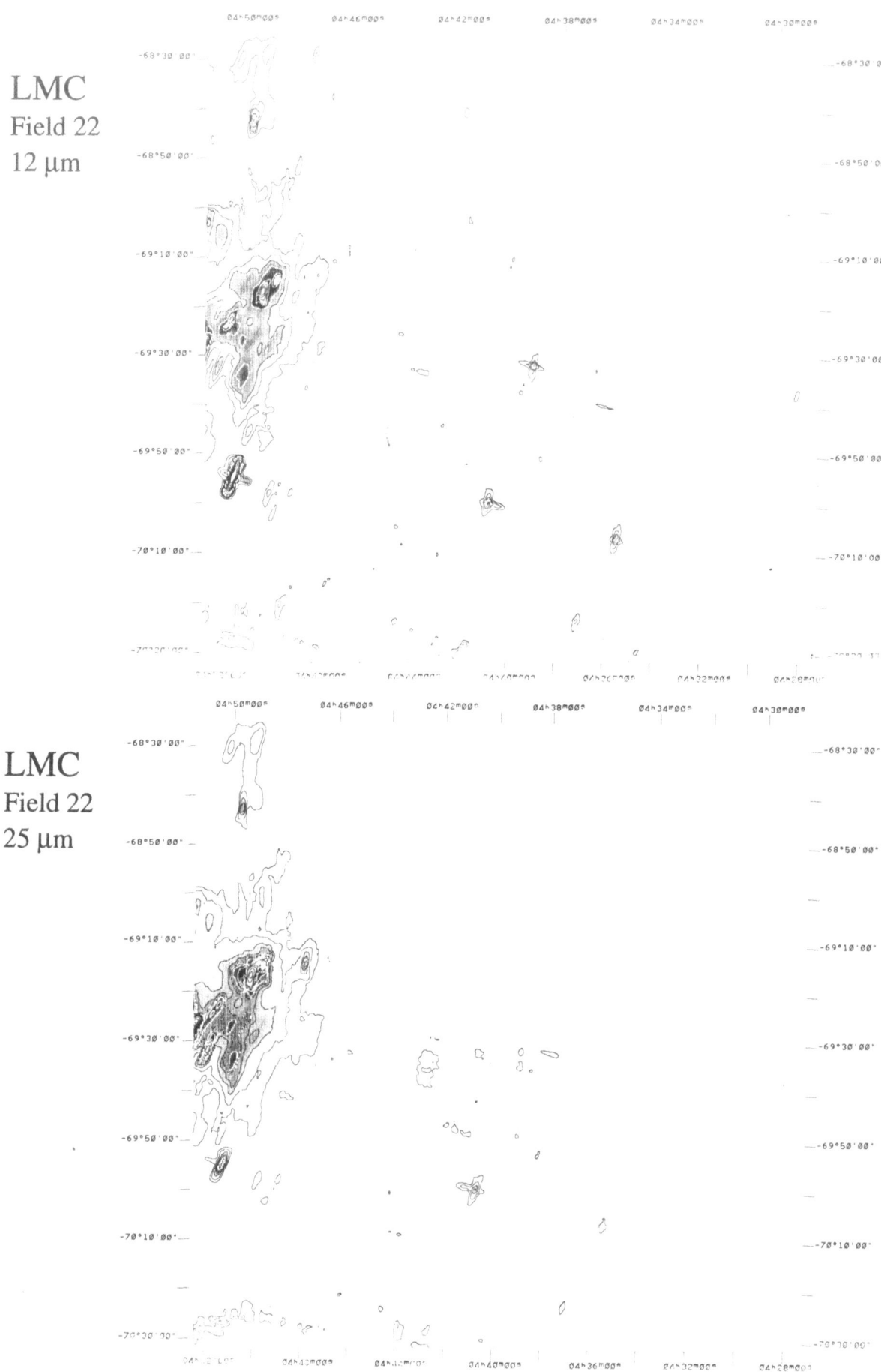

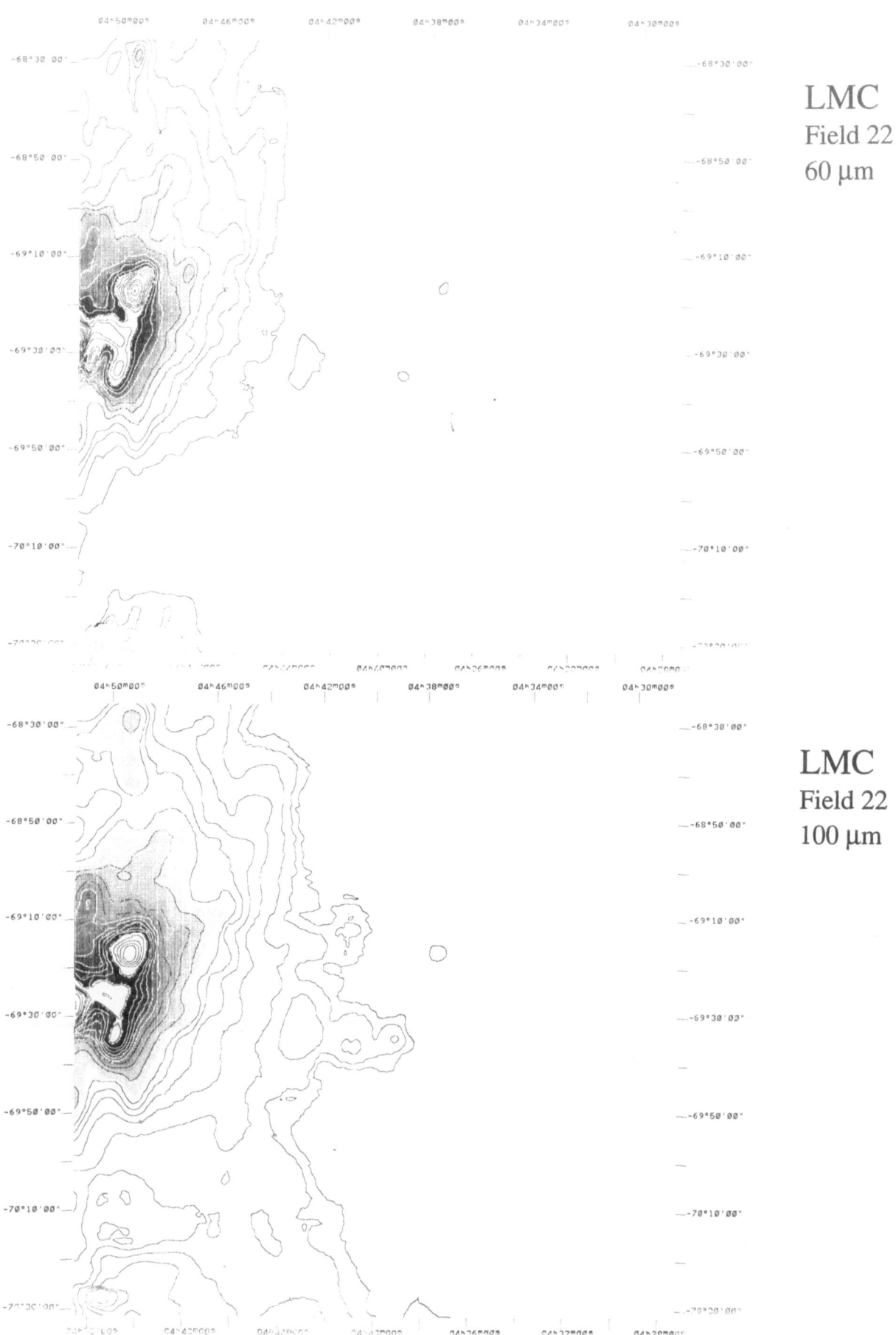

LMC
Field 22
60 µm

LMC
Field 22
100 µm

LMC
Field 23
12 μm

LMC
Field 23
25 μm

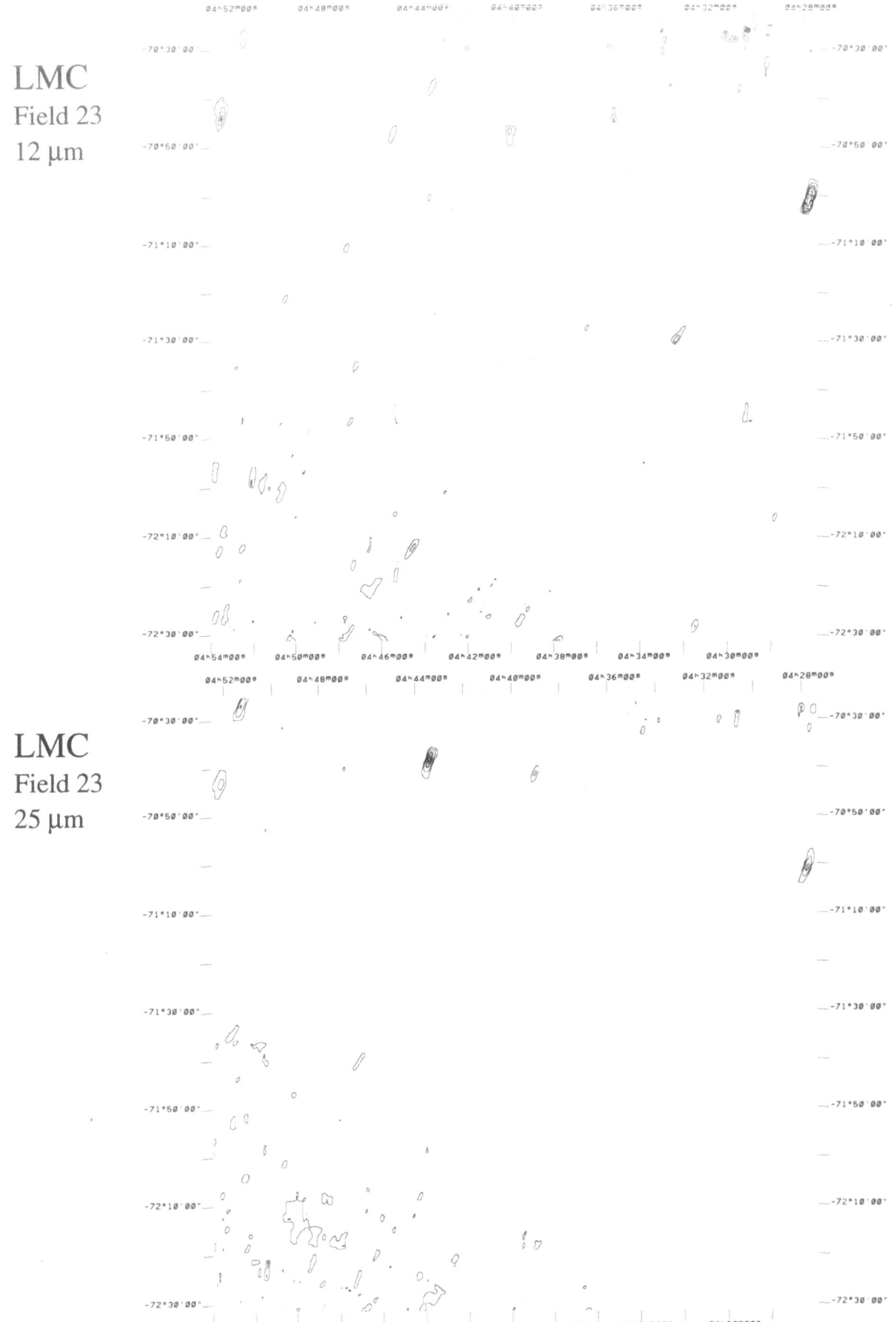

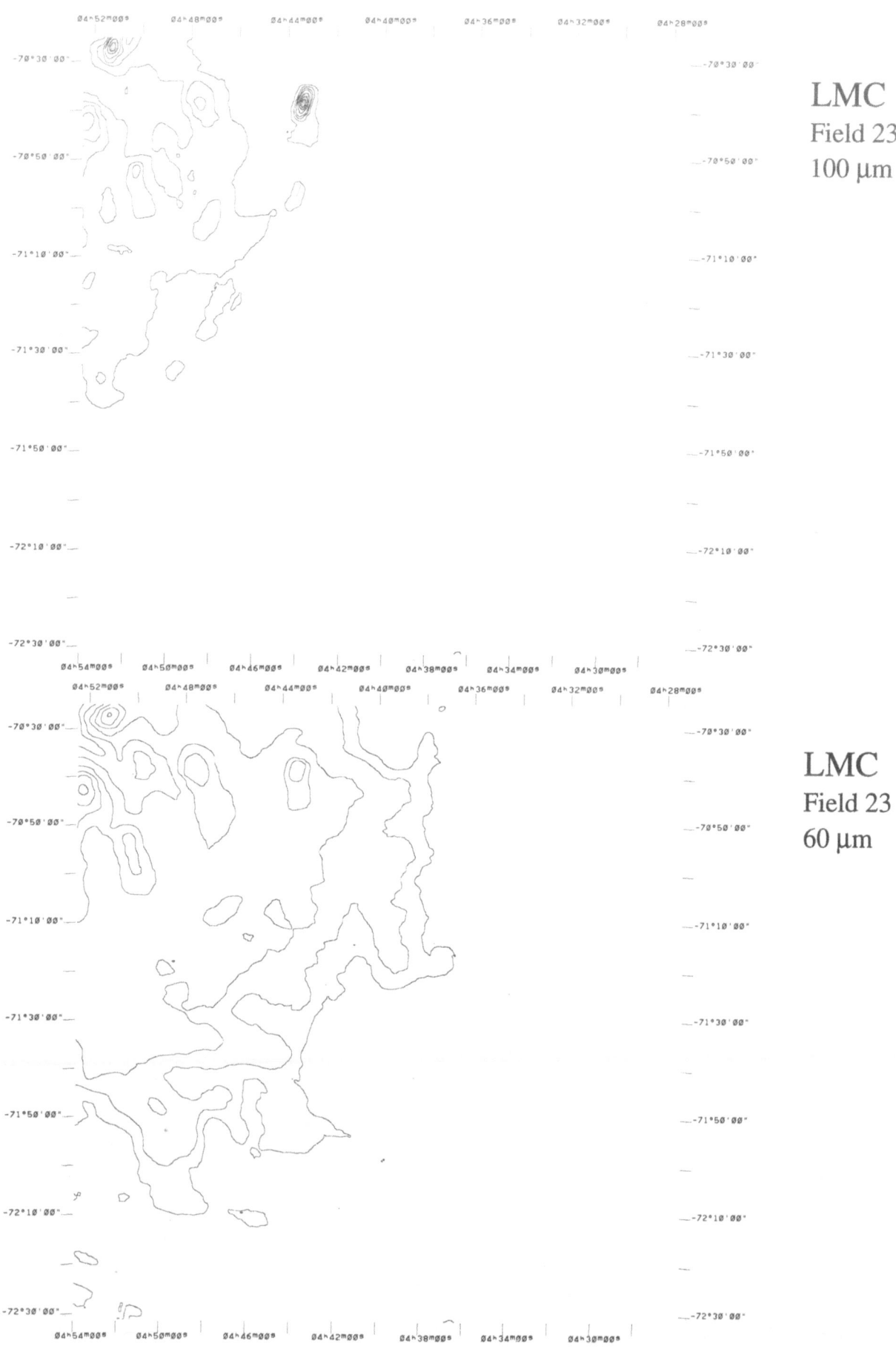

LMC
Field 23
100 μm

LMC
Field 23
60 μm

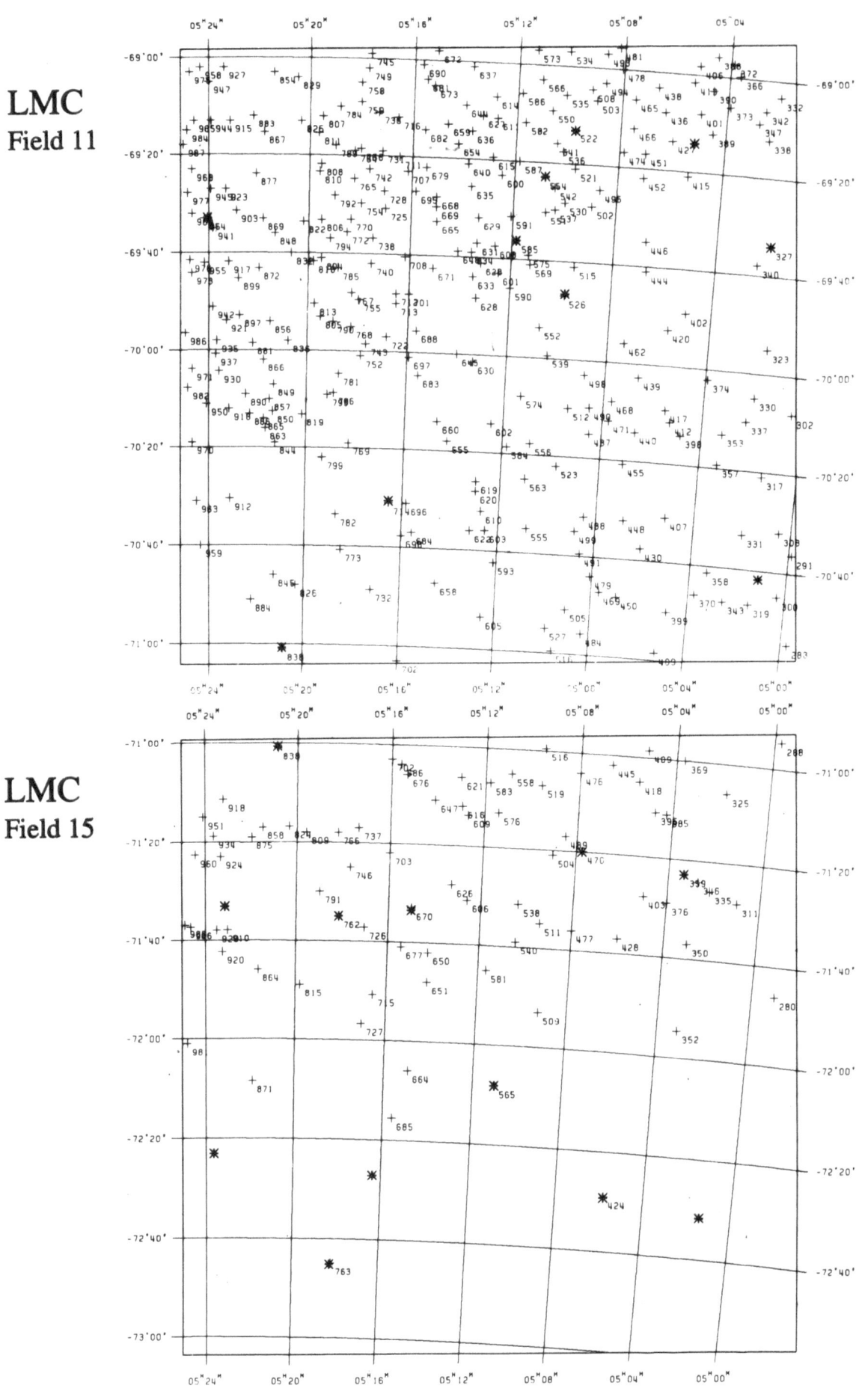

LMC
Field 11

LMC
Field 15

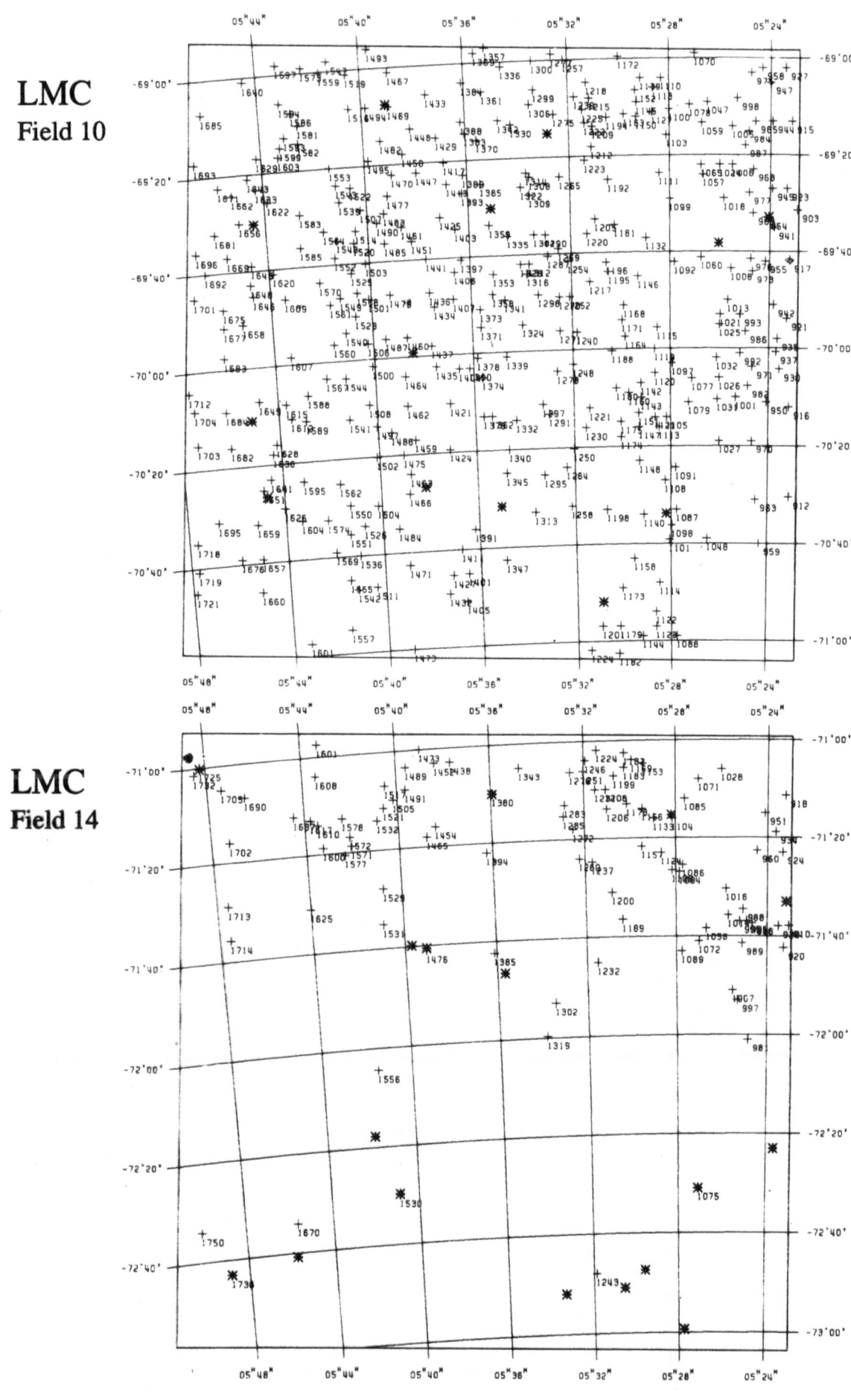

LMC
Field 10

LMC
Field 14

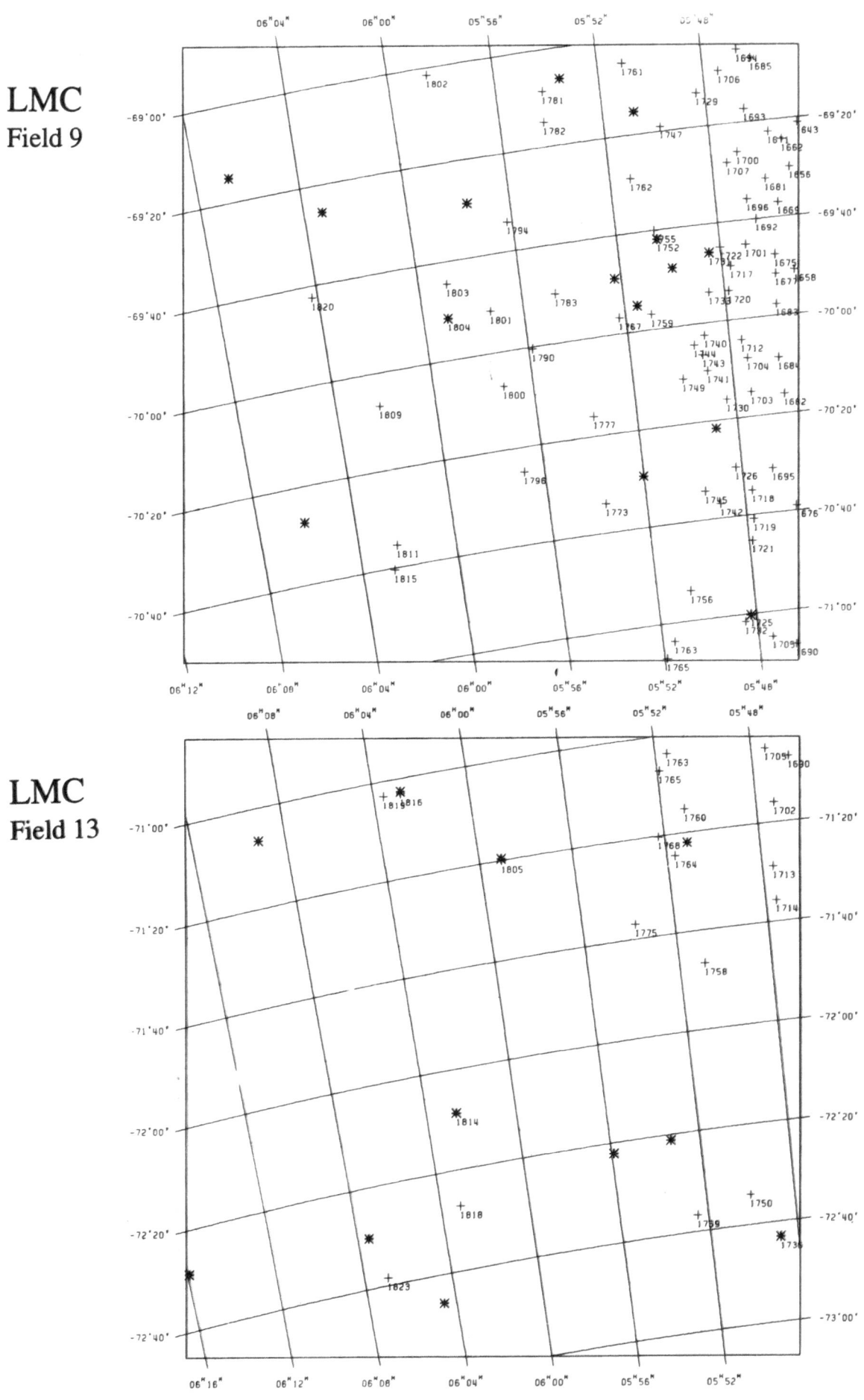

LMC
Field 9

LMC
Field 13

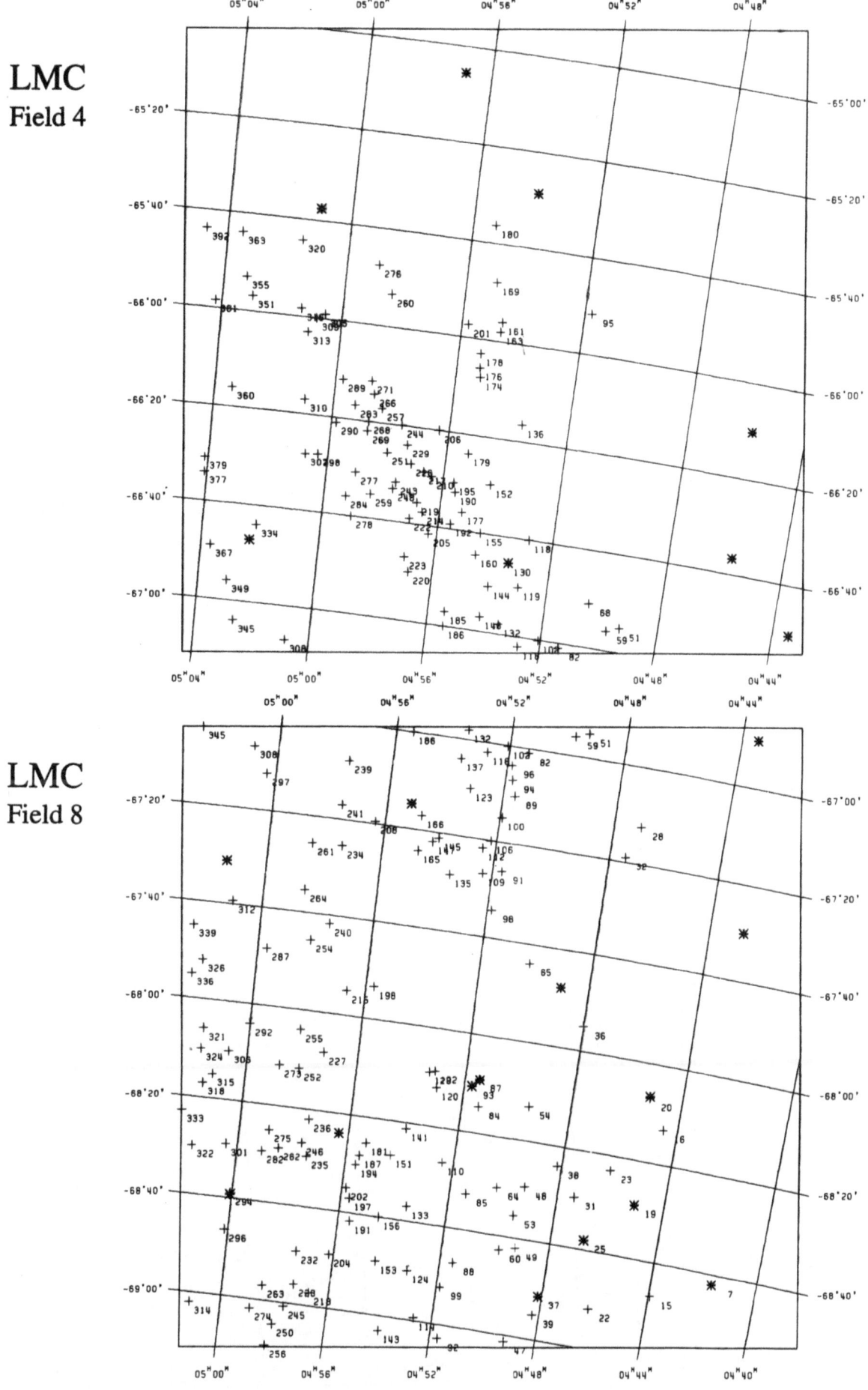

LMC
Field 4

LMC
Field 8

LMC
Field 3

LMC
Field 7

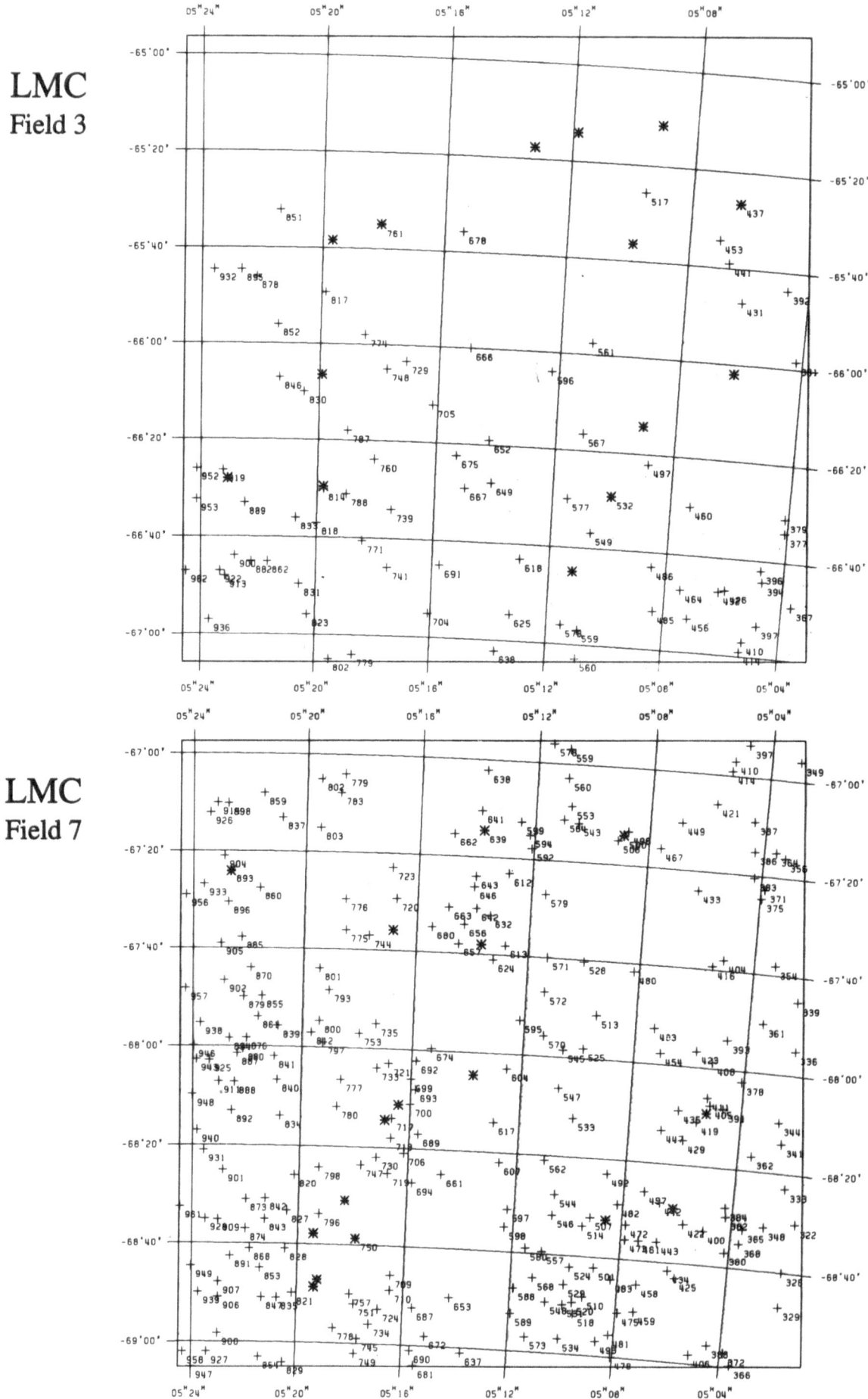

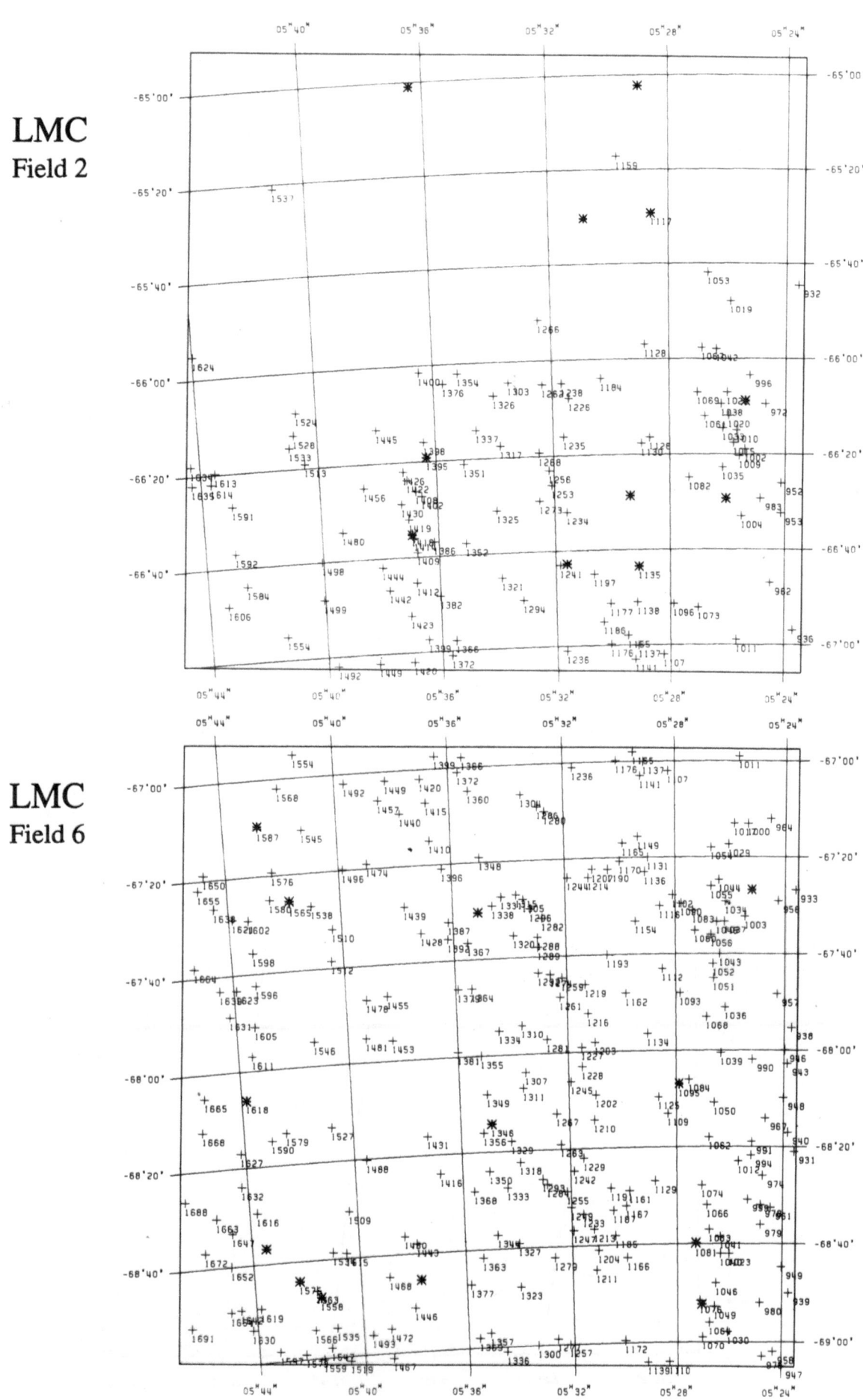

LMC
Field 2

LMC
Field 6

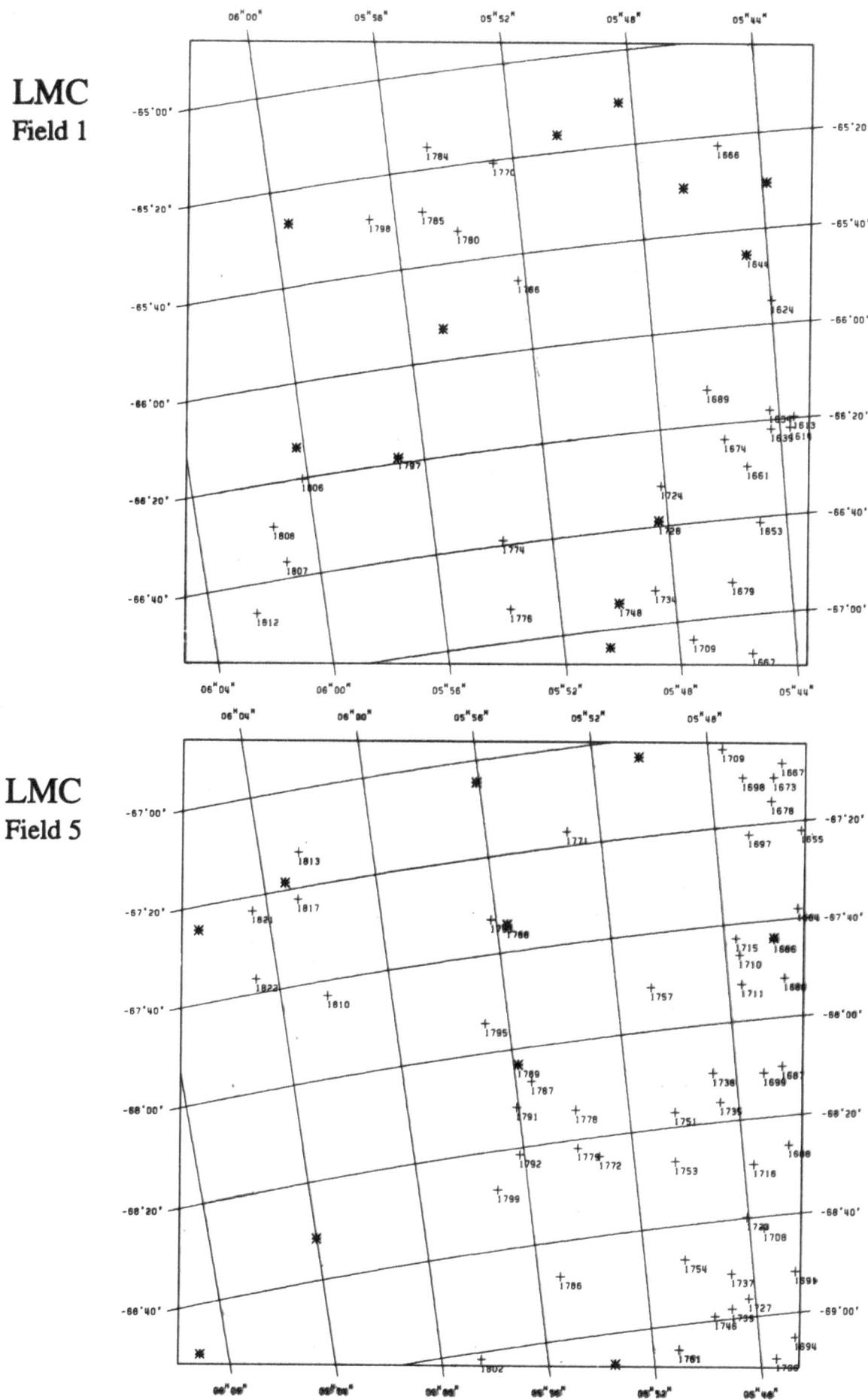

LMC
Field 1

LMC
Field 5

SMC
Field 5

SMC
Field 6

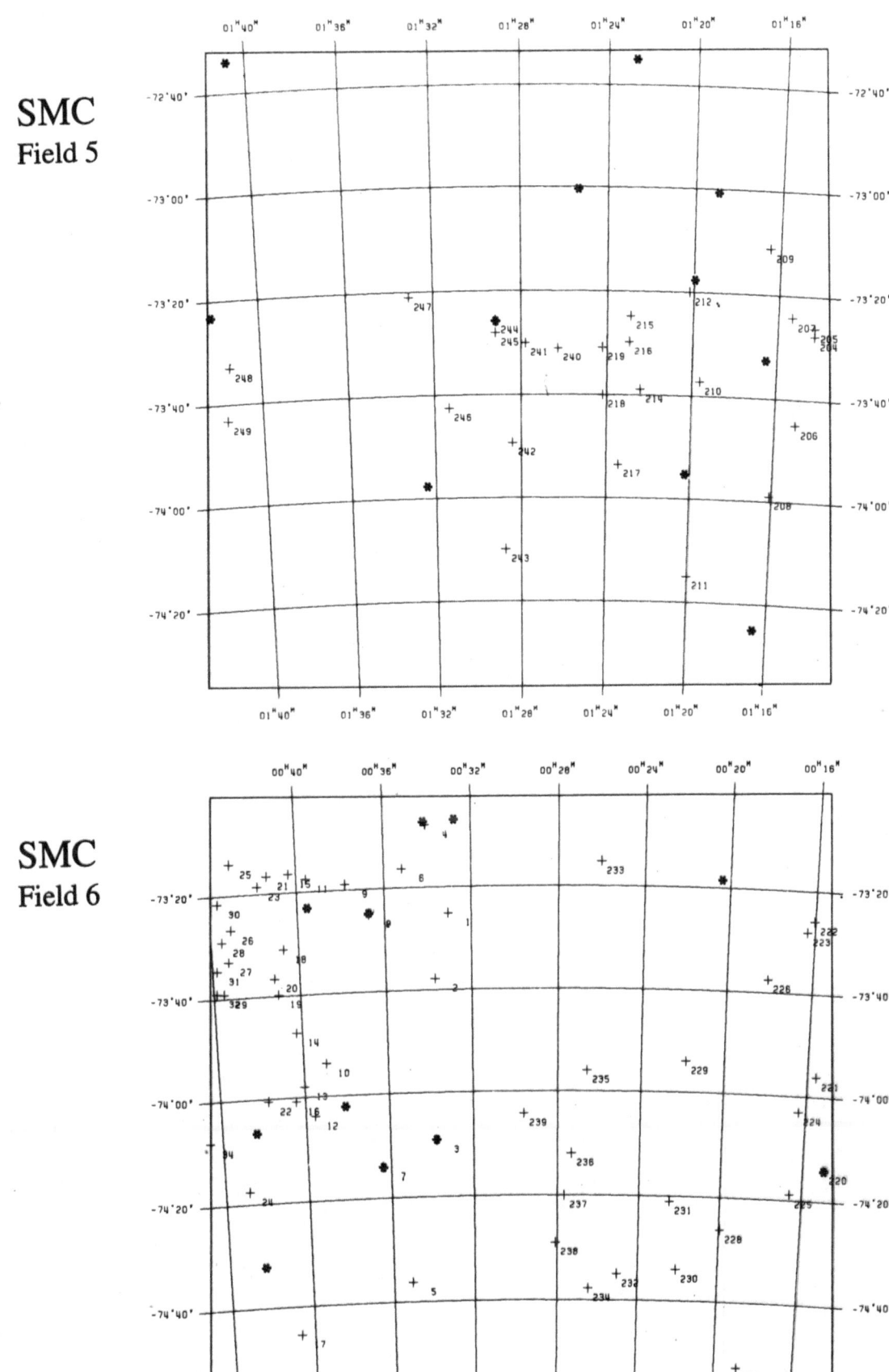

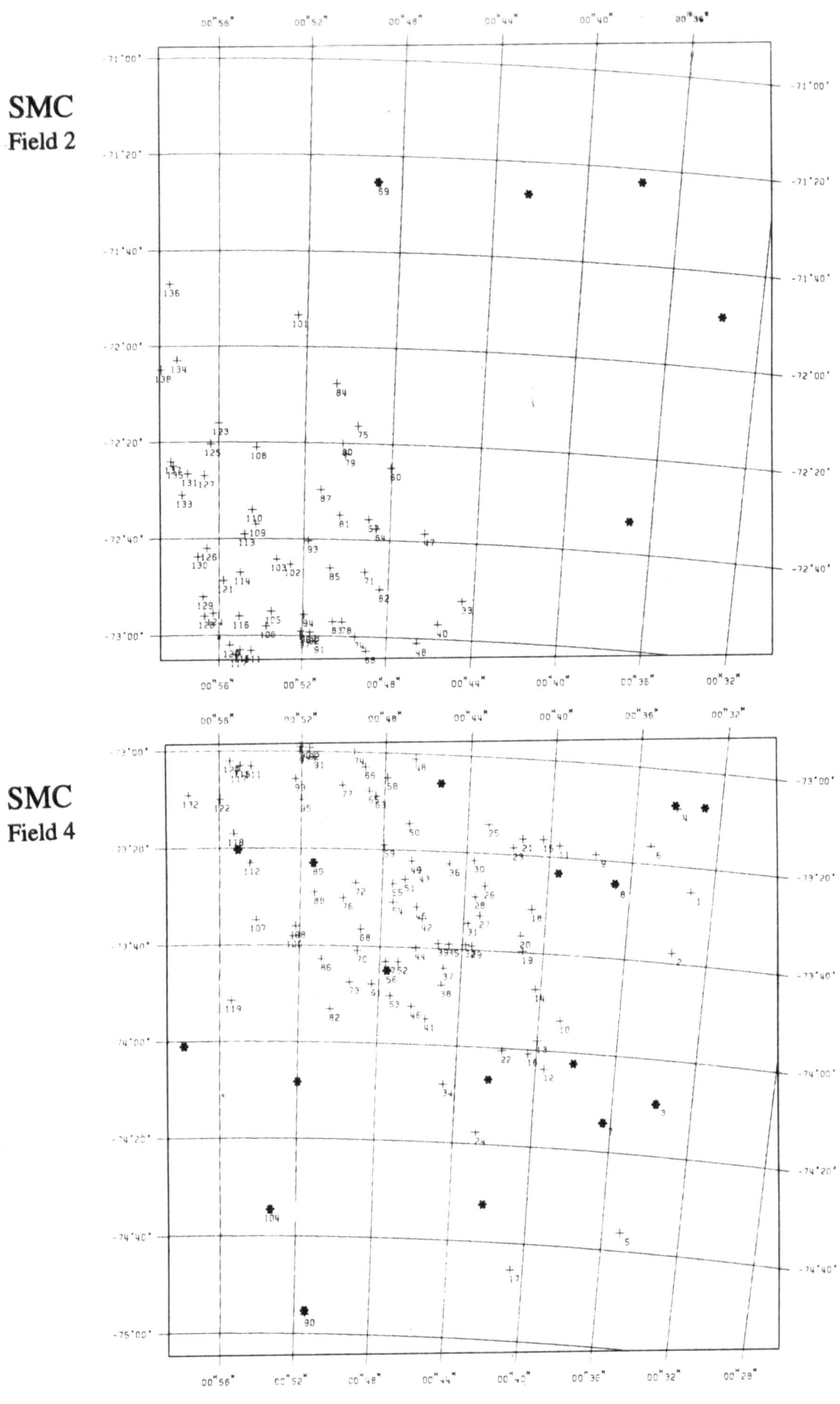

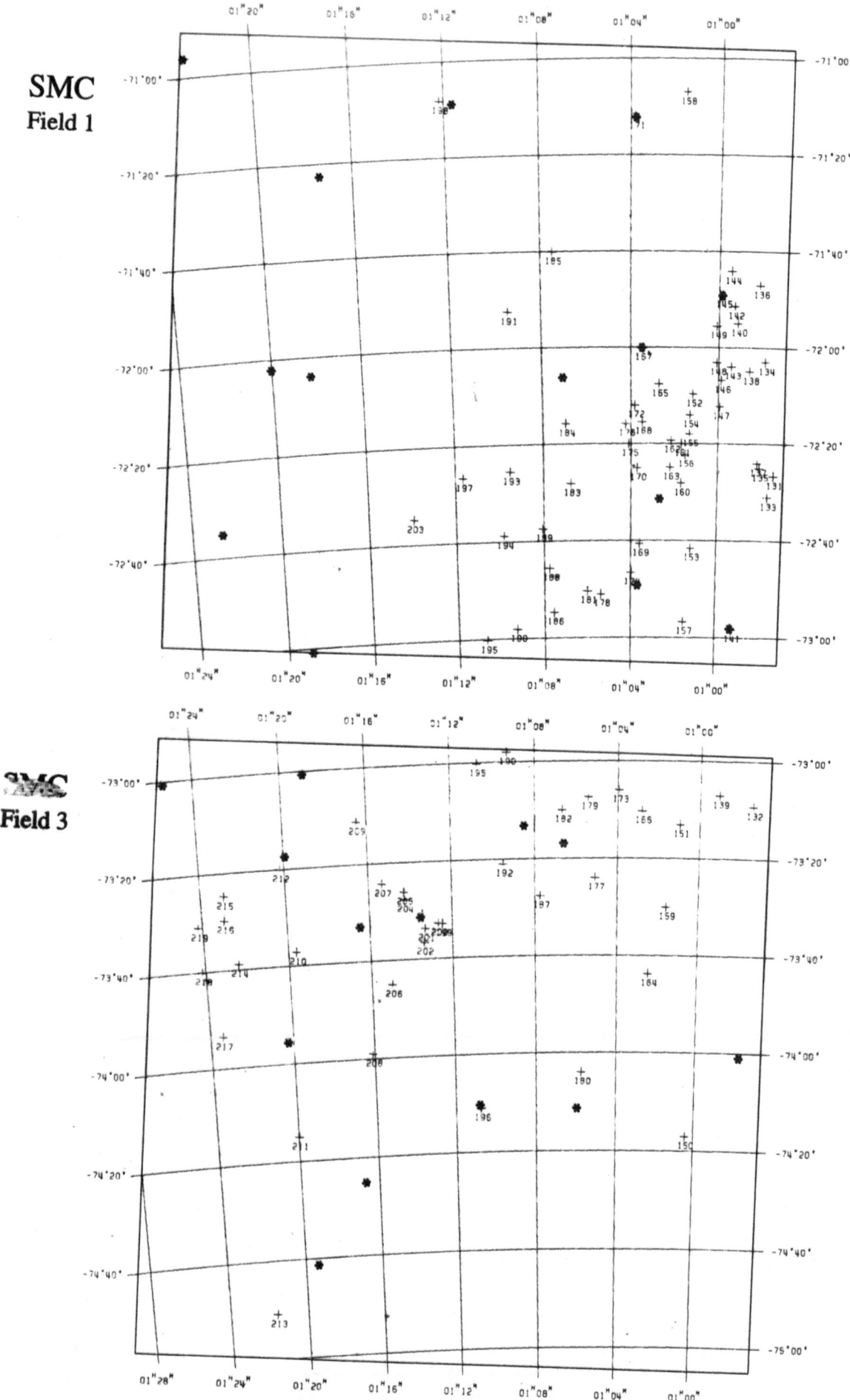

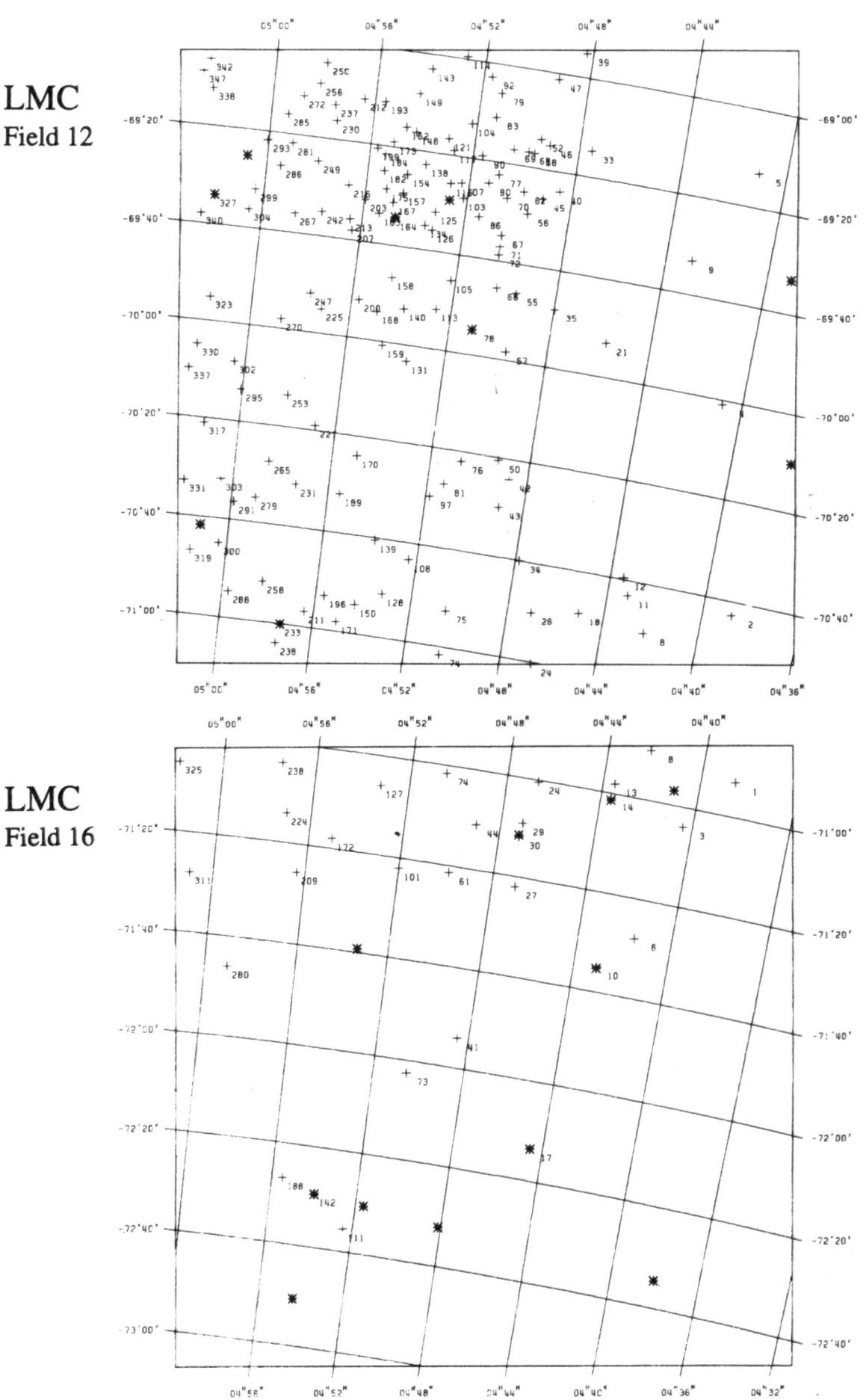

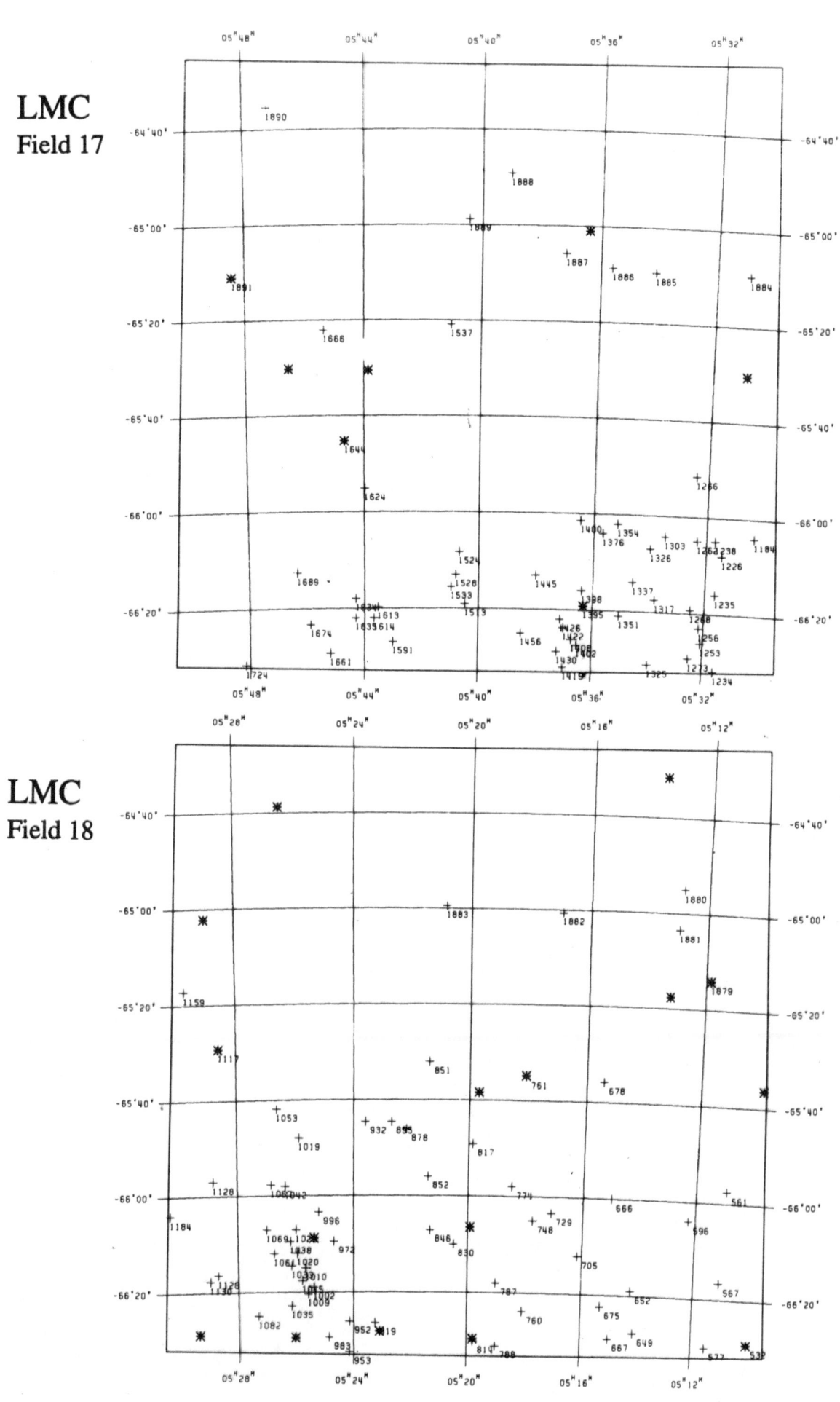

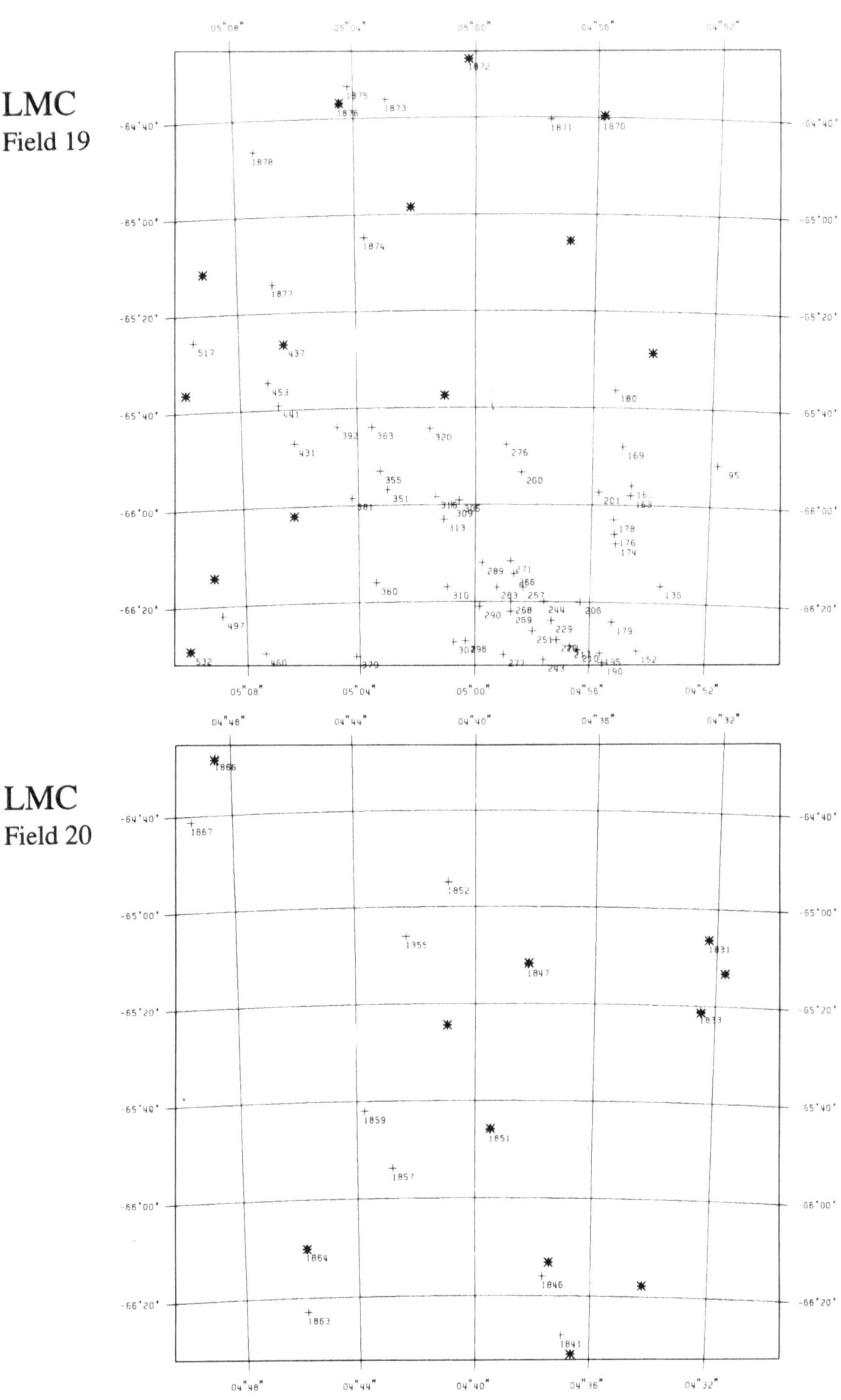

LMC
Field 19

LMC
Field 20

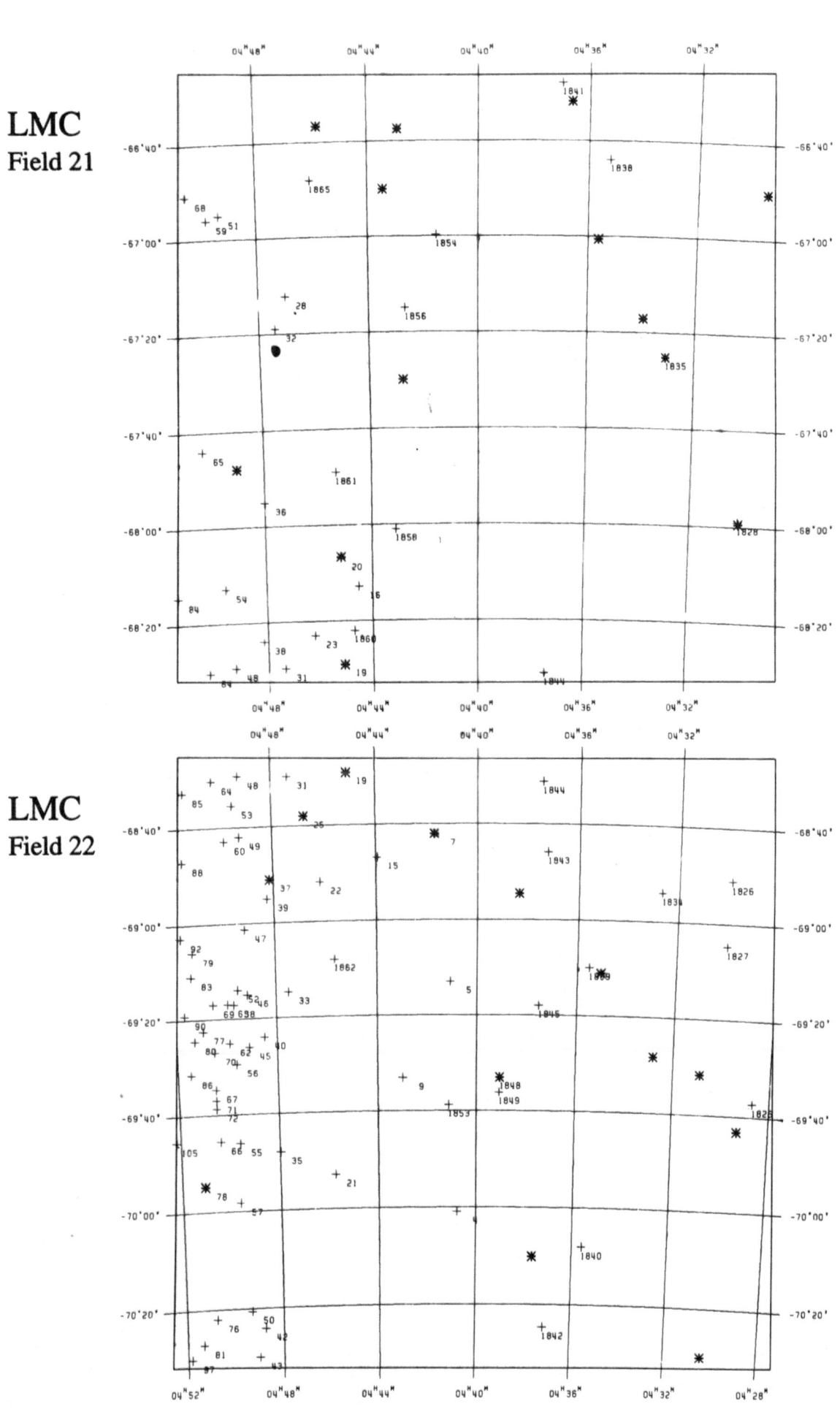

LMC
Field 21

LMC
Field 22

LMC
Field 23

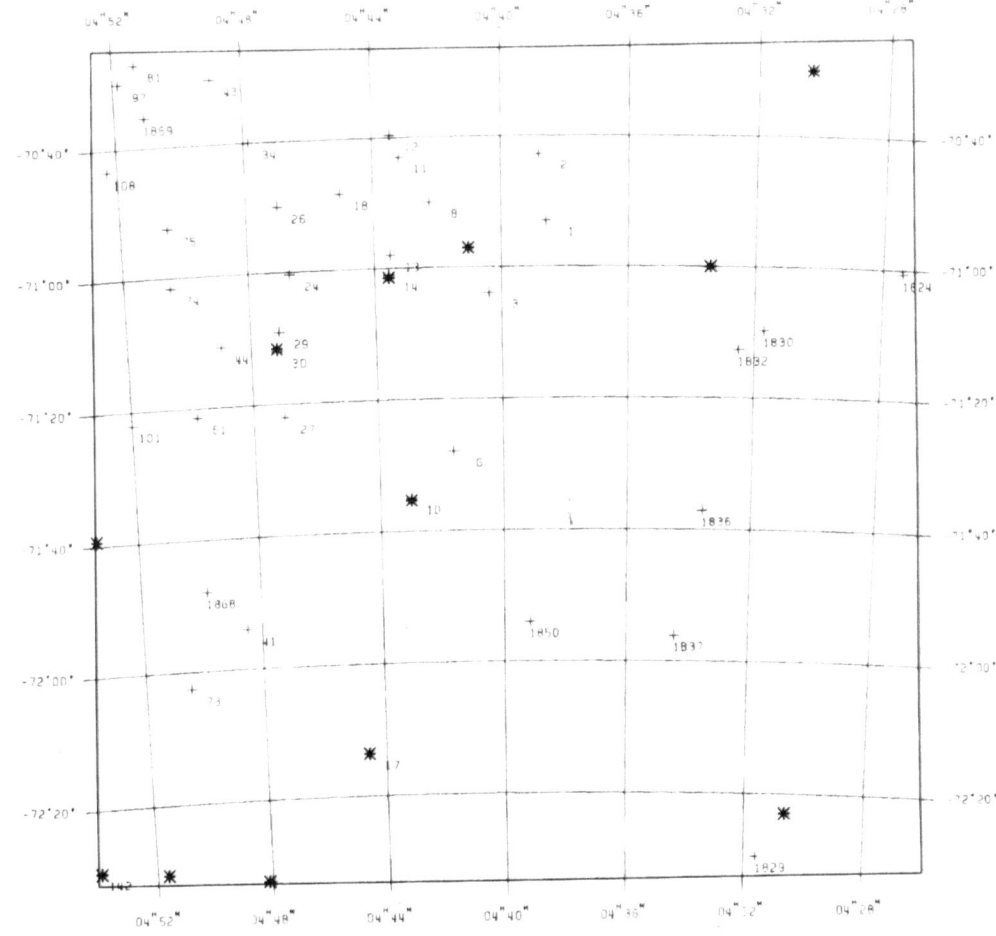